THE LOEB CLASSICAL LIBRARY

FOUNDED BY JAMES LOEB 1911

EDITED BY

JEFFREY HENDERSON

EDITOR EMERITUS

G. P. GOOLD

ARISTOTLE

VIII

LCL 288

ARISTOTLE

ON THE SOUL
PARVA NATURALIA
ON BREATH

WITH AN ENGLISH TRANSLATION BY

W. S. HETT

HARVARD UNIVERSITY PRESS
CAMBRIDGE, MASSACHUSETTS
LONDON, ENGLAND

First published 1936
Revised and reprinted 1957
Reprinted 1964, 1975, 1986, 1995, 2000

ISBN 0-674-99318-7

Printed in Great Britain by St Edmundsbury Press Ltd,
Bury St Edmunds, Suffolk, on acid-free paper.
Bound by Hunter & Foulis Ltd, Edinburgh, Scotland.

CONTENTS

CONTENTS

INTRODUCTION

ARISTOTLE, the man, we know; the works of Aristotle, the philosopher, present almost insuperable problems. We know as much of the life of Aristotle as of most of Hellas' great men, but we find it difficult, if not impossible, to regard his work as his contemporaries regarded it. His writing was considered a model of Greek prose style at its best, but no one who reads his works as they have come down to us could subscribe to this view. In the treatises comprised in this volume, the philosopher is sometimes lucid and bald, sometimes involved and obscure, and sometimes even unintelligible, but very rarely brilliant in style.

Yet we need not necessarily blame Aristotle for this. In days when there was no copyright corruption of the text was fatally easy, and if we are to believe Strabo, the works of Aristotle were peculiarly unfortunate in this respect. According to that author, Aristotle left his library to Theophrastus, who handed it on to others. It passed through a good many hands, including those of one Apellicon, who is delightfully described by Strabo as " a bookworm, rather than a philosopher," by whom the gaps in the manuscript caused by damp and moth were filled in " not well." Later on still, when Sulla captured Athens, it was transferred to Rome, and edited by inferior clerks under the direction of

INTRODUCTION

booksellers who made no attempt to collate the different copies. Small wonder if we find them often obscure and lacking in style. There is, however, another view of the works which have come down to us bearing the name of Aristotle. Cicero refers to them as *commentarii*, and it is quite possible that the present volume contains nothing but lecture notes compiled either by Aristotle himself or his pupils. This would account for the unevenness of exposition. Some points are argued in full, some are only briefly outlined. It might also account for the numerous cross-references to other treatises, some of which cannot be traced.

Yet, in spite of obscurity and baldness of style, it would be quite wrong to suppose that these treatises are valueless. An attempt has been made in the introductions to the separate treatises to show where this value lies for us to-day.

The modern reader studying Greek philosophy is confronted at the outset by a difficulty which must be boldly faced. Aristotle received his training in the school of Plato, and from him inherited many of his fundamental beliefs. Plato had come to regard the world of sense as unreal, and only the world as apprehended by the mind, detached as far as possible from sense impressions, as real. Aristotle repeatedly shows this Platonic influence in his handling of philosophical questions. He pursues an argument in the direction in which Logic leads him, quite unmoved by any apparent absurdity in the conclusions at which he arrives. Thus in this volume Aristotle will often be found to reach conclusions, which at first sight seem only fantastic ; but such are always worth careful and unbiased consideration.

INTRODUCTION

This collection of treatises belongs to subjects on the borderline between bodily and mental. Aristotle was the son of a doctor, and himself a biologist, who believed in experiment and dissection as a means of collecting evidence. Thus his views on the soul are influenced by his physiology. Yet he never falls into the meshes of materialism, and appears quite certain that the body cannot possibly explain the mind. His method is analytical, his logic, within the limitations imposed by his age and personal character, is ruthless, and his knowledge is encyclopaedic. His arguments and conclusions should assuredly serve to stimulate, even where they fail to satisfy.

Finally, although his conception of a divine Creator is nowhere clearly worked out, Aristotle believes implicitly in design and purpose in the universe, and lays it down as axiomatic that no account of any part of the body or function of the mind can be considered adequate, unless it shows the purpose which it serves in the scheme of creation.

The Text

The present text is based upon that of Bekker's Berlin edition of 1831, and as a rule critical notes are only added where his readings have been rejected. The manuscripts are as follows :

E	Parisiensis 1853	10th cent.	402 a 1 —464 b 18.
L	Vaticanus 253	14th ,,	424 b 22—486 b 4.
M	Urbinas 37	,, ,,	436 a 1 —470 b 5.
N	Vaticanus 258	,, ,,	?
P	Vaticanus 1339	12/13th ,,	412 a 13—424 b 19,
			436 a 1 —449 b 3,
			464 b 19—486 b 4.

Q	Marcianus 200	12th cent.	481 a 1 —486 b 4.
S	Laurentianus 81.1	12/13th ,,	402 a 1 — ,,
T	Vaticanus 256	14th ,,	,, —435 b 25.
U	Vaticanus 260	13th ,,	,, —464 b 18.
V	Vaticanus 266	14th ,,	,, —480 b 30.
W	Vaticanus 1026	? ,,	,, —435 b 25.
X	Ambrosianus H 50	? ,,	,, — ,,
Y	Vaticanus 261	13/14th ,,	436 a 1 —464 b 18.
Z	Oxoniensis C.C.C. 108 (formerly W. A. 27)	9/10th ,,	464 b 19—486 b 4.
Ba	Bekkeri Palatinus Vaticanus 162	15/16th ,,	481 a 1 — ,,

The more important of these form two distinct groups preserving different traditions.[a] The earlier (Π) consists of EVY, the later (Φ) of LNSU. Of the other mss., M generally follows Π, while PQTWX seem more akin to Φ. Z (which is supplementary and related to E) has independent value for the *De Respiratione* and *De Spiritu*, and Ba occasionally helps in establishing the text of the latter.

Light is also thrown upon the text by the treatises of Alexander Aphrodisiensis (2nd/3rd cent.), the paraphrases of Themistius (4th) and Sophonias (14th cent.), the commentaries of Simplicius and Philoponus (6th) and of Michael of Ephesus (11th cent.) ; and by two Latin versions, the anonymous Vetus Translatio (12th cent.) and the Nova Translatio by William of Moerbeke (13th cent.).

The following are the principal modern editions of the works included in this volume :

De Anima (text only) : W. Biehl, Leipzig, 1896 ; (text, translation and commentary) : R. D. Hicks, Cambridge, 1907.

---

[a] See R. Mugnier, *Les Manuscrits des " Parva Naturalia" d'Aristote, Mélanges Desrousseaux*, Paris, 1937.

INTRODUCTION

Parva Naturalia (text only): W. Biehl, Leipzig, 1898.

De Sensu and *De Memoria* (text, translation and commentary): G. R. T. Ross, Cambridge, 1906.

De Somno et Vigilia (with the Latin versions and the commentary of Theodorus Metochites): H. J. Drossaart Lulofs, Leyden, 1943.

De Spiritu (text only, with *De Animalium Motione* and *De Animalium Incessu*): W. W. Jaeger, Leipzig, 1913.

───────

I have found it impossible to express in detail my indebtedness to those many scholars who have edited or revised the text of Aristotle's *De Anima* and *Parva Naturalia*, but all who have worked in this field must be conscious how great that debt inevitably is. In the translation it has been my aim to make Aristotle's meaning as clear as possible, even if this has involved contraction or (more commonly) expansion of the Greek. I have one other debt to acknowledge: my thanks are due to Messrs. R. & R. Clark's reader, who has dealt so ably with the German text (in spite of the comparative poorness of the type and the irritating contractions) that my own task of proof correction has been reduced to a minimum.

It may assist the general reader to add an explanation of some of the technical terms used by Aristotle in this work.

> δυνάμει ("potential"). A man is "potentially" (*e.g.*) a thinking being, but he is not always thinking. When this capacity is actually operating, Aristotle calls him ἐνεργείᾳ ("actually") a thinking being.

ἐντελέχεια ("actuality" or "actualization"). This
term frequently overlaps ἐνέργεια, but in its
most exact sense it means more. It is the
perfect realization of all that any creature or
power is capable of becoming.

εἶδος ("form"), often contrasted with ὕλη
("matter"). Matter in itself is formless;
e.g. the marble from which a statue is carved.
When it received εἶδος, it becomes τόδε τι
("particular thing").

κατὰ συμβεβηκός ("accidentally"), τὰ συμβεβηκότα
("contingent attributes"). A table may be
green, but its greenness is only "accidental";
that is to say, if it were (*e.g.*) brown, it would
be just as much a table as before. But there
are some qualities which belong to its οὐσία
("essence," "essential nature"). We assume
this to be true of all objects of sense, and
although it is always difficult and usually impossible to determine exactly what these
qualities are their sum total constitutes the
τὸ τί ἦν εἶναι ("that which makes it what it
is").

νοῦς ("mind"); the most general word, including
frequently both the percipient and the intellectual faculties. It is sometimes sub-
divided by Aristotle into νοῦς πρακτικός (the
mind as applied to producing results) and the
νοῦς θεωρητικός (the mind regarded as purely
contemplative).

πάσχω ("be acted upon" or "affected"). Both in
the physical and intellectual worlds influences
of different kinds are at work. If such an
influence operates the object of it is said

πάσχειν. The noun πάθος ("affection") is to be interpreted in this sense.

φαντασία ("imagination"). The word is used by Aristotle in two senses: (1) sometimes it operates in the presence of the sensible object and thus "interprets" the object to the mind; (2) sometimes it operates in the absence of the sensible object, and then is either a form of memory or what we call "pure imagination."

INTRODUCTION

THE *De Anima* is a discussion of the main problems connected with the soul or vital principle of living things : what it is, how it is related to the body, and how it functions. The approach is more metaphysical and the treatment more abstract than might have been expected from a biologist ; indeed Aristotle does not seem to have contributed much to the subject from his own observation. His criticism of earlier views is almost entirely logical, and the novelty of his own doctrine consists mainly (1) in a more careful analysis of the vital functions, (2) in the application of the actuality-potentiality relation to explain the interaction of soul and body, sense and sensible, mind and intelligible. Even when he comes to discuss the operation of the several senses he does little more than generalize from preconceived opinions, and it is not until we reach the next " treatise," *On Sense and Sensible Objects*, that we find any serious attempt to argue from observed phenomena. The fact is that the *De Anima* is intended to outline a systematic general theory, unencumbered by superfluous detail because it is required to exhibit the maximum internal coherence and symmetry.

Book I is purely introductory. After stressing the importance and difficulty of the subject, and indicating the chief problems which have to be investigated, Aristotle proceeds in his usual way to consider earlier theories about the nature of the soul. These are shown to agree generally in attributing to it the characteristics of motion and sensation, and in regarding it as either incorporeal or at any rate the least corporeal of all substances.

The criticisms which Aristotle then makes are rather misleading. Instead of admitting the partial truth of these traditional views, he subjects them to a fusillade of arguments apparently designed to discredit them altogether. The process is highly dialectical; there is much repetition and digression; and although it was doubtless effective in convincing Aristotle's students that the whole subject needed reexamination, it must have left their minds in some confusion—that is, if our text fairly represents his discourses. In fact, however, the views which are really the objects of his attack are in each case more narrowly specific than he at first implies. They are (1) that the soul has a spatial motion of its own which it imparts mechanically to the body, (2) that the soul perceives external objects because it is composed of the same physical elements as they are, and (3) that the soul is not entirely incorporeal. (The third, of course, really underlies the other two.) All three are duly shown to involve contradictions and absurdities. Two other theories—one Pythagorean, that the soul is a " harmony " or adjustment of physical constituents, and the other Platonic, that soul pervades the material universe—are also discredited. The ground is now cleared for Aristotle's own treatment of the subject, which may be summarized as follows.

Relation of Soul to Body. Book II begins with a general definition of soul as the first actualization of a potentially living organism. The definition is hardly satisfactory on Aristotle's own principles, since in a sense it contains the term to be defined, " soul " being virtually equivalent to " life " or " vital principle." Nevertheless it is highly significant. It clearly makes the soul incorporeal ; it states a relation between soul and body ; and it distinguishes organic from inorganic matter. The doctrine implied in the definition is this. All matter as found in nature has a definite potentiality, and some matter is capable of being vitalized as an organism. The specific form or actuality of this matter is vitality or " soul." In so far as the matter attains to this form, *i.e.*, has life in it, it is actualized in the first or widest sense ; but it is not completely actualized unless the organism is exercising its vital faculties.

Faculties of Soul. These are not separate " parts " of the soul ; they are logically distinguishable aspects of vitality, corresponding to different grades of existence or to different activities in the same grade. The Nutritive faculty (which enables an organism to feed, grow, and reproduce) is shared by all living things. The Sensitive is confined to animals and higher beings ; the Intellective to man " and anything higher than man." The other faculties—imaginative, appetitive, locomotive, etc.—are more irregularly distributed and resist tabulation ; they either fall under the sensitive or are shared by it with the intellective faculty.

Nutrition is not discussed at any length and needs no comment here.

Sensation.—The senses are in themselves mere potentialities. To be actualized they must be acted upon by sensible objects. This actualization (in which the subject's sensitivity and the object's sensibility become identified) is sensation. It requires a medium (*e.g.*, transparency, air, flesh). This medium

4

is activated by the sense-object and transmits a stimulus to the sense-organ, thereby assimilating it to the sense-object. Before stimulation the sense-organ must be in a neutral state, since any determinate quality would interfere with its receptivity. Every sense is therefore a sort of mean or balance between contraries ; which explains why sense is injured or even destroyed by over-stimulation.

Aristotle describes the operation of the several senses separately, but the details call for little comment, except that the medium-theory has to be forced somewhat to make it cover touch and taste. The treatment is intended to be comprehensive ; particular problems are left for discussion in the *De Sensu* or other sections of the *Parva Naturalia*.

Book III is chiefly concerned with the other vital faculties, but some accessories to the theory of sensation overflow into its opening chapters. The connecting link is formed by the problem : What is it that unifies (or distinguishes) the data of sense ?

Sensus Communis. The solution given is that there is a common sense-faculty (located in or near the heart : *De Iuventute* 469 a 10) which receives and co-ordinates the stimuli passed on to it from the various sense-organs. This same faculty also directly perceives the " common sensibles " (*i.e.*, those attributes, such as shape, size, number, etc., which are perceptible by more than one sense), among which Aristotle includes movement and time (erroneously, since perception of these involves calculation). It also accounts for our consciousness of sensation, and it is responsible for the process of imagination.

Imagination is distinct from sensation on the one hand and intellect on the other ; but it is closely bound up with both and forms the link between them.

5

It prolongs and records the former (thus making memory and recollection possible) ; it also supplies the latter with illustrations, as it were, to accompany the intelligible forms which are the objects of thought. It is moreover a prerequisite for Appetite (*q.v.*).

Thought. Thinking is a process analogous to sensation. Mind, though insusceptible to alteration, is receptive of form : *i.e.*, like sense, it is potentially the same as its object, and has no characteristic of its own except pure receptivity. When it has fully realized its capacity for thinking it attains to a state of intellectivity in which it can think " itself "—*i.e.,* any of the objects with which it has become identified in previous actualizations.

As in every other department of reality there are two factors, a passive or indeterminate and an active or formative, so Mind too must be partly passive, partly active. Active mind is separable from body, impassive, unmixed and divine. The precise implications of this doctrine of active mind have always been doubtful. On the one hand the relation of active to passive mind in the individual soul is not explained ; on the other it is not made clear whether the human active mind is identical with, derived from or merely analogous to the Divine Mind which is the First Mover or God (*Metaphysics* Λ vii., 1072 b 23 ff.).

Locomotion and Appetite. There remains the problem of how soul moves body. Locomotion in animals is caused by the appetitive faculty, which is found in all grades of life except plants and the lowest animals. Desirable objects (*viz.*, those which offer a " good " which is attainable by action) through imagination stimulate the appetite, and this in turn moves the joints of the animal. The material instrument by which it effects this motion is not explicitly

6

mentioned here. It is the σύμφυτον πνεῦμα or " connatural spirit," a kind of hot air (analogous to the fiery ether of the upper cosmos), which, passing along the blood-vessels, pervades the whole body and causes local changes of temperature with consequent expansion or contraction. The whole process is excellently analysed by Dr. A. L. Peck in Appendix B to his *Generation of Animals* in this series.

In rational beings there is a complication. Besides the purely animal craving for the immediate or apparent " good " apprehended by sensation through sensitive imagination, there is also the rational will which seeks the ultimate or real " good " apprehended by intellect through calculative imagination. Hence the frequent clash between will and appetite in the narrower sense of mere desire.

Aristotle does not seem to have been much interested in will and emotion. They receive some incidental discussion in the *Ethics* and *Rhetoric*, but nothing comparable to the detailed treatment which is given in the *Parva Naturalia* to some of the particular problems of sense and consciousness.

However, so far as the *De Anima* is concerned, he has upon the whole carried out his undertaking to give a tolerably coherent account of " soul " and its faculties in relation to body. It is easy, in the light of our fuller information, to gibe at the crudeness of his theories and the falsity of some of his assumptions. It would be fairer to recognize the skill with which he constructed a reasoned system out of wholly inadequate material, and to reflect that much later psychologists have frequently made far worse use of better opportunities.

ΑΡΙΣΤΟΤΕΛΟΥΣ
ΠΕΡΙ ΨΥΧΗΣ

Α

402 a   I. Τῶν καλῶν καὶ τιμίων τὴν εἴδησιν ὑπολαμβάνοντες, μᾶλλον δ' ἑτέραν ἑτέρας ἢ κατ' ἀκρίβειαν
ἢ τῷ βελτιόνων τε καὶ θαυμασιωτέρων εἶναι, δι'
ἀμφότερα ταῦτα τὴν τῆς ψυχῆς ἱστορίαν εὐλόγως
5 ἂν ἐν πρώτοις τιθείημεν. δοκεῖ δὲ καὶ πρὸς
ἀλήθειαν ἅπασαν ἡ γνῶσις αὐτῆς μεγάλα συμβάλ-
λεσθαι, μάλιστα δὲ πρὸς τὴν φύσιν· ἔστι γὰρ οἷον
ἀρχὴ τῶν ζῴων. ἐπιζητοῦμεν δὲ θεωρῆσαι καὶ
γνῶναι τήν τε φύσιν αὐτῆς καὶ τὴν οὐσίαν, εἶθ'
ὅσα συμβέβηκε περὶ αὐτήν· ὧν τὰ μὲν ἴδια πάθη
10 τῆς ψυχῆς εἶναι δοκεῖ, τὰ δὲ δι' ἐκείνην καὶ τοῖς
ζῴοις ὑπάρχειν. πάντῃ δὲ πάντως ἐστὶ τῶν χαλε-
πωτάτων λαβεῖν τινὰ πίστιν περὶ αὐτῆς. καὶ γὰρ
ὄντος κοινοῦ τοῦ ζητήματος πολλοῖς ἑτέροις, λέγω
δὲ τοῦ περὶ τὴν οὐσίαν καὶ τοῦ τί ἐστι, τάχ' ἂν
τῳ δόξειε μία τις εἶναι μέθοδος κατὰ πάντων περὶ
15 ὧν βουλόμεθα γνῶναι τὴν οὐσίαν, ὥσπερ καὶ τῶν
κατὰ συμβεβηκὸς ἰδίων ἀπόδειξις,[1] ὥστε ζητητέον

---

[1] ἀπόδειξιν SUWX, Bekker.

ἂν εἴη τὴν μέθοδον ταύτην. εἰ δὲ μή ἐστι μία
τις καὶ κοινὴ μέθοδος περὶ τὸ τί ἐστιν, ἔτι χαλεπώτερον γίνεται τὸ πραγματευθῆναι· δεήσει γὰρ
λαβεῖν περὶ ἕκαστον τίς ὁ τρόπος, ἐὰν δὲ φανερὸν
20 ᾖ πότερον ἀπόδειξίς τίς ἐστιν ἢ διαίρεσις ἢ καί
τις ἄλλη μέθοδος, ἔτι¹ πολλὰς ἀπορίας ἔχει καὶ
πλάνας, ἐκ τίνων δεῖ ζητεῖν· ἄλλαι γὰρ ἄλλων
ἀρχαί, καθάπερ ἀριθμῶν καὶ ἐπιπέδων.

Πρῶτον δ' ἴσως ἀναγκαῖον διελεῖν ἐν τίνι τῶν
γενῶν καὶ τί ἐστι, λέγω δὲ πότερον τόδε τι καὶ
25 οὐσία ἢ ποιὸν ἢ ποσὸν ἢ καί τις ἄλλη τῶν διαιρεθεισῶν κατηγοριῶν, ἔτι δὲ πότερον τῶν ἐν δυνάμει
ὄντων ἢ μᾶλλον ἐντελέχειά τις· διαφέρει γὰρ οὔ τι
402 b σμικρόν. σκεπτέον δὲ καὶ εἰ μεριστὴ ἢ ἀμερής,
καὶ πότερον ὁμοειδὴς ἅπασα ψυχὴ ἢ οὔ· εἰ δὲ μὴ
ὁμοειδής, πότερον εἴδει διαφέρουσιν ἢ γένει. νῦν
μὲν γὰρ οἱ λέγοντες καὶ ζητοῦντες περὶ ψυχῆς
5 περὶ τῆς ἀνθρωπίνης μόνης ἐοίκασιν ἐπισκοπεῖν.
εὐλαβητέον δ' ὅπως μὴ λανθάνῃ πότερον εἷς ὁ
λόγος αὐτῆς ἐστι, καθάπερ ζῴου, ἢ καθ' ἕκαστον
ἕτερος, οἷον ἵππου, κυνός, ἀνθρώπου, θεοῦ· τὸ δὲ
ζῷον τὸ καθόλου ἤτοι οὐθέν ἐστιν ἢ ὕστερον.
ὁμοίως δὲ κἂν εἴ τι κοινὸν ἄλλο κατηγοροῖτο. ἔτι
δ' εἰ μὴ πολλαὶ ψυχαὶ ἀλλὰ μόρια, πότερον δεῖ

---

¹ μέθοδος. ἔτι δὲ TUVW, Bekker.

ᵃ *Division.* Plato used and perhaps invented this method
of forming a concept. If, for instance, you divide the term
" living creature " into " footed " and " footless," and
" footed " again into " biped " and " quadruped," and so
on, you will after many such divisions and subdivisions
form a clear concept of the meaning of " living creature."
A. objects (*Post. An.* 11, c. 7, 92 b 5) to the method on several
grounds, one being that such division presupposes the
existence of the concept.

must be discovered ; but if there is no one common method of finding the essential nature, our handling of the subject becomes still more difficult. For we shall be obliged to establish the proper method in each individual case ; and even if it is patent whether syllogistic demonstration, or division,[a] or some other method is the right one, there is still room for confusion and error as to the premisses from which we must start the inquiry ; for the premisses of all subjects are not the same ; for instance those of arithmetic and those of plane geometry are different.

Perhaps our first business is to determine to which of the genera the soul belongs, and what it is ; I mean whether it is a particular thing, *i.e.*, a substance, or whether it is a quality, or quantity, or belongs to any other of our pre-established categories, and furthermore, whether it has potential or actual existence. For this makes no small difference. In the second place we must inquire whether it has parts or not, and whether every soul is of the same kind or not ; and if not, whether the difference is one of species or of genus. For speakers and inquirers about the soul seem to-day to confine their inquiries to the soul of man. But one must be careful not to evade the question whether one definition of " soul " is enough, as we can give one definition of " living creature," or whether there must be a different one in each case ; that is, one of the horse, one of the dog, one of man, and one of God, and whether the words " living creature " as a common term have no meaning, or logically come later. This question can of course be raised about any common term. Further, supposing that there are not several souls, but only parts of one, are we to inquire first

To what category does the soul belong?

11

10 ζητεῖν πρότερον τὴν ὅλην ψυχὴν ἢ τὰ μόρια.
χαλεπὸν δὲ καὶ τούτων διορίσαι ποῖα πέφυκεν
ἕτερα ἀλλήλων, καὶ πότερον τὰ μόρια χρὴ ζητεῖν
πρότερον ἢ τὰ ἔργα αὐτῶν, οἷον τὸ νοεῖν ἢ τὸν
νοῦν καὶ τὸ αἰσθάνεσθαι ἢ τὸ αἰσθητικόν· ὁμοίως
δὲ καὶ ἐπὶ τῶν ἄλλων. εἰ δὲ τὰ ἔργα πρότερον,
15 πάλιν ἄν τις ἀπορήσειεν εἰ τὰ ἀντικείμενα πρότερον
τούτων ζητητέον, οἷον τὸ αἰσθητὸν τοῦ αἰσθητικοῦ
καὶ τὸ νοητὸν τοῦ νοῦ.[1] ἔοικε δ' οὐ μόνον τὸ τί
ἐστι γνῶναι χρήσιμον εἶναι πρὸς τὸ θεωρῆσαι τὰς
αἰτίας τῶν συμβεβηκότων ταῖς οὐσίαις, ὥσπερ
ἐν τοῖς μαθήμασι τί τὸ εὐθὺ καὶ καμπύλον ἢ τί
20 γραμμὴ καὶ ἐπίπεδον πρὸς τὸ κατιδεῖν πόσαις
ὀρθαῖς αἱ τοῦ τριγώνου γωνίαι ἴσαι, ἀλλὰ καὶ
ἀνάπαλιν τὰ συμβεβηκότα συμβάλλεται μέγα μέρος
πρὸς τὸ εἰδέναι τὸ τί ἐστιν· ἐπειδὰν γὰρ ἔχωμεν
ἀποδιδόναι κατὰ τὴν φαντασίαν περὶ τῶν συμβεβηκότων, ἢ πάντων ἢ τῶν πλείστων, τότε καὶ
25 περὶ τῆς οὐσίας ἕξομεν λέγειν κάλλιστα· πάσης
γὰρ ἀποδείξεως ἀρχὴ τὸ τί ἐστιν, ὥστε καθ' ὅσους
403 a τῶν ὁρισμῶν μὴ συμβαίνει τὰ συμβεβηκότα γνωρίζειν, ἀλλὰ μηδ' εἰκάσαι περὶ αὐτῶν εὐμαρές, δῆλον
ὅτι διαλεκτικῶς εἴρηνται καὶ κενῶς ἅπαντες.

'Απορίαν δ' ἔχει καὶ τὰ πάθη τῆς ψυχῆς, πότερόν
ἐστι πάντα κοινὰ καὶ τοῦ ἔχοντος ἢ ἐστί τι καὶ

---

[1] νοῦ EVX, Biehl, Hicks : νοητοῦ S : νοητικοῦ cet.

into the soul as a whole or the several parts? Here and how shall we approach the analysis? again it is difficult to determine which parts differ in nature from one another, and whether we should inquire first into the parts of the soul, or their functions; for example, into the thinking or into that which thinks, into sensation or into that which feels; and a similar difficulty arises with all the other parts. If we are to take the functions first, a further problem may arise; whether we should consider the objects corresponding to them before the parts themselves —I mean, the thing felt before the part of the soul which feels it, and the thing thought before the mind which thinks it. A knowledge of a thing's essential nature is of course a valuable assistance towards the examination of the causes of its attributes; for instance, in mathematics, to know the meaning of " straight," " curved," " line," and " plane figure " helps to determine the number of right-angles to which the angles of a triangle are equal. But the converse is also true; the attributes contribute materially to the knowledge of what a thing is. For when we are in a position to expound all or most of the attributes as presented to us, we shall also be best qualified to speak about the essence. For the starting-point of every demonstration is the statement of the subject's essential nature, and definitions which do not enable us to know the attributes, or even to make a tolerable guess about them, are clearly laid down merely for argument's sake and are utterly valueless.

The affections of the soul present a further diffi- Relations of soul and body. culty—Are they all shared also by that which contains the soul, or is any of them peculiar to the soul

⁵ τῆς ψυχῆς ἴδιον αὐτῆς· τοῦτο γὰρ λαβεῖν μὲν
ἀναγκαῖον, οὐ ῥᾴδιον δέ. φαίνεται δὲ τῶν πλείστων
οὐθὲν ἄνευ σώματος πάσχειν οὐδὲ ποιεῖν, οἷον
ὀργίζεσθαι, θαρρεῖν, ἐπιθυμεῖν, ὅλως αἰσθάνεσθαι.
μάλιστα δ' ἔοικεν ἴδιον τὸ νοεῖν· εἰ δ' ἐστὶ καὶ
τοῦτο φαντασία τις ἢ μὴ ἄνευ φαντασίας, οὐκ
¹⁰ ἐνδέχοιτ' ἂν οὐδὲ τοῦτ' ἄνευ σώματος εἶναι. εἰ
μὲν οὖν ἐστί τι τῶν τῆς ψυχῆς ἔργων ἢ παθημάτων
ἴδιον, ἐνδέχοιτ' ἂν αὐτὴν χωρίζεσθαι· εἰ δὲ μη-
θέν ἐστιν ἴδιον αὐτῆς, οὐκ ἂν εἴη χωριστή, ἀλλὰ
καθάπερ τῷ εὐθεῖ, ᾗ εὐθύ, πολλὰ συμβαίνει, οἷον
¹⁵ ἅπτεσθαι τῆς χαλκῆς σφαίρας κατὰ στιγμήν, οὐ
μέντοι γ' ἅψεται οὕτω¹ χωρισθὲν τὸ εὐθύ· ἀχώριστον γάρ, εἴπερ ἀεὶ μετὰ σώματός τινός ἐστιν.
ἔοικε δὲ καὶ τὰ τῆς ψυχῆς πάθη πάντα εἶναι μετὰ
σώματος, θυμός, πραότης, φόβος, ἔλεος, θάρσος,
ἔτι χαρὰ καὶ τὸ φιλεῖν τε καὶ μισεῖν· ἅμα γὰρ
²⁰ τούτοις πάσχει τι τὸ σῶμα. μηνύει δὲ τὸ ποτὲ
μὲν ἰσχυρῶν καὶ ἐναργῶν παθημάτων συμβαινόντων
μηδὲν παροξύνεσθαι ἢ φοβεῖσθαι, ἐνίοτε δ' ὑπὸ
μικρῶν καὶ ἀμαυρῶν κινεῖσθαι, ὅταν ὀργᾷ τὸ σῶμα
καὶ οὕτως ἔχῃ ὥσπερ ὅταν ὀργίζηται. ἔτι δὲ
τοῦτο μᾶλλον φανερόν· μηθενὸς γὰρ φοβεροῦ συμβαίνοντος ἐν τοῖς πάθεσι γίνονται τοῖς τοῦ φοβου-
²⁵ μένου. εἰ δ' οὕτως ἔχει, δῆλον ὅτι τὰ πάθη λόγοι
ἔνυλοί εἰσιν, ὥστε οἱ ὅροι τοιοῦτοι· οἷον τὸ ὀργίζεσθαι κίνησίς τις τοῦ τοιουδὶ σώματος ἢ μέρους

---

¹ οὕτω E, Bonitz, Hicks : τούτου.

14

itself? This question must be faced, but its solution
is not easy. In most cases it seems that none of the
affections, whether active or passive, can exist apart
from the body. This applies to anger, courage,
desire and sensation generally, though possibly
thinking is an exception. But if this too is a kind
of imagination, or at least is dependent upon
imagination, even this cannot exist apart from the
body. If then any function or affection of the soul
is peculiar to it, it can be separated from the body;
but if there is nothing peculiar to the soul it cannot
be separated. In the same way there are many attributes belonging to the straight, *qua* straight, as, for
instance, that a straight line touches a bronze sphere
at a point, yet if separated, the straightness will not
so touch. It is in fact inseparable, if it is always
associated with some body. Probably all the affections of the soul are associated with the body—anger,
gentleness, fear, pity, courage and joy, as well as
loving and hating; for when they appear the body
is also affected. There is good evidence for this.
Sometimes no irritation or fear is expressed, though
the provocations are strong and obvious; and conversely, small and obscure causes produce movement,
when the body is disposed to anger, and when it is
in an angry mood. And here is a still more obvious
proof. There are times when men show all the
symptoms of fear without any cause of fear being
present. If this is the case, then clearly the affections
of the soul are formulae expressed in matter. Their
definitions therefore must be in harmony with this;
for instance, anger must be defined as a movement
of a body, or of a part or faculty of a body, in a particular state roused by such a cause, with such an

end in view. This at once makes it the business of the natural philosopher to inquire into the soul, either generally, or at least in this special aspect. But the natural philosopher and the logician will in every case offer different definitions, *e.g.*, in answer to the question what is anger. The latter will call it a craving for retaliation, or something of the sort ; the former will describe it as a surging of the blood and heat round the heart. The one is describing the matter, the other the form or formula of the essence. For what he states is the formula of the thing, and if it is to exist, it must appear in appropriate matter. To illustrate this : the formula of a house is a covering to protect from damage by wind, rain and heat. But another will mean by a house stones, bricks and timber ; and another again will mean the form expressed in these materials to achieve these objects. Now which of these is really the natural philosopher ? The man who ignores the formula and is only concerned with the matter, or the man who is only concerned with the formula ? Probably the man who bases his concept on both. What then are we to say of the other two ? Surely there is no one who treats of the affections of matter which are inseparable, or regarded as inseparable.[a] The natural philosopher's concern is with all the functions and affections of a given body, *i.e.*, of matter in a given state ; any attribute not of this kind is the business of another ; in some subjects it is the business of the expert, the carpenter, it may be, or the physician ; but inseparables in so far as they are not affections of the body in such a state, that is, in the abstract,

---

[a] *Sc.*, from their matter.

403 b

15 σώματος πάθη καὶ ἐξ ἀφαιρέσεως, ὁ μαθηματικός,
ᾗ δὲ κεχωρισμένα, ὁ πρῶτος φιλόσοφος.

Ἀλλ' ἐπανιτέον ὅθεν ὁ λόγος. ἐλέγομεν δ' ὅτι
τὰ πάθη τῆς ψυχῆς ἀχώριστα τῆς φυσικῆς ὕλης
τῶν ζῴων, ᾗ δὴ τοιαῦθ' ὑπάρχει, θυμὸς καὶ φόβος,
καὶ οὐχ ὥσπερ γραμμὴ καὶ ἐπίπεδον.

20 II. Ἐπισκοποῦντας δὲ περὶ ψυχῆς ἀναγκαῖον ἅμα
διαποροῦντας περὶ ὧν εὐπορεῖν δεῖ, προελθόντας
τὰς τῶν προτέρων δόξας συμπαραλαμβάνειν ὅσοι
τι περὶ αὐτῆς ἀπεφήναντο, ὅπως τὰ μὲν καλῶς
εἰρημένα λάβωμεν, εἰ δέ τι μὴ καλῶς, τοῦτ' εὐ-
25 λαβηθῶμεν. ἀρχὴ δὲ τῆς ζητήσεως προθέσθαι τὰ
μάλιστα δοκοῦνθ' ὑπάρχειν αὐτῇ κατὰ φύσιν. τὸ
ἔμψυχον δὴ τοῦ ἀψύχου δυοῖν μάλιστα διαφέρειν
δοκεῖ, κινήσει τε καὶ τῷ αἰσθάνεσθαι· παρειλή-
φαμεν δὲ καὶ παρὰ τῶν προγενεστέρων σχεδὸν
δύο ταῦτα περὶ ψυχῆς. φασὶ γὰρ ἔνιοι καὶ μάλιστα
καὶ πρώτως ψυχὴν εἶναι τὸ κινοῦν, οἰηθέντες δὲ
30 τὸ μὴ κινούμενον αὐτὸ μὴ ἐνδέχεσθαι κινεῖν ἕτερον,
τῶν κινουμένων τι τὴν ψυχὴν ὑπέλαβον εἶναι.

404 a ὅθεν Δημόκριτος μὲν πῦρ τι καὶ θερμόν φησιν
αὐτὴν εἶναι· ἀπείρων γὰρ ὄντων σχημάτων καὶ
ἀτόμων τὰ σφαιροειδῆ πῦρ καὶ ψυχὴν λέγει, οἷον
ἐν τῷ ἀέρι τὰ καλούμενα ξύσματα, ἃ φαίνεται ἐν
ταῖς διὰ τῶν θυρίδων ἀκτῖσιν, ὧν τὴν πανσπερμίαν

ᵃ *i.e.* the Metaphysician.

are the province of the mathematician, and in so far as they are separable are the sphere of the First Philosopher.[a]

But we must now return to the point from which our digression started. We were saying that the affections of the soul, such as anger and fear, are inseparable from the matter of living things in which their nature is manifested, and are not separable like a line or a plane.

II. In our inquiry about the soul we shall have to raise problems for which we must find a solution, and in our progress we must take with us for comparison the theories expounded by our predecessors, in order that we may adopt those which are well stated, and be on our guard against any which are unsatisfactory. But our inquiry must begin by laying down in advance those things which seem most certainly to belong to the soul by nature. There are two qualities in which that which has a soul seems to differ radically from that which has not ; these are movement and sensation. We have practically accepted these two distinguishing characteristics of the soul from our predecessors. Some say that capacity to produce movement is first and foremost the characteristic of the soul. But because they believe that nothing can produce movement which does not itself move, they have supposed that the soul is one of the things which move. On this supposition Democritus argues that the soul is a sort of fire or heat. For forms and atoms being countless, he calls the spherical ones fire and soul, and likens them to the (so-called) motes in the air, which can be seen in the sunbeams passing through our windows ; the aggregate of these particles he calls the elements of which all nature is

404 a

5 στοιχεῖα λέγει τῆς ὅλης φύσεως. ὁμοίως δὲ καὶ
Λεύκιππος. τούτων δὲ τὰ σφαιροειδῆ ψυχήν, διὰ
τὸ μάλιστα διὰ παντὸς δύνασθαι διαδύνειν τοὺς
τοιούτους ῥυσμούς, καὶ κινεῖν τὰ λοιπὰ κινούμενα
καὶ αὐτά, ὑπολαμβάνοντες τὴν ψυχὴν εἶναι τὸ
παρέχον τοῖς ζῴοις τὴν κίνησιν. διὸ καὶ τοῦ ζῆν
10 ὅρον εἶναι τὴν ἀναπνοήν· συνάγοντος γὰρ τοῦ
περιέχοντος τὰ σώματα, καὶ ἐκθλίβοντος τῶν
σχημάτων τὰ παρέχοντα τοῖς ζῴοις τὴν κίνησιν
διὰ τὸ μηδ' αὐτὰ ἠρεμεῖν μηδέποτε, βοήθειαν
γίγνεσθαι θύραθεν ἐπεισιόντων ἄλλων τοιούτων ἐν
τῷ ἀναπνεῖν· κωλύειν γὰρ αὐτὰ καὶ τὰ ἐνυπάρ-
15 χοντα ἐν τοῖς ζῴοις ἐκκρίνεσθαι, συνανείργοντα τὸ
συνάγον καὶ πηγνύον· καὶ ζῆν δὲ ἕως ἂν δύνωνται
τοῦτο ποιεῖν. ἔοικε δὲ καὶ τὸ παρὰ τῶν Πυθαγορείων λεγόμενον τὴν αὐτὴν ἔχειν διάνοιαν·
ἔφασαν γάρ τινες αὐτῶν ψυχὴν εἶναι τὰ ἐν τῷ ἀέρι
ξύσματα, οἱ δὲ τὸ ταῦτα κινοῦν. περὶ δὲ τούτων
20 εἴρηται, διότι συνεχῶς φαίνεται κινούμενα, κἂν ᾖ
νηνεμία παντελής. ἐπὶ ταὐτὸ δὲ φέρονται καὶ ὅσοι
λέγουσι τὴν ψυχὴν τὸ αὐτὸ κινοῦν· ἐοίκασι γὰρ
οὗτοι πάντες ὑπειληφέναι τὴν κίνησιν οἰκειότατον
εἶναι τῇ ψυχῇ, καὶ τὰ μὲν ἄλλα πάντα κινεῖσθαι
διὰ τὴν ψυχήν, ταύτην δ' ὑφ' ἑαυτῆς, διὰ τὸ μηθὲν
25 ὁρᾶν κινοῦν ὃ μὴ καὶ αὐτὸ κινεῖται.

Ὁμοίως δὲ καὶ Ἀναξαγόρας ψυχὴν εἶναι λέγει
τὴν κινοῦσαν, καὶ εἴ τις ἄλλος εἴρηκεν ὡς τὸ πᾶν

20

composed. And Leucippus adopts a similar position. It is the spherical atoms which they call the soul, because such shapes can most readily pass through anything, and can move other things by virtue of their own motion; for they suppose that the soul is that which imparts motion to living things. Hence they consider also that respiration is the essential condition of life; for the surrounding atmosphere exerts pressure upon bodies and thus forces out the atoms which produce movement in living things, because they themselves are never at rest. The resulting shortage is reinforced from outside, when other similar atoms enter in the act of breathing; for they prevent the atoms which are in the bodies at the time from escaping by checking the compressive and solidifying action of the surrounding atmosphere; and animals can live just as long as they are competent to do this. The theory handed down from the Pythagoreans seems to entail the same view; for some of them have declared that the soul is identical with the particles in the air, and others with what makes these particles move. These particles have found their place in the theory because they can be seen perpetually in motion even when the air is completely calm. Those who say that the soul is that which moves itself tend towards the same view. For they all seem to assume that movement is the distinctive characteristic of the soul, and that everything else owes its movement to the soul, which they suppose to be self-moved, because they see nothing producing movement which does not itself move.

In the same way Anaxagoras (and so too anyone Anaxagoras. else who has held that mind set everything in motion) says that the soul is the producer of movement,

21

though not quite as Democritus taught. The latter actually identified soul and mind; for he believed that truth is subjective. Hence he regards Homer's description of Hector in his swooning as " lying thinking other thoughts " as accurate.[a] He does not **then** employ the term mind as denoting a faculty concerned with the truth, but identifies the soul and the mind.

Anaxagoras is less precise in his dealing with the subject; for on many occasions he speaks of mind as responsible for what is right and correct, but at others he says that this is the soul : for mind he regards as existing in all living things, great and small, noble and base ; but mind in the sense of intelligence does not appear to belong to all living things alike, nor even to all men.

Those then who have interpreted the soul in terms of motion have regarded the soul as most capable of producing movement. But those who have referred it to cognition and perception regard the soul as the first beginning of all things—some regarding this first beginning as plural and some as singular. Empedocles, for instance, thought that the soul was composed of all the elements, and yet considered each of these to be a soul. He says : *Empedocles.*

> By Earth we see Earth, by Water Water,
> By Air the divine Air, by Fire destroying Fire,
> Love by Love, and Strife by bitter Strife.

In the same way, in the *Timaeus*, Plato constructs the soul out of the elements. For he maintains that " like " can only be known by " like," and that from these first beginnings grow the things which we perceive. A similar definition is laid down in his

404 b

20 λεγομένοις διωρίσθη, αὐτὸ μὲν τὸ ζῷον ἐξ αὐτῆς
τῆς τοῦ ἑνὸς ἰδέας καὶ τοῦ πρώτου μήκους καὶ
πλάτους καὶ βάθους, τὰ δ' ἄλλα ὁμοιοτρόπως.
ἔτι δὲ καὶ ἄλλως, νοῦν μὲν τὸ ἕν, ἐπιστήμην δὲ
τὰ δύο (μοναχῶς γὰρ ἐφ' ἕν), τὸν δὲ τοῦ ἐπιπέδου
ἀριθμὸν δόξαν, αἴσθησιν δὲ τὸν τοῦ στερεοῦ· οἱ
25 μὲν γὰρ ἀριθμοὶ τὰ εἴδη αὐτὰ καὶ αἱ ἀρχαὶ ἐλέγοντο,
εἰσὶ δ' ἐκ τῶν στοιχείων. κρίνεται δὲ τὰ πράγματα
τὰ μὲν νῷ, τὰ δ' ἐπιστήμῃ, τὰ δὲ δόξῃ, τὰ δ'
αἰσθήσει· εἴδη δ' οἱ ἀριθμοὶ οὗτοι τῶν πραγμάτων.

Ἐπεὶ δὲ καὶ κινητικὸν ἐδόκει ἡ ψυχὴ εἶναι καὶ
γνωριστικόν, οὕτως ἔνιοι συνέπλεξαν ἐξ ἀμφοῖν
30 ἀποφηνάμενοι τὴν ψυχὴν ἀριθμὸν κινοῦνθ' ἑαυτόν.
διαφέρονται δὲ περὶ τῶν ἀρχῶν, τίνες καὶ πόσαι,
μάλιστα μὲν οἱ σωματικὰς ποιοῦντες τοῖς ἀσω-
405 a μάτοις,[1] τούτοις δ' οἱ μίξαντες καὶ ἀπ' ἀμφοῖν τὰς
ἀρχὰς ἀποφηνάμενοι. διαφέρονται δὲ καὶ περὶ τὸ
πλῆθος· οἱ μὲν γὰρ μίαν οἱ δὲ πλείους λέγουσιν.
ἑπομένως δὲ τούτοις καὶ τὴν ψυχὴν ἀποδιδόασιν·

---

[1] τοῖς ἀσωμάτους comm. vett.: τὰς ἀσωμάτους X: τοῖς
ἀσωμάτοις vulgo.

---

^a This difficult passage gives Aristotle's interpretation of
Plato's theory of Ideas, as applied to the origin of the universe. We know of no treatise of Plato *About Philosophy*,
but tradition ascribes the reference to some lecture notes of
Plato to which Aristotle had access. According to this
theory Pure Knowledge has some object corresponding to it;
this object is not the world of Sense, but the world of Ideas.
All the sensible world is thus but an imperfect copy of this
world of Ideas. There is, for instance, in the world of Ideas
an Idea Beauty. Objects in the world of Sense are beautiful
only in so far as they are copies of this. But the world of
Ideas includes the Ideas of numbers and, according to the
theory Aristotle is discussing, from these are derived some

comments *About Philosophy*, where he maintains that the living universe is derived from the idea of the One and from the primary length, breadth and depth ; and everything else in the same way. But he also gives another account, that mind is One and knowledge Two (for there is only one straight line from one point to another) ; and the number of the plane (Three) is opinion, and the number of the cube (Four) is sensation. For numbers are alleged to be identical with the forms themselves and ultimate principles, but they are composed of the elements.[a] The sensible world is apprehended in some cases by mind, in others by knowledge, in others again by opinion, and in others by sensation ; and these numbers are the forms of things.

But since the soul appears to contain an element which produces movement and one which produces knowledge, so some thinkers have constructed it from both, explaining the soul as a number moving itself. But men differ about the first principles of things, both as to their nature and quantity, especially those who make them corporeal from those who make them incorporeal, and from both these differ those who combine the two and explain the ultimate principles as compounded of both. They differ again about the number, some alleging that there is one, and others more than one. The account they give of the soul in each case follows their conclusions ;

of our concepts. One was not considered by the Greeks as a number, but only as the fountain of all numbers. So the three dimensions are derived from the numbers 2, 3, and 4 respectively. As " like " is known by " like," there are similarly in the soul faculties corresponding to these. So mind corresponds to One, Knowledge to Two, and so on.

405 a

5 τό τε γὰρ κινητικὸν τὴν φύσιν τῶν πρώτων ὑπ-
ειλήφασιν, οὐκ ἀλόγως. ὅθεν ἔδοξέ τισι πῦρ εἶναι·
καὶ γὰρ τοῦτο λεπτομερέστατόν τε καὶ μάλιστα
τῶν στοιχείων ἀσώματον, ἔτι δὲ κινεῖταί τε καὶ
κινεῖ τὰ ἄλλα πρώτως. Δημόκριτος δὲ καὶ γλα-
φυρωτέρως εἴρηκεν ἀποφηνάμενος διὰ τί τούτων
ἑκάτερον· ψυχὴν μὲν γὰρ εἶναι ταυτὸ καὶ νοῦν,
10 τοῦτο δ' εἶναι τῶν πρώτων καὶ ἀδιαιρέτων σωμάτων, κινητικὸν δὲ διὰ μικρομέρειαν καὶ τὸ
σχῆμα· τῶν δὲ σχημάτων εὐκινητότατον τὸ σφαιρο-
ειδὲς λέγει· τοιοῦτον δ' εἶναι τόν τε νοῦν καὶ τὸ
πῦρ.

Ἀναξαγόρας δ' ἔοικε μὲν ἕτερον λέγειν ψυχήν
τε καὶ νοῦν, ὥσπερ εἴπομεν καὶ πρότερον, χρῆται
15 δ' ἀμφοῖν ὡς μιᾷ φύσει, πλὴν ἀρχήν γε τὸν νοῦν
τίθεται μάλιστα πάντων· μόνον γοῦν φησιν αὐτὸν
τῶν ὄντων ἁπλοῦν εἶναι καὶ ἀμιγῆ τε καὶ καθαρόν.
ἀποδίδωσι δ' ἄμφω τῇ αὐτῇ ἀρχῇ, τό τε γινώσκειν
καὶ τὸ κινεῖν, λέγων νοῦν κινῆσαι τὸ πᾶν. ἔοικε
20 δὲ καὶ Θαλῆς ἐξ ὧν ἀπομνημονεύουσι κινητικόν τι
τὴν ψυχὴν ὑπολαβεῖν, εἴπερ τὸν λίθον ἔφη ψυχὴν
ἔχειν, ὅτι τὸν σίδηρον κινεῖ. Διογένης δ' ὥσπερ
καὶ ἕτεροί τινες ἀέρα,[a] τοῦτον οἰηθεὶς πάντων
λεπτομερέστατον εἶναι καὶ ἀρχήν· καὶ διὰ τοῦτο
γινώσκειν τε καὶ κινεῖν τὴν ψυχήν, ᾗ μὲν πρῶτόν
ἐστι, καὶ ἐκ τούτου τὰ λοιπά, γινώσκειν, ᾗ δὲ
25 λεπτότατον, κινητικὸν εἶναι. καὶ Ἡράκλειτος δὲ
τὴν ἀρχὴν εἶναί φησι ψυχήν, εἴπερ τὴν ἀναθυμίασιν,

---

[a] *i.e.* " the Magnesian stone," or, as we call it, the magnet.

for they consider that which by its own nature produces movement to be a primary reality ; which is not unreasonable. And so some have thought the soul to be fire ; for this is composed of the finest particles, and of all the elements is the nearest to incorporeal, and it also in a primary sense moves and causes movement in other things. Democritus has explained with greater precision why each of these two things is so ; for he identifies the soul and the mind. This, he says, consists of primary and indivisible bodies, and its power of producing movement is due to the smallness of its parts, and its shape ; for he calls the spherical the most easily moved of all shapes ; and this characteristic is shared by mind and fire. Democritus.

Anaxagoras indeed seems to regard soul and mind as different, as we have said before, but he treats them both as of one nature, except that he regards mind as above all things the ultimate principle ; at any rate, he speaks of it as the only existing thing which is simple, unmixed, and pure. But he assigns both the power of knowing and of moving to the same principle when he says that mind set everything moving. Thales, too, to judge from what is recorded of his views, seems to suppose that the soul is in a sense the cause of movement, since he says that a stone [a] has a soul because it causes movement to iron. Diogenes and some others think that the soul is air, regarding this as composed of the finest particles, and as an ultimate principle ; for this reason he believes that the soul both knows, and causes movement ; it knows because it is primary and from it all else comes ; it causes movement because of its extreme tenuity. Heracleitus also calls the first principle soul, as the emanation from which Anaxagoras. Thales. Diogenes. Heracleitus.

he constructs all other things ; it is most incorporeal and in ceaseless flux : he, like many others, supposed that a thing moving can only be known by something which moves, and that all that exists is in motion. Alcmaeon's suppositions about the soul are Alcmaeon. somewhat similar to these ; for he says it is immortal, because it resembles immortal things, and that this characteristic is due to its perpetual motion ; for things divine, the moon, the sun, the stars, and the whole heavens, are in a state of perpetual motion. Some of the less exact thinkers, like Hippo, have Hippo. declared the soul to be water. This belief seems to arise from the fact that the seed of all animals is moist. For he rebuts those who say that the soul is blood, on the ground that the seed is not blood ; and seed, he says, is primary soul. Others, like Critias, have Critias. imagined the soul to be blood, because they have supposed that sensation is the peculiar characteristic of the soul, and that this is due to the nature of blood. In fact each of the elements in turn has found a supporter, except earth ; but this no one has suggested except in so far as one [a] has said that the soul is composed of, or is identical with, all the elements.

But all, or almost all, distinguish the soul by three of its attributes, movement, sensation, and incorporeality ; and each of these is referred back to the first principles. So those who define it by the power of knowing describe it as an element, or as derived from the elements, all arguing with one [b] exception on similar lines ; for they say that " like " is known by " like " ; for since everything is known by the soul, they construct it of all the principles. Those, then, who allege that there is only one cause, and but one

405 b

ψυχὴν ἓν τιθέασιν, οἷον πῦρ ἢ ἀέρα· οἱ δὲ πλείους
λέγοντες τὰς ἀρχὰς καὶ τὴν ψυχὴν πλείω ποιοῦσιν.

20 Ἀναξαγόρας δὲ μόνος ἀπαθῆ φησὶν εἶναι τὸν νοῦν,
καὶ κοινὸν οὐθὲν οὐθενὶ τῶν ἄλλων ἔχειν. τοιοῦτος
δ' ὢν πῶς γνωριεῖ καὶ διὰ τίν' αἰτίαν, οὔτ' ἐκεῖνος
εἴρηκεν οὔτ' ἐκ τῶν εἰρημένων συμφανές ἐστιν.
ὅσοι δ' ἐναντιώσεις ποιοῦσιν ἐν ταῖς ἀρχαῖς, καὶ
τὴν ψυχὴν ἐκ τῶν ἐναντίων συνιστᾶσιν· οἱ δὲ

25 θάτερον τῶν ἐναντίων, οἷον θερμὸν ἢ ψυχρὸν ἤ τι
τοιοῦτον ἄλλο, καὶ τὴν ψυχὴν ὁμοίως ἕν τι τούτων
τιθέασιν. διὸ καὶ τοῖς ὀνόμασιν ἀκολουθοῦσιν, οἱ
μὲν τὸ θερμὸν λέγοντες, ὅτι διὰ τοῦτο καὶ τὸ ζῆν
ὠνόμασται, οἱ δὲ τὸ ψυχρὸν διὰ τὴν ἀναπνοὴν καὶ
τὴν κατάψυξιν καλεῖσθαι ψυχήν. τὰ μὲν οὖν παρα-

30 δεδομένα περὶ ψυχῆς, καὶ δι' ἃς αἰτίας λέγουσιν
οὕτω, ταῦτ' ἐστίν.

III. Ἐπισκεπτέον δὲ πρῶτον μὲν περὶ κινήσεως·
ἴσως γὰρ οὐ μόνον ψεῦδός ἐστι τὸ τὴν οὐσίαν

406 a αὐτῆς τοιαύτην εἶναι οἵαν φασὶν οἱ λέγοντες ψυχὴν
εἶναι τὸ κινοῦν ἑαυτὸ ἢ δυνάμενον κινεῖν, ἀλλ' ἔν
τι τῶν ἀδυνάτων τὸ ὑπάρχειν αὐτῇ κίνησιν. ὅτι
μὲν οὖν οὐκ ἀναγκαῖον τὸ κινοῦν καὶ αὐτὸ κινεῖσθαι,
πρότερον εἴρηται· διχῶς δὲ κινουμένου παντός (ἢ

5 γὰρ καθ' ἕτερον ἢ καθ' αὑτό· καθ' ἕτερον δὲ
λέγομεν, ὅσα κινεῖται τῷ ἐν κινουμένῳ εἶναι, οἷον
πλωτῆρες· οὐ γὰρ ὁμοίως κινοῦνται τῷ πλοίῳ· τὸ
μὲν γὰρ καθ' αὑτὸ κινεῖται, οἱ δὲ τῷ ἐν κινουμένῳ

element, also make the soul one element, such as
fire or air ; but those who believe in more than one
first principle make the soul also plural.   Anaxagoras is alone in his belief that mind cannot be acted
upon, and that it has nothing in common with anything else.   But how mind, being thus constructed,
can ever recognize anything, and by what agency,
he does not explain, nor is it clear from his expressed
views.   All those who assume pairs of contrary
opposites among their first principles also construct
the soul from contraries ; while those who suppose
the first principle to be one of a pair of contraries such
as hot and cold or the like, similarly also suppose
the soul to be one of these.   Thus they appeal to
etymology also ; those who identify the soul with
heat derive ζῆν (to live) from ζεῖν (to boil), but those
who identify it with cold maintains that soul (ψυχή)
is so called after the cooling process (κατάψυξις) associated with respiration.   These, then, are the traditional views about the soul and the grounds upon
which they are held.

III. In the first place we must investigate the _{Does the}
question of movement. For perhaps it is not merely ^{soul move?}
untrue that the essence of the soul is such as those
describe it to be who say that the soul moves or can
move itself, but it may be quite impossible that movement should be characteristic of the soul at all.   We
have said before that it is not necessary that that
which produces movement should itself move.   But
everything may be moved in two senses (directly and
indirectly. We call movement indirect, when a thing
moves because it is in something which moves ; for
instance the passengers in a ship. For they do not
move in the same sense as the ship moves ; for the
ship moves directly, but they move only by being in

εἶναι. δῆλον δ' ἐπὶ τῶν μορίων· οἰκεία μὲν γάρ
ἐστι κίνησις ποδῶν βάδισις, αὕτη δὲ καὶ ἀνθρώπων·
10 οὐχ ὑπάρχει δὲ τοῖς πλωτῆρσι τότε), διχῶς δὲ
λεγομένου τοῦ κινεῖσθαι, νῦν ἐπισκοποῦμεν περὶ
τῆς ψυχῆς εἰ καθ' αὑτὴν κινεῖται καὶ μετέχει
κινήσεως.

Τεσσάρων δὲ κινήσεων οὐσῶν, φορᾶς ἀλλοιώσεως
φθίσεως αὐξήσεως, ἢ μίαν τούτων κινοῖτ' ἂν ἢ
πλείους ἢ πάσας. εἰ δὲ κινεῖται μὴ κατὰ συμ-
15 βεβηκός, φύσει ἂν ὑπάρχοι κίνησις αὐτῇ· εἰ δὲ
τοῦτο, καὶ τόπος· πᾶσαι γὰρ αἱ λεχθεῖσαι κινήσεις
ἐν τόπῳ. εἰ δ' ἐστὶν ἡ οὐσία τῆς ψυχῆς τὸ κινεῖν
ἑαυτήν, οὐ κατὰ συμβεβηκὸς αὐτῇ τὸ κινεῖσθαι
ὑπάρξει, ὥσπερ τῷ λευκῷ ἢ τῷ τριπήχει· κινεῖται
20 γὰρ καὶ ταῦτα, ἀλλὰ κατὰ συμβεβηκός· ᾧ γὰρ
ὑπάρχουσιν, ἐκεῖνο κινεῖται, τὸ σῶμα. διὸ καὶ οὐκ
ἔστι τόπος αὐτῶν· τῆς δὲ ψυχῆς ἔσται, εἴπερ φύσει
κινήσεως μετέχει.

Ἔτι δ' εἰ φύσει κινεῖται, κἂν βίᾳ κινηθείη· κἂν
εἰ βίᾳ, καὶ φύσει. τὸν αὐτὸν δὲ τρόπον ἔχει καὶ
περὶ ἠρεμίας· εἰς ὃ γὰρ κινεῖται φύσει, καὶ ἠρεμεῖ
25 ἐν τούτῳ φύσει· ὁμοίως δὲ καὶ εἰς ὃ κινεῖται βίᾳ,
καὶ ἠρεμεῖ ἐν τούτῳ βίᾳ. ποῖαι δὲ βίαιοι τῆς
ψυχῆς κινήσεις ἔσονται καὶ ἠρεμίαι, οὐδὲ πλάττειν

something which moves. And this becomes obvious if we consider the parts of the body. For the movement proper to the feet is walking, that is the movement natural to human beings ; and at the moment the passengers are not exhibiting this kind of motion). Movement then having two different senses, we are at present inquiring whether the soul moves and has a share in direct movement.

Now there are four kinds of movement : (1) change of position, (2) change of state, (3) decay and (4) growth ; if then the soul moves, it must have one, or more than one, or all of these kinds of movement. And if the movement of the soul is not accidental, then movement must belong to it by nature ; if this is so, it must have position in space, for all the kinds of movement mentioned are in space. But if it is the essence of the soul to move itself, then movement will not belong to it by accident, as it does for instance to the quality of whiteness, or to a length of three cubits ; these are liable to be moved, but only accidentally, and merely because the body to which they belong is moved. For this reason they have no position in space. But the soul must have position in space, if of its own nature it participates in movement. *Four kinds of movement.*

Again, if it moves naturally, it must also be movable by force ; and conversely if it is movable by force, then it must also move naturally. And the same thing is true about its rest ; for it comes to rest by nature at the point to which it is moved by nature ; and similarly it rests by force in the place to which it is moved by force. But what these enforced movements of the soul and enforced rests can be is not easy to explain, even if we are prepared to allow our *Difficulties of attributing movement to the soul.*

βουλομένοις ῥᾴδιον ἀποδοῦναι. ἔτι δ' εἰ μὲν ἄνω
κινήσεται, πῦρ ἔσται, εἰ δὲ κάτω, γῆ· τούτων γὰρ
τῶν σωμάτων αἱ κινήσεις αὗται. ὁ δ' αὐτὸς λόγος
30 καὶ περὶ τῶν μεταξύ.

   Ἔτι δ' ἐπεὶ φαίνεται κινοῦσα τὸ σῶμα, ταύτας
εὔλογον κινεῖν τὰς κινήσεις ἃς καὶ αὐτὴ κινεῖται.
εἰ δὲ τοῦτο, καὶ ἀντιστρέψασιν εἰπεῖν ἀληθὲς ὅτι
406 b ἦν τὸ σῶμα κινεῖται, ταύτην καὶ αὐτή. τὸ δὲ
σῶμα κινεῖται φορᾷ· ὥστε καὶ ἡ ψυχὴ μεταβάλλοι
ἂν κατὰ τὸ σῶμα ἢ ὅλη ἢ κατὰ μόρια μεθισταμένη.
εἰ δὲ τοῦτ' ἐνδέχεται, καὶ ἐξελθοῦσαν εἰσιέναι
5 πάλιν ἐνδέχοιτ' ἄν· τούτῳ δ' ἔποιτ' ἂν τὸ ἀν-
ίστασθαι τὰ τεθνεῶτα τῶν ζῴων.

   Τὴν δὲ κατὰ συμβεβηκὸς κίνησιν κἂν ὑφ' ἑτέρου
κινοῖτο· ὠσθείη γὰρ ἂν βίᾳ τὸ ζῷον. οὐ δεῖ δὲ ᾧ
τὸ ὑφ' ἑαυτοῦ κινεῖσθαι ἐν τῇ οὐσίᾳ, τοῦθ' ὑπ'
ἄλλου κινεῖσθαι, πλὴν εἰ μὴ κατὰ συμβεβηκός,
ὥσπερ οὐδὲ τὸ καθ' αὑτὸ ἀγαθὸν ἢ δι' αὑτό, τὸ
10 μὲν δι' ἄλλο εἶναι, τὸ δ' ἑτέρου ἕνεκεν. τὴν δὲ
ψυχὴν μάλιστα φαίη τις ἂν ὑπὸ τῶν αἰσθητῶν
κινεῖσθαι, εἴπερ κινεῖται.

   Ἀλλὰ μὴν καὶ εἰ κινεῖ γε αὐτὴ αὑτήν, καὶ αὐτὴ
κινοῖτ' ἄν, ὥστ' εἰ πᾶσα κίνησις ἔκστασίς ἐστι
τοῦ κινουμένου ᾗ κινεῖται, καὶ ἡ ψυχὴ ἐξίσταιτ'
ἂν ἐκ τῆς οὐσίας, εἰ μὴ κατὰ συμβεβηκὸς αὑτὴν

Many of the things we want we want for the sake of
something else ; *e.g.* we may want money for the sake of
health, health for the sake of our work in the world, and so
on.   But ultimately we must reach a " good " which we want

fancies free play. Again, if the soul moves upwards it will be fire, and if downwards, earth; for these two movements belong respectively to these two bodies; and the same argument will apply to movements intermediate between " up " and " down."

Moreover, since the soul can be seen to move the body, it is reasonable to suppose that it imparts to it the same movements that it has itself; and if this is so, then it is true to assert conversely that the soul has the same movements as the body. Now the body moves by change of position; and therefore the soul must change position in the same manner as the body, either as a complete whole or in respect of its parts. But, if this is possible, it would also be possible for a soul which has left the body to enter in again; and upon this would follow the possibility of resurrection for animals which are dead.

Indirect movement of the soul may indeed be caused by something external to it; the living creature may be pushed by force. But that which has self-movement as part of its essence cannot be moved by anything else except incidentally: just as that which is good in itself is not good because of anything else, and that which is good for its own sake is not good for the sake of anything else.[a] But one would be inclined to assert that the soul, if it is moved at all, is most likely to be moved by sensible objects.

Moreover, if the soul moves itself, it is also itself moved, so that, if all movement is a displacement of that which is moved *qua* moved, then the soul must depart from its essential nature, if it does not move

for its own sake. Similarly the essential self-movement of the soul is not to be explained as imparted indirectly from without.

15 κινεῖ, ἀλλ' ἔστιν ἡ κίνησις τῆς οὐσίας αὐτῆς καθ'
αὑτήν.

Ἔνιοι δὲ καὶ κινεῖν φασὶ τὴν ψυχὴν τὸ σῶμα ἐν
ᾧ ἐστὶν ὡς αὐτὴ κινεῖται, οἷον Δημόκριτος, παραπλησίως λέγων Φιλίππῳ τῷ κωμῳδοδιδασκάλῳ·
φησὶ γὰρ τὸν Δαίδαλον κινουμένην ποιῆσαι τὴν
ξυλίνην Ἀφροδίτην, ἐγχέαντ' ἄργυρον χυτόν.
20 ὁμοίως δὲ καὶ Δημόκριτος λέγει· κινουμένας γάρ
φησι τὰς ἀδιαιρέτους σφαίρας διὰ τὸ πεφυκέναι
μηδέποτε μένειν, συνεφέλκειν καὶ κινεῖν τὸ σῶμα
πᾶν. ἡμεῖς δ' ἐρωτήσομεν εἰ καὶ ἠρέμησιν ποιεῖ
ταὐτὰ ταῦτα. πῶς δὲ ποιήσει, χαλεπὸν ἢ καὶ
ἀδύνατον εἰπεῖν. ὅλως δ' οὐχ οὕτω φαίνεται
25 κινεῖν ἡ ψυχὴ τὸ ζῷον, ἀλλὰ διὰ προαιρέσεώς τινος
καὶ νοήσεως.

Τὸν αὐτὸν δὲ τρόπον καὶ ὁ Τίμαιος φυσιολογεῖ
τὴν ψυχὴν κινεῖν τὸ σῶμα· τῷ γὰρ κινεῖσθαι αὐτὴν
καὶ τὸ σῶμα κινεῖν διὰ τὸ συμπεπλέχθαι πρὸς
αὐτό. συνεστηκυῖαν γὰρ ἐκ τῶν στοιχείων, καὶ
μεμερισμένην κατὰ τοὺς ἁρμονικοὺς ἀριθμούς,
30 ὅπως αἴσθησίν τε σύμφυτον ἁρμονίας ἔχῃ καὶ τὸ
πᾶν φέρηται συμφώνους φοράς, τὴν εὐθυωρίαν εἰς
κύκλον κατέκαμψεν· καὶ διελὼν ἐκ τοῦ ἑνὸς δύο
407 a κύκλους δισσαχῇ συνημμένους πάλιν τὸν ἕνα
διεῖλεν εἰς ἑπτὰ κύκλους, ὡς οὔσας τὰς τοῦ οὐρανοῦ
φορὰς τὰς τῆς ψυχῆς κινήσεις.

Πρῶτον μὲν οὖν οὐ καλῶς τὸ λέγειν τὴν ψυχὴν
μέγεθος εἶναι· τὴν γὰρ τοῦ παντὸς δῆλον ὅτι
τοιαύτην εἶναι βούλεται οἷόν ποτ' ἐστὶν ὁ καλού-

---

[a] The point of the comparison is that both offer a purely
external and mechanical explanation of movement.

itself accidentally, but movement is part of its very essence.

Some say that the soul moves its body exactly as it is moved itself. Such is the view of Democritus, arguing in the vein of Philippus the comic dramatist; for he tells us that Daedalus made his wooden Aphrodite move by pouring in quicksilver.[a] Democritus speaks in a similar strain; for he says that the spherical atoms, as they move because it is their nature never to remain still, draw the whole body with them and so move it. But we shall ask whether these same atoms also produce rest. How they can do so, it is difficult, if not impossible, to say. In general the living creature does not appear to be moved by the soul in this way, but by some act of mind or will.

In the same way Plato's *Timaeus* [b] also gives a physical account of how the soul moves the body; he thinks that the soul moves the body by its own movement, owing to their intimate inter-connexion. For first the Creator fashioned it out of all the elements, and divided it according to the harmonic ratios, in order that it might have innate perception of harmony and the universe might move by harmonic movements; then he bent the straight line into the form of a circle, and, having divided the one circle into two, meeting at two points, he again divided one of these into seven. Thus Plato identifies the movements of the soul with the spatial movements of the heavenly bodies. *Plato's view that the soul causes movement of the body.*

(1) Now to say that the soul is a spatial magnitude is unsound; for he clearly means " the soul of the world " to be some such thing as what is called mind; *Objections to Plato's view.*

---

[b] Plato, *Timaeus*, pp. 33 *sqq.*

407 a

5 μενος νοῦς· οὐ γὰρ δὴ οἷόν γ᾽ ἡ αἰσθητική, οὐδ᾽
οἷον ἡ ἐπιθυμητική· τούτων γὰρ ἡ κίνησις οὐ
κυκλοφορία. ὁ δὲ νοῦς εἷς καὶ συνεχής, ὥσπερ
καὶ ἡ νόησις· ἡ δὲ νόησις τὰ νοήματα· ταῦτα δὲ
τῷ ἐφεξῆς ἕν, ὡς ἀριθμός, ἀλλ᾽ οὐχ ὡς τὸ μέγεθος.
διόπερ οὐδ᾽ ὁ νοῦς οὕτω συνεχής, ἀλλ᾽ ἤτοι ἀμερὴς
10 ἢ οὐχ ὡς μέγεθός τι συνεχής· πῶς γὰρ δὴ καὶ
νοήσει μέγεθος ὤν; πότερον[1] ὁτῳοῦν μορίῳ τῶν
αὑτοῦ; μορίων δ᾽ ἤτοι κατὰ μέγεθος ἢ κατὰ στιγμήν, εἰ δεῖ καὶ τοῦτο μόριον εἰπεῖν. εἰ μὲν οὖν
κατὰ στιγμήν, αὗται δ᾽ ἄπειροι, δῆλον ὡς οὐδέποτε
διέξεισιν, εἰ δὲ κατὰ μέγεθος, πολλάκις ἢ ἀπειράκις
15 νοήσει τὸ αὐτό. φαίνεται δὲ καὶ ἅπαξ ἐνδεχόμενον.
εἰ δ᾽ ἱκανὸν θιγεῖν ὁτῳοῦν τῶν μορίων, τί δεῖ
κύκλῳ κινεῖσθαι ἢ καὶ ὅλως μέγεθος ἔχειν; εἰ δ᾽
ἀναγκαῖον νοῆσαι τῷ ὅλῳ κύκλῳ θιγόντα, τίς ἐστιν
ἡ τοῖς μορίοις θίξις; ἔτι δὲ πῶς νοήσει τὸ μεριστὸν
ἀμερεῖ καὶ τὸ ἀμερὲς μεριστῷ; ἀναγκαῖον δὲ τὸν
20 νοῦν εἶναι τὸν κύκλον τοῦτον. νοῦ μὲν γὰρ κίνησις
νόησις, κύκλου δὲ περιφορά. εἰ οὖν ἡ νόησις περιφορά, καὶ νοῦς ἂν εἴη ὁ κύκλος, οὗ ἡ τοιαύτη περιφορὰ [νόησις].[2] ἀεὶ δὲ δὴ τί νοήσει;[3] δεῖ γάρ, εἴπερ
ἀΐδιος ἡ περιφορά· τῶν μὲν γὰρ πρακτικῶν νοήσεων
25 ἔστι πέρατα (πᾶσαι γὰρ ἑτέρου χάριν), αἱ δὲ θεωρη-
τικαὶ τοῖς λόγοις ὁμοίως ὁρίζονται· λόγος δὲ πᾶς

---

[1] ὤν ; πότερον ὁτῳοῦν τῶν μορίων E[1], Biehl : ὧν ὁτῳοῦν τῶν
μορίων Bekker.
[2] Torstrik.
[3] δὴ τί νοήσει ; Simplicius, Torstrik, Biehl, Rodier : δή τι
νοήσει·

it is nothing like either the perceptive or desiderative faculty ; for their movements are not circular. But mind is one and continuous in the same sense as the process of thinking ; thinking consists of thoughts. But the unity of these is one of succession, like that of numbers, whereas the unity of spatial magnitudes is not. So also the mind is not continuous in this sense, but it either has no parts, or at any rate is not continuous as a magnitude. For, if it is a magnitude, how can it think ? With any one of its parts indifferently ? The parts must be regarded either as magnitudes or as points, if one can call a point a part. In the latter case, since the points are infinite in number, the mind can obviously never exhaust them ; in the former, it will think the same thoughts very many or an infinite number of times. But it is clear that it is also capable of thinking a thought once only. (2) If it is sufficient for it to touch with any one of its parts, why should it move in a circle, or have magnitude at all ? But if it can only think when its whole circle is in contact, what does the contact of its parts mean ? (3) Again, how can it think that which has parts with that which has not, or that which has not with that which has ? The mind must be identical with this circle ; for the movement of the mind is thinking, and the movement of a circle is revolution. If then thinking is revolution, then the circle whose revolution is of this kind must be mind. But what can it be which mind always thinks ?—as it must if the revolution is eternal. All practical thinking has limits (for it always has an object in view), and speculation is bounded like the verbal formulae which express it. Every such formula is a definition or a demonstration.

407 a

ὁρισμὸς ἢ ἀπόδειξις· αἱ δ' ἀποδείξεις καὶ ἀπ'
ἀρχῆς, καὶ ἔχουσί πως τέλος τὸν συλλογισμὸν ἢ
τὸ συμπέρασμα· εἰ δὲ μὴ περατοῦνται, ἀλλ' οὐκ
ἀνακάμπτουσί γε πάλιν ἐπ' ἀρχήν, προσλαμβά-
νουσαι δ' ἀεὶ μέσον καὶ ἄκρον εὐθυποροῦσιν· ἡ
30 δὲ περιφορὰ πάλιν ἐπ' ἀρχὴν ἀνακάμπτει. οἱ δ'
ὁρισμοὶ πάντες πεπερασμένοι. ἔτι εἰ ἡ αὐτὴ περιφορὰ πολλάκις, δεήσει πολλάκις νοεῖν τὸ αὐτό.
ἔτι δ' ἡ νόησις ἔοικεν ἠρεμήσει τινὶ καὶ ἐπιστάσει
μᾶλλον ἢ κινήσει· τὸν αὐτὸν δὲ τρόπον καὶ ὁ
συλλογισμός. ἀλλὰ μὴν οὐδὲ μακάριόν γε τὸ μὴ
407 b ῥᾴδιον ἀλλὰ βίαιον· εἰ δ' ἐστὶν ἡ κίνησις αὐτῆς μὴ
οὐσία, παρὰ φύσιν ἂν κινοῖτο. ἐπίπονον δὲ καὶ τὸ
μεμῖχθαι τῷ σώματι μὴ δυνάμενον ἀπολυθῆναι, καὶ
προσέτι φευκτόν, εἴπερ βέλτιον τῷ νῷ μὴ μετὰ
5 σώματος εἶναι, καθάπερ εἴωθέ τε λέγεσθαι καὶ
πολλοῖς συνδοκεῖ. ἄδηλος δὲ καὶ τοῦ κύκλῳ
φέρεσθαι τὸν οὐρανὸν ἡ αἰτία· οὔτε γὰρ τῆς ψυχῆς
ἡ οὐσία αἰτία τοῦ κύκλῳ φέρεσθαι, ἀλλὰ κατὰ
συμβεβηκὸς οὕτω κινεῖται, οὔτε τὸ σῶμα αἴτιον,
ἀλλ' ἡ ψυχὴ μᾶλλον ἐκείνῳ. ἀλλὰ μὴν οὐδ' ὅτι
10 βέλτιον λέγεται· καίτοι γ' ἐχρῆν διὰ τοῦτο τὸν
θεὸν κύκλῳ ποιεῖν φέρεσθαι τὴν ψυχήν, ὅτι βέλτιον
αὐτῇ τὸ κινεῖσθαι τοῦ μένειν, κινεῖσθαι δ' οὕτως
ἢ ἄλλως. ἐπεὶ δ' ἐστὶν ἡ τοιαύτη σκέψις ἑτέρων
λόγων οἰκειοτέρα, ταύτην μὲν ἀφῶμεν τὸ νῦν.

Ἐκεῖνο δὲ ἄτοπον συμβαίνει καὶ τούτῳ τῷ λόγῳ
15 καὶ τοῖς πλείστοις τῶν περὶ ψυχῆς· συνάπτουσι
γὰρ καὶ τιθέασιν εἰς σῶμα τὴν ψυχήν, οὐθὲν προσ-

Demonstrations both start from a beginning, and have in a sense an end, *viz.*, the inference or conclusion. Even if they do not arrive at a conclusion, at least they do not return again to the beginning, but advance in a straight line by means of additional middle or extreme terms. But circular movement is for ever returning to its starting-point. Definitions, too, are all finite. Again, if the same revolution recurs frequently, the mind must frequently think the same thing. (4) Again, thinking seems more like a state of rest or a halting than a movement ; and the same thing is true of the syllogism. (5) Furthermore, that which moves not easily but only by force cannot be happy ; and if the soul's movement is not part of its essence, it will be moved unnaturally. (6) Again, the inescapable association of the mind with the body would be wearisome ; such a conception must be rejected, if it is true that it is better for the mind to be without the body, as is usually said and widely accepted. (7) Again, the reason why the heavens should move in a circle is obscure. For the essence of the soul is not the cause of this circular movement—it only moves in this way by accident ; nor is the body ; on the contrary, the soul causes the body's movement. Nor is there any suggestion that this circular movement is better, and yet God should surely have made the soul move in a circle for this very reason, that movement is a better condition for it than rest, and this movement better than any other. But since this inquiry belongs more properly to another subject, let us leave it for the present.

But there is one absurd feature both in this argument, and in most of those about the soul. Men associate the soul with and place it in the body, with-

407 b

διορίσαντες διὰ τίν' αἰτίαν καὶ πῶς ἔχοντος τοῦ
σώματος. καίτοι δόξειεν ἂν τοῦτ' ἀναγκαῖον εἶναι·
διὰ γὰρ τὴν κοινωνίαν τὸ μὲν ποιεῖ τὸ δὲ πάσχει
καὶ τὸ μὲν κινεῖται τὸ δὲ κινεῖ, τούτων δ' οὐθὲν
20 ὑπάρχει πρὸς ἄλληλα τοῖς τυχοῦσιν. οἱ δὲ μόνον
ἐπιχειροῦσι λέγειν ποῖόν τι ἡ ψυχή, περὶ δὲ τοῦ
δεξομένου σώματος οὐθὲν ἔτι προσδιορίζουσιν,
ὥσπερ ἐνδεχόμενον κατὰ τοὺς Πυθαγορικοὺς
μύθους τὴν τυχοῦσαν ψυχὴν εἰς τὸ τυχὸν ἐνδύεσθαι
σῶμα· δοκεῖ γὰρ ἕκαστον ἴδιον ἔχειν εἶδος καὶ
μορφήν. παραπλήσιον δὲ λέγουσιν ὥσπερ εἴ τις
25 φαίη τὴν τεκτονικὴν εἰς αὐλοὺς ἐνδύεσθαι· δεῖ γὰρ
τὴν μὲν τέχνην χρῆσθαι τοῖς ὀργάνοις, τὴν δὲ
ψυχὴν τῷ σώματι.

IV. Καὶ ἄλλη δέ τις δόξα παραδέδοται περὶ ψυχῆς,
πιθανὴ μὲν πολλοῖς οὐδεμιᾶς ἧττον τῶν λεγομένων,
λόγους δ' ὥσπερ εὐθύνας δεδωκυῖα καὶ τοῖς ἐν
30 κοινῷ γινομένοις λόγοις· ἁρμονίαν γάρ τινα αὐτὴν
λέγουσι· καὶ γὰρ τὴν ἁρμονίαν κρᾶσιν καὶ σύνθεσιν
ἐναντίων εἶναι, καὶ τὸ σῶμα συγκεῖσθαι ἐξ ἐναντίων.

Καίτοι γε ἡ μὲν ἁρμονία λόγος τίς ἐστι τῶν
μιχθέντων ἢ σύνθεσις, τὴν δὲ ψυχὴν οὐδέτερον οἷόν
τ' εἶναι τούτων. ἔτι δὲ τὸ κινεῖν οὐκ ἔστιν ἁρ-
408 a μονίας, ψυχῇ δὲ πάντες ἀπονέμουσι τοῦτο μάλισθ'
ὡς εἰπεῖν. ἁρμόζει δὲ μᾶλλον καθ' ὑγιείας λέγειν
ἁρμονίαν, καὶ ὅλως τῶν σωματικῶν ἀρετῶν, ἢ
κατὰ ψυχῆς. φανερώτατον δ' εἴ τις ἀποδιδόναι
5 πειραθείη τὰ πάθη καὶ τὰ ἔργα τῆς ψυχῆς ἁρμονίᾳ
τινί· χαλεπὸν γὰρ ἐφαρμόζειν. ἔτι δ' εἰ λέγομεν[1]
τὴν ἁρμονίαν εἰς δύο ἀποβλέποντες, κυριώτατα μὲν

---

[1] λέγομεν E¹STVX : λέγοιμεν.

42

out specifying why this is so, and how the body is conditioned ; and yet this would seem to be essential. For it is by this association that the one acts and the other is acted upon, that the one moves and the other is moved ; and no such mutual relation is found in haphazard combinations. But these thinkers only try to explain what is the nature of the soul, without adding any details about the body which is to receive it ; as though it were possible, as the Pythagorean stories suggest, for any soul to find its way into any body, ⟨which is absurd,⟩ for we can see that every body has its own peculiar shape or form. Such a theory is like suggesting that carpentry can find its way into flutes ; each craft must employ its own tools, and each soul its own body.

IV. There is another traditional theory about the soul, which many find the most credible of all current theories, and which has been approved by the verdict of public opinion. It is said that the soul is a harmony of some kind ; for, they argue, a harmony is a blend or composition of contraries, and the body is composed of contraries. *Is the soul a harmony or proportion ?*

But (1) a harmony is a fixed proportion or composition of the ingredients blended, and the soul cannot be either of these things. (2) Again, it is no part of harmony to cause movement, yet almost everyone ascribes this to the soul as its chief characteristic. (3) It seems more in accord with the facts to connect harmony with health or generally with good conditions of the body than with the soul. This will become quite obvious if one tries to attribute the soul's experences and actions to some sort of harmony ; for it is difficult to make them fit. (4) Again, we use the word harmony in two different senses : most properly *Objections to this theory.*

τῶν μεγεθῶν ἐν τοῖς ἔχουσι κίνησιν καὶ θέσιν τὴν
σύνθεσιν αὐτῶν, ἐπειδὰν οὕτω συναρμόζωσιν ὥστε
μηδὲν συγγενὲς παραδέχεσθαι, ἐντεῦθεν δὲ καὶ τὸν
τῶν μεμιγμένων λόγον.

10 Οὐδετέρως μὲν οὖν εὔλογον, ἡ δὲ σύνθεσις τῶν
τοῦ σώματος μερῶν λίαν εὐεξέταστος· πολλαί τε
γὰρ αἱ συνθέσεις τῶν μερῶν καὶ πολλαχῶς· τίνος
οὖν ἢ πῶς ὑπολαβεῖν τὸν νοῦν χρὴ σύνθεσιν εἶναι,
ἢ καὶ τὸ αἰσθητικὸν ἢ ὀρεκτικόν; ὁμοίως δὲ
ἄτοπον καὶ τὸν λόγον τῆς μίξεως εἶναι τὴν ψυχήν·
15 οὐ γὰρ τὸν αὐτὸν ἔχει λόγον ἡ μίξις τῶν στοιχείων
καθ' ἣν σὰρξ καὶ καθ' ἣν ὀστοῦν. συμβήσεται οὖν
πολλάς τε ψυχὰς ἔχειν καὶ κατὰ πᾶν τὸ σῶμα,
εἴπερ πάντα μὲν ἐκ τῶν στοιχείων μεμιγμένων, ὁ
δὲ τῆς μίξεως λόγος ἁρμονία καὶ ψυχή. ἀπαιτή-
σειε δ' ἄν τις τοῦτό γε καὶ παρ' Ἐμπεδοκλέους·
20 ἕκαστον γὰρ αὐτῶν λόγῳ τινί φησιν εἶναι· πότερον
οὖν ὁ λόγος ἐστὶν ἡ ψυχή, ἢ μᾶλλον ἕτερόν τι
οὖσα ἐγγίνεται τοῖς μέλεσιν; ἔτι δὲ πότερον ἡ
φιλία τῆς τυχούσης αἰτία μίξεως ἢ τῆς κατὰ τὸν
λόγον; καὶ αὕτη πότερον ὁ λόγος ἐστὶν ἢ παρὰ
τὸν λόγον ἕτερόν τι; ταῦτα μὲν οὖν ἔχει τοιαύτας
25 ἀπορίας· εἰ δ' ἐστὶν ἕτερον ἡ ψυχὴ τῆς μίξεως, τί
δή ποτε ἅμα τῷ σαρκὶ εἶναι ἀναιρεῖται καὶ τῷ
τοῖς ἄλλοις μορίοις τοῦ ζῴου; πρὸς δὲ τούτοις

---

[a] Or " composition."
[b] Love is Empedocles' personification of the force of
attraction which accounts for the combination of elements
into a whole, as its opposite repulsion (Strife) accounts for
the separation of things into separate wholes.

of spatial magnitudes, to mean compaction *a* in the case of things which have movement and position, when they cohere in such a way that they do not admit the intrusion of anything homogeneous ; but in a derivative sense we also use the word to mean the ratio in which constituents are mixed.

In neither of these senses can harmony be reasonably identified with soul, but the view that the soul is a composition of parts of the body is easily refuted. For the compositions of parts are many, and take place in many ways. Of which of the parts, then, are we to suppose that the mind or the perceptive or appetitive faculty is a composition, and how is such a composition effected ? But the view that the soul is a harmony in the sense of a ratio of mixture is equally absurd. For the mixture of the elements which go to make the flesh has not the same ratio as that which makes the bone. It will follow, then, that there are many souls distributed all over the body, since every part of it is a mixture of the elements and the ratio of each mixture is a harmony, *i.e.* a soul. One might put this question to Empedocles (for he says that each part of the body owes its distinctive nature to the ratio of its mixture) : is this ratio the soul, or is the soul something distinct, which develops in the limbs ? Again, does his principle of Love *b* cause any random mixture, or only a mixture in the right ratio ? And is Love this ratio, or is it some other thing distinct from the ratio ? Such are the difficulties which these theories present. And if the soul is a different thing from the mixture, why is it destroyed at the same time as that which constitutes the flesh and the other parts of the living animal ? Besides this, if each of the parts has not a separate

408 a

εἴπερ μὴ ἕκαστον τῶν μορίων ψυχὴν ἔχει, εἰ μή
ἐστιν ἡ ψυχὴ ὁ λόγος τῆς μίξεως, τί ἐστιν ὃ
φθείρεται τῆς ψυχῆς ἀπολειπούσης;

"Ὅτι μὲν οὖν οὔθ' ἁρμονίαν οἷόν τ' εἶναι τὴν
30 ψυχὴν οὔτε κύκλῳ περιφέρεσθαι, δῆλον ἐκ τῶν
εἰρημένων. κατὰ συμβεβηκὸς δὲ κινεῖσθαι, καθάπερ εἴπομεν, ἔστι καὶ κινεῖν ἑαυτήν, οἷον κινεῖσθαι
μὲν ἐν ᾧ ἐστί, τοῦτο δὲ κινεῖσθαι ὑπὸ τῆς ψυχῆς·
ἄλλως δ' οὐχ οἷόν τε κινεῖσθαι κατὰ τόπον αὐτήν.

Εὐλογώτερον δ' ἀπορήσειεν ἄν τις περὶ αὐτῆς ὡς
408 b κινουμένης, εἰς τὰ τοιαῦτα ἀποβλέψας. φαμὲν γὰρ
τὴν ψυχὴν λυπεῖσθαι χαίρειν θαρρεῖν φοβεῖσθαι,
ἔτι δὲ ὀργίζεσθαί τε καὶ αἰσθάνεσθαι καὶ διανοεῖ-
σθαι· ταῦτα δὲ πάντα κινήσεις εἶναι δοκοῦσιν.
5 ὅθεν οἰηθείη τις ἂν αὐτὴν κινεῖσθαι· τὸ δ' οὐκ ἔστιν
ἀναγκαῖον· εἰ γὰρ καὶ ὅτι μάλιστα τὸ λυπεῖσθαι ἢ
χαίρειν ἢ διανοεῖσθαι κινήσεις εἰσί, καὶ ἕκαστον
κινεῖσθαι τούτων, τὸ δὲ κινεῖσθαί ἐστιν ὑπὸ τῆς
ψυχῆς, οἷον τὸ ὀργίζεσθαι ἢ φοβεῖσθαι τὸ[1] τὴν καρδίαν ὡδὶ κινεῖσθαι, τὸ δὲ διανοεῖσθαι ἢ τὸ τοῦτο[2]
10 ἴσως ἢ ἕτερόν τι (τούτων δὲ συμβαίνει τὰ μὲν
κατὰ φορὰν τινῶν κινουμένων, τὰ δὲ κατ' ἀλ-
λοίωσιν. ποῖα δὲ καὶ πῶς, ἕτερός ἐστι λόγος).
τὸ δὲ λέγειν ὀργίζεσθαι τὴν ψυχὴν ὅμοιον κἂν εἴ
τις λέγοι τὴν ψυχὴν ὑφαίνειν ἢ οἰκοδομεῖν· βέλτιον
γὰρ ἴσως μὴ λέγειν τὴν ψυχὴν ἐλεεῖν ἢ μανθάνειν ἢ
15 διανοεῖσθαι, ἀλλὰ τὸν ἄνθρωπον τῇ ψυχῇ. τοῦτο
δὲ μὴ ὡς ἐν ἐκείνῃ τῆς κινήσεως οὔσης, ἀλλ' ὁτὲ
μὲν μέχρι ἐκείνης, ὁτὲ δ' ἀπ' ἐκείνης, οἷον ἡ μὲν

---

[1] τὸ V, Bonitz : τῷ.   [2] τὸ τοῦτο Bonitz : τοιοῦτον.

soul of its own, and if the soul is not the ratio of the mixture, what is it which perishes when the soul leaves the body ?

It is clear from what has been said that the soul cannot be a harmony, nor can it revolve in a circle. It is, however, possible, as we have said, that it may be moved, and even move itself, incidentally (*e.g.*, that which contains it may be moved, and be moved by the soul) ; but in no other sense can it move in space.

The following considerations suggest even more reasonable criticisms of the theory that the soul moves. We say that the soul grieves, rejoices, is courageous, or afraid, and also grows angry, perceives and thinks ; all these seem to be movements ; hence one might suppose that the soul is moved ; but this is not a necessary inference. Let us grant that grief, joy and thinking are all movements, *i.e.*, that each of them is a process of being moved ; let us further admit that the movement is caused by the soul—*e.g.*, that anger and fear are particular movements of the heart, and that thinking is a movement of this or of something else, some of these processes involving change of place and others change of quality in certain parts (of what parts and under what conditions need not be considered now) : still to say that the soul gets angry is as if one were to say that the soul weaves or builds a house. Probably it is better not to say that the soul pities, or learns, or thinks, but to say rather that the soul is the instrument whereby man does these things ; that is to say, that the movement does not take place in the soul, but sometimes penetrates to it, and sometimes starts from it. For instance perception starts from particular objects and reaches

408 b

αἴσθησις ἀπὸ τωνδί, ἡ δ' ἀνάμνησις ἀπ' ἐκείνης
ἐπὶ τὰς ἐν τοῖς αἰσθητηρίοις κινήσεις ἢ μονάς.

20 Ὁ δὲ νοῦς ἔοικεν ἐγγίνεσθαι οὐσία τις οὖσα,
καὶ οὐ φθείρεσθαι. μάλιστα γὰρ ἐφθείρετ' ἂν ὑπὸ
τῆς ἐν τῷ γήρᾳ ἀμαυρώσεως, νῦν δ' ἴσως ὅπερ
ἐπὶ τῶν αἰσθητηρίων συμβαίνει· εἰ γὰρ λάβοι ὁ
πρεσβύτης ὄμμα τοιονδί, βλέποι ἂν ὥσπερ καὶ ὁ
νέος. ὥστε τὸ γῆρας οὐ τῷ τὴν ψυχήν τι πεπον-
θέναι, ἀλλ' ἐν ᾧ, καθάπερ ἐν μέθαις καὶ νόσοις.
25 καὶ τὸ νοεῖν δὴ καὶ τὸ θεωρεῖν μαραίνεται ἄλλου
τινὸς ἔσω φθειρομένου, αὐτὸ δὲ ἀπαθές ἐστιν. τὸ
δὲ διανοεῖσθαι καὶ φιλεῖν ἢ μισεῖν οὐκ ἔστιν ἐκείνου
πάθη, ἀλλὰ τουδὶ τοῦ ἔχοντος ἐκεῖνο, ᾗ ἐκεῖνο
ἔχει. διὸ καὶ τούτου φθειρομένου οὔτε μνημονεύει
οὔτε φιλεῖ· οὐ γὰρ ἐκείνου ἦν, ἀλλὰ τοῦ κοινοῦ, ὃ
ἀπόλωλεν· ὁ δὲ νοῦς ἴσως θειότερόν τι καὶ ἀπαθές
30 ἐστιν. ὅτι μὲν οὖν οὐχ οἷόν τε κινεῖσθαι τὴν
ψυχήν, φανερὸν ἐκ τούτων· εἰ δ' ὅλως μὴ κινεῖται,
δῆλον ὡς οὐδ' ὑφ' ἑαυτῆς.

Πολὺ δὲ τῶν εἰρημένων ἀλογώτατον τὸ λέγειν
ἀριθμὸν εἶναι τὴν ψυχὴν κινοῦνθ' ἑαυτόν· ὑπάρχει
γὰρ αὐτοῖς ἀδύνατα·ᵃ πρῶτα μὲν τὰ ἐκ τοῦ κινεῖσθαι
409 a συμβαίνοντα, ἰδίᾳ δ' ἐκ τοῦ λέγειν αὐτὴν ἀριθμόν·
πῶς γὰρ χρὴ νοῆσαι μονάδα κινουμένην, καὶ ὑπὸ
τίνος, καὶ πῶς, ἀμερῆ καὶ ἀδιάφορον οὖσαν; εἰ
γάρ ἐστι κινητικὴ καὶ κινητή, διαφέρειν δεῖ. ἔτι

ᵃ This is the theory of Xenocrates, a contemporary of
Aristotle, who succeeded Speusippus as head of the
" Academy."

the soul ; recollection starts from the soul and extends to the movements or resting points in the sense organs.

But mind seems to be an independent substance Does the engendered in us, and to be imperishable. If it could soul perish? be destroyed the most probable cause would be the feebleness of old age, but, in fact, probably the same thing occurs as in the sense organs ; for if an old man could acquire the right kind of eye, he would see as a young man sees. Hence old age is due to an affection, not of the soul, but only of that in which the soul resides, as in the case in drunkenness and disease. Thus the power of thought and speculation decays because something else within perishes, but itself it is unaffected. Thinking, loving and hating, are affections not of the mind, but rather of the individual which possesses the mind, in so far as it does so. Memory and love fail when this perishes ; for they were never part of the mind, but of the whole entity which has perished. Presumably the mind is something more divine, and is unaffected. It is then obvious from these considerations that the soul cannot be moved ; and, if it cannot be moved at all, it is obviously not moved by itself.

But of all the unreasonable theories about the soul The soul as the most unreasonable is that which calls the soul a a self-mov-number which moves itself.[a] In this theory there are ing number. inherent impossibilities, first those which are implied by the theory of the soul's being moved, and also special ones which follow from calling the soul a number. For (1) how can one conceive of a unit moving ? by what is it moved, and in what way, being as it is without parts or differences? For if it can cause and suffer movement it must have differences.

δ' ἐπεί φασι κινηθεῖσαν γραμμὴν ἐπίπεδον ποιεῖν,
5 στιγμὴν δὲ γραμμήν, καὶ αἱ τῶν μονάδων κινήσεις
γραμμαὶ ἔσονται· ἡ γὰρ στιγμὴ μονάς ἐστι θέσιν
ἔχουσα· ὁ δ' ἀριθμὸς τῆς ψυχῆς ἤδη πού ἐστι καὶ
θέσιν ἔχει. ἔτι δ' ἀριθμοῦ μὲν ἐὰν ἀφέλῃ τις
ἀριθμὸν ἢ μονάδα, λείπεται ἄλλος ἀριθμός· τὰ δὲ
φυτὰ καὶ τῶν ζῴων πολλὰ διαιρούμενα ζῇ, καὶ
10 δοκεῖ τὴν αὐτὴν ψυχὴν ἔχειν τῷ εἴδει. δόξειε δ'
ἂν οὐθὲν διαφέρειν μονάδας λέγειν ἢ σωμάτια
μικρά· καὶ γὰρ ἐκ τῶν Δημοκρίτου σφαιρίων ἐὰν
γένωνται στιγμαί, μόνον δὲ μένῃ τὸ ποσόν, ἔσται
τι ἐν αὐτῷ τὸ μὲν κινοῦν τὸ δὲ κινούμενον, ὥσπερ
ἐν τῷ συνεχεῖ· οὐ γὰρ διὰ τὸ μεγέθει διαφέρειν ἢ
15 μικρότητι συμβαίνει τὸ λεχθέν, ἀλλ' ὅτι ποσόν.
διὸ ἀναγκαῖον εἶναί τι τὸ κινῆσον τὰς μονάδας.
εἰ δ' ἐν τῷ ζῴῳ τὸ κινοῦν ἡ ψυχή, καὶ ἐν τῷ
ἀριθμῷ, ὥστε οὐ τὸ κινοῦν καὶ τὸ κινούμενον ἡ
ψυχή, ἀλλὰ τὸ κινοῦν μόνον. ἐνδέχεται δὲ [1]δὴ
πῶς μονάδα ταύτην εἶναι;[1] δεῖ γὰρ ὑπάρχειν τινὰ
20 αὐτῇ διαφορὰν πρὸς τὰς ἄλλας· στιγμῆς δὲ μονα-
δικῆς τίς ἂν εἴη διαφορὰ πλὴν θέσις; εἰ μὲν οὖν
εἰσιν ἕτεραι αἱ ἐν τῷ σώματι μονάδες καὶ αἱ
στιγμαί, ἐν τῷ αὐτῷ ἔσονται αἱ μονάδες· καθέξει
γὰρ χώραν στιγμῆς. καίτοι τί κωλύει ἐν τῷ
αὐτῷ εἶναι, εἰ δύο, καὶ ἀπείρους; ὧν γὰρ ὁ τόπος
25 ἀδιαίρετος, καὶ αὐτά. εἰ δ' αἱ ἐν τῷ σώματι
στιγμαὶ ὁ ἀριθμὸς ὁ τῆς ψυχῆς, ἢ εἰ ὁ ἐκ τῶν ἐν
τῷ σώματι στιγμῶν ἀριθμὸς ἡ ψυχή, διὰ τί οὐ

---

[1] δή πως . . . εἶναι. Bekker.

(2) Again, since they say that a moving line describes a surface, and a moving point a line, the movements of the soul's units will be lines. For a point is a unit having position; and the number of the soul is *ipso facto* somewhere, and has position. (3) Now, if one subtracts a number or unit from a number, another number is left. But plants and many animals continue to live even when divided, and seem to retain in these fragments a soul specifically the same as before. It would seem to make no difference whether we speak of units or of minute particles; for if we suppose Democritus's spherical atoms to be converted into points and to retain nothing but their quantitative nature, there will still be in each of them something which moves and something which is moved, just as in a continuum. For what we have mentioned does not occur through any difference of size in the atoms, but because they possess quantity. There must, then, be something to give movement to the units. But if that which produces movement in the animal is the soul, then it is also so in the number, so that the soul is not both that which produces movement and that which is moved, but only that which produces movement. But how can this possibly be a unit? Such a unit must differ inherently from the others. But what difference can a unit which is a point exhibit, except position? If then the soul-units in the body are different from the points in the body, the former will be in the same place as the latter, for each will occupy the place of a point. And yet if two units can be in the same place, why not an infinite number? for things which occupy an indivisible space are themselves indivisible. But if the bodily points are identical with the units of the soul number, or if the number of bodily points is the soul, why do not

all bodies have a soul ? For there appear to be points —infinitely many, indeed—in all of them. And again how is it possible to separate the points and free them from the bodies, if lines cannot be resolved into points ?

V. In effect, as we have said, this theory in one respect repeats the view of those who suppose the soul to be a body of fine particles, and in another, just as when Democritus states that the body is moved by the soul, it has an absurdity of its own. For if the soul exists in every part of the sentient body then there must be two bodies in the same place, if the soul is a body. And those who say that the soul is a number must believe that there are many points in one point, or else that every body has a soul, unless the number engendered in the body is different and distinct from the points already present in the body. And it follows that the living creature is moved by the number, just as we have already said that Democritus accounted for its movement ; for what difference does it make whether we call them small spheres, or large units, or generally moving units ? For in either case we can only account for the movement of the living creature by the movement of these particles. *Xenocrates' theory further criticized.*

These are some of the difficulties in the view which combines movement and number, and there are many others of a like nature ; for this combination, so far from being a definition of the soul, cannot even be one of its attributes. And this will become clear to anyone, if he tries on this theory to give an explanation of the affections and functions of the soul, such as calculations, perceptions, pleasures, pains, and so on ; for, as we have said before, on these lines it is not easy even to conjecture an explanation. *Conclusion.*

53

409 b

Τριῶν δὲ τρόπων παραδεδομένων καθ' οὓς
20 ὁρίζονται τὴν ψυχήν, οἱ μὲν τὸ κινητικώτατον
ἀπεφήναντο τῷ κινεῖν ἑαυτό, οἱ δὲ σῶμα τὸ λεπτομερέστατον ἢ τὸ ἀσωματώτατον τῶν ἄλλων.
ταῦτα δὲ τίνας ἀπορίας τε καὶ ὑπεναντιώσεις ἔχει,
διεληλύθαμεν σχεδόν. λείπεται δ' ἐπισκέψασθαι
πῶς λέγεται τὸ ἐκ τῶν στοιχείων αὐτὴν εἶναι.
25 λέγουσι μὲν γάρ, ἵν' αἰσθάνηταί τε τῶν ὄντων
καὶ ἕκαστον γνωρίζῃ, ἀναγκαῖον δὲ συμβαίνειν
πολλὰ καὶ ἀδύνατα τῷ λόγῳ· τίθενται γὰρ γνωρίζειν
τῷ ὁμοίῳ τὸ ὅμοιον, ὥσπερ ἂν εἰ τὴν ψυχὴν τὰ
πράγματα τιθέντες. οὐκ ἔστι δὲ μόνα ταῦτα,
πολλὰ δὲ καὶ ἕτερα, μᾶλλον δ' ἴσως ἄπειρα τὸν
30 ἀριθμόν, τὰ ἐκ τούτων. ἐξ ὧν μὲν οὖν ἐστιν
ἕκαστον τούτων, ἔστω γινώσκειν τὴν ψυχὴν καὶ
αἰσθάνεσθαι· ἀλλὰ τὸ σύνολον τίνι γνωριεῖ ἢ
αἰσθήσεται, οἷον τί θεὸς ἢ ἄνθρωπος ἢ σὰρξ ἢ
410 a ὀστοῦν; ὁμοίως δὲ καὶ ἄλλο ὁτιοῦν τῶν συνθέτων·
οὐ γὰρ ὁπωσοῦν ἔχοντα τὰ στοιχεῖα τούτων ἕκαστον, ἀλλὰ λόγῳ τινὶ καὶ συνθέσει, καθάπερ φησὶ
καὶ Ἐμπεδοκλῆς τὸ ὀστοῦν·

ἡ δὲ χθὼν ἐπίηρος ἐν εὐστέρνοις χοάνοισιν
5 τὼ¹ δύο τῶν ὀκτὼ μερέων λάχε νήστιδος αἴγλης,
τέσσαρα δ' Ἡφαίστοιο· τὰ δ' ὀστέα λεύκ'
ἐγένοντο.

οὐδὲν οὖν ὄφελος εἶναι τὰ στοιχεῖα ἐν τῇ ψυχῇ, εἰ
μὴ καὶ οἱ λόγοι ἐνέσονται καὶ ἡ σύνθεσις· γνωριεῖ
γὰρ ἕκαστον τὸ ὅμοιον, τὸ δ' ὀστοῦν ἢ τὸν ἄνθρω-

---

¹ τὼ Torstrik, Biehl¹, Rodier, Diels : τῶν E¹STUXy, comm.
vett., Biehl ² : τὰ E²V, Bekker.

Three methods of defining the soul have come down to us ; some have regarded it as the principal cause of movement, because it moves itself ; others have described the soul as composed of the finest particles, or as the least corporeal of all bodies. We have pretty well exhausted the difficulties and contradictions which these two definitions involve. But it remains to see what is meant by saying that the soul is composed of the elements. This theory is intended to account for the soul's perception and cognition of everything that is, but the theory necessarily involves many impossibilities ; its supporters assume that like is recognized by like, as though they thus identified the soul with the things it knows. But these elements are not the only things existing ; there are many—to be more exact, infinitely many—other things, composed of the elements. Granted that the soul might know and perceive the elements of which each of these things is composed ; yet by what will it perceive and know a composite whole : e.g., what god, man, flesh, or bone is ? and similarly any other compound whole ; for such wholes do not consist of the elements arranged at random, but in a certain ratio and with some principle of composition, as Empedocles says in his description of bone :

" The kindly earth in broad-bosomed crucibles got two of the eight parts from the gleam of moisture, and four from Hephaestus ; and bones come into being all white."

It is then no use for the elements to exist in the soul, unless the ratios and the principle of composition also exist in it ; for each element will recognize its like, but there will be nothing in the soul to recognize

410 a
10 πον οὐθέν, εἰ μὴ καὶ ταῦτ' ἐνέσται. τοῦτο δ' ὅτι
ἀδύνατον, οὐθὲν δεῖ λέγειν· τίς γὰρ ἂν ἀπορήσειεν
εἰ ἔνεστιν ἐν τῇ ψυχῇ λίθος ἢ ἄνθρωπος; ὁμοίως
δὲ καὶ τὸ ἀγαθὸν καὶ τὸ μὴ ἀγαθόν. τὸν αὐτὸν
δὲ τρόπον καὶ περὶ τῶν ἄλλων.

Ἔτι δὲ πολλαχῶς λεγομένου τοῦ ὄντος (σημαίνει
γὰρ τὸ μὲν τόδε τι, τὸ δὲ ποσὸν ἢ ποιὸν ἢ καί
15 τινα ἄλλην τῶν διαιρεθεισῶν κατηγοριῶν) πότερον
ἐξ ἁπάντων ἔσται ἡ ψυχὴ ἢ οὔ; ἀλλ' οὐ δοκεῖ
κοινὰ πάντων εἶναι στοιχεῖα. ἆρ' οὖν ὅσα τῶν
οὐσιῶν ἐκ τούτων μόνον; πῶς οὖν γινώσκει καὶ
τῶν ἄλλων ἕκαστον; ἢ φήσουσιν ἑκάστου γένους
εἶναι στοιχεῖα καὶ ἀρχὰς ἰδίας, ἐξ ὧν τὴν ψυχὴν
20 συνεστάναι; ἔσται ἄρα ποσὸν καὶ ποιὸν καὶ οὐσία.
ἀλλ' ἀδύνατον ἐκ τῶν τοῦ ποσοῦ στοιχείων οὐσίαν
εἶναι καὶ μὴ ποσόν. τοῖς δὴ λέγουσιν ἐκ πάντων
ταῦτά τε καὶ τοιαῦθ' ἕτερα συμβαίνει. ἄτοπον δὲ
καὶ τὸ φάναι μὲν ἀπαθὲς εἶναι τὸ ὅμοιον ὑπὸ τοῦ
ὁμοίου, αἰσθάνεσθαι δὲ τὸ ὅμοιον τοῦ ὁμοίου καὶ
25 γινώσκειν τῷ ὁμοίῳ τὸ ὅμοιον. τὸ δ' αἰσθάνεσθαι
πάσχειν τι καὶ κινεῖσθαι τιθέασιν· ὁμοίως δὲ καὶ
τὸ νοεῖν τε καὶ γινώσκειν.

Πολλὰς δ' ἀπορίας καὶ δυσχερείας ἔχοντος τοῦ
λέγειν, καθάπερ Ἐμπεδοκλῆς, ὡς τοῖς σωματικοῖς
στοιχείοις ἕκαστα γνωρίζεται καὶ πρὸς τὸ ὅμοιον,
30 μαρτυρεῖ τὸ νῦν λεχθέν· ὅσα γὰρ ἔνεστιν ἐν τοῖς
410 b τῶν ζῴων σώμασιν ἁπλῶς γῆς, οἷον ὀστᾶ νεῦρα
τρίχες, οὐθενὸς αἰσθάνεσθαι δοκεῖ, ὥστ' οὐδὲ τῶν
ὁμοίων· καίτοι προσῆκεν. ἔτι δ' ἑκάστῃ τῶν

---

[a] So that the soul can know other categories besides that of
substance.   [b] Sc., in the soul.

bone, for instance, or man, unless they too exist in it. But it is unnecessary to say that this is impossible. For who could seriously ask whether there is a stone or a man in the soul ? The same argument applies to good and not-good ; and so with all the rest.

Again, the word " being " has many senses ; it is applied to substance, quantity, quality, or any other of the categories which we have distinguished. Will the soul consist of all of these or not ? The categories cannot surely all have common elements. Does the soul then consist only of those elements which compose substances ? How then is it to know each of the other categories ? Will they maintain that every genus has its peculiar elements and principles, and that the soul is composed of all of these ? [a] In that case it will be quantity, quality, and substance. But it is impossible that that which is composed of the elements of quantity should be a substance and not a quantity. Those who say that the soul is composed of all the elements are confronted with these and similar difficulties. It is also unreasonable to say on the one hand that like is not acted on by like, and on the other that like perceives and recognizes like by like ; but they regard perceiving as a form of being acted upon and moved, and similarly with thinking and knowing. *Difficulties in the theory.*

There are many obscurities and difficulties in saying, as Empedocles does, that each class of things is known by its corporeal elements, and by reference to its like,[b] as is further testified by this fresh argument For in the bodies of living creatures all the parts which are composed simply of earth, such as bone, sinews, and hair, seem to have no perception at all, and so cannot perceive their like ; and yet on this theory they should do so. Again, in each of these *Problems arising from Empedocles' theory.*

first principles there will be more ignorance than understanding ; for each will know one thing, but will be ignorant of many, in fact of everything else. On Empedocles' view at least it follows that God must be most unintelligent ; for He alone will be ignorant of one of these elements, namely strife, whereas mortal creatures will know them all ; for each individual is composed of them all. In general also, why have not all existing things a soul, since everything is an element, or composed of an element, or of more than one, or of all ? For each of them must know one thing, or some things, or all things. There would be a further difficulty in deciding what is the unifying principle, for the elements correspond to matter, and the force, whatever it is, which combines them is supreme ; but it is impossible that anything should be superior to and control the soul, or (*a fortiori*) the mind ; for it is reasonable to suppose that the mind is by nature original and dominant, but they say that the elements are the first of all existing things.

All those too who describe the soul as composed of the elements, because it knows and perceives existing things, and equally those who call it the chief cause of motion, fail to offer an explanation which will cover every soul. For not everything which has sensation has movement also ; for instance some living things seem to be stationary in space ; and yet this seems to be the only kind of movement which the soul imparts to the living creature. The same difficulty arises for those who construct the mind and the perceptive faculty out of the elements ; for plants seem to live without sharing in locomotion or in perception, and many living animals have no power

The definition will not cover all the facts.

25 διάνοιαν οὐκ ἔχειν. εἰ δέ τις καὶ ταῦτα παρα-
χωρήσειε, καὶ θείη τὸν νοῦν μέρος τι τῆς ψυχῆς,
ὁμοίως δὲ καὶ τὸ αἰσθητικόν, οὐδ' ἂν οὕτω λέγοιεν
καθόλου περὶ πάσης ψυχῆς οὐδὲ περὶ ὅλης οὐδὲ
μιᾶς. τοῦτο δὲ πέπονθε καὶ ὁ ἐν τοῖς Ὀρφικοῖς
ἔπεσι καλουμένοις λόγος· φησὶ γὰρ τὴν ψυχὴν ἐκ
30 τοῦ ὅλου εἰσιέναι ἀναπνεόντων, φερομένην ὑπὸ τῶν
ἀνέμων. οὐχ οἷόν τε δὴ τοῖς φυτοῖς τοῦτο συμβαί-
411 a νειν οὐδὲ τῶν ζῴων ἐνίοις, εἴπερ μὴ πάντα ἀναπνέουσιν. τοῦτο δὲ λέληθε τοὺς οὕτως ὑπειληφό-
τας. εἴ τε δεῖ τὴν ψυχὴν ἐκ τῶν στοιχείων ποιεῖν,
οὐθὲν δεῖ ἐξ ἁπάντων· ἱκανὸν γὰρ θάτερον μέρος
τῆς ἐναντιώσεως ἑαυτό τε κρίνειν καὶ τὸ ἀντικεί-
5 μενον. καὶ γὰρ τῷ εὐθεῖ καὶ αὐτὸ καὶ τὸ καμπύλον
γινώσκομεν· κριτὴς γὰρ ἀμφοῖν ὁ κανών, τὸ δὲ
καμπύλον οὔθ' ἑαυτοῦ οὔτε τοῦ εὐθέος. καὶ ἐν τῷ
ὅλῳ δέ τινες αὐτὴν μεμῖχθαί φασιν, ὅθεν ἴσως καὶ
Θαλῆς ᾠήθη πάντα πλήρη θεῶν εἶναι. τοῦτο δ'
10 ἔχει τινὰς ἀπορίας· διὰ τίνα γὰρ αἰτίαν ἐν μὲν τῷ
ἀέρι ἢ τῷ πυρὶ οὖσα ἡ ψυχὴ οὐ ποιεῖ ζῷον, ἐν δὲ
τοῖς μικτοῖς, καὶ ταῦτα βελτίων ἐν τούτοις εἶναι
δοκοῦσα; ἐπιζητήσειε γὰρ ἄν τις καὶ διὰ τίν'
αἰτίαν ἡ ἐν τῷ ἀέρι ψυχὴ τῆς ἐν τοῖς ζῴοις
βελτίων ἐστὶ καὶ ἀθανατωτέρα. συμβαίνει δ'
ἀμφοτέρως ἄτοπον καὶ παράλογον· καὶ γὰρ τὸ
15 λέγειν ζῷον τὸ πῦρ ἢ τὸν ἀέρα τῶν παραλογω-
τέρων ἐστί, καὶ τὸ μὴ λέγειν ζῷα ψυχῆς ἐνούσης
ἄτοπον. ὑπολαβεῖν δ' ἐοίκασιν εἶναι τὴν ψυχὴν

of thinking. But supposing one were to let this too pass, and assume that the mind is part of the soul, and similarly the perceptive faculty : not even so would their account hold good generally of every soul, or of the whole of any one soul. The theory in the so-called poems of Orpheus presents the same difficulty ; for this theory alleges that the soul, borne by the winds, enters from the universe into animals when they breathe. Now this cannot happen to plants, nor to some animals, since they do not all breathe : a point which has escaped those who support this theory. And if we are to construct the soul out of the elements, it is quite unnecessary that it should be composed of all the elements ; for only one of a pair of contraries is needed to discern both itself and its opposite. For instance, by that which is straight we discern both straight and crooked ; for the carpenter's rule is the test of both, but the crooked tests neither itself nor the straight. Some think that the soul pervades the whole universe, whence perhaps came Thales' view that everything is full of gods. But this theory contains certain difficulties ; for why does not the soul make an animal when it is in air or in fire, but only when it is in a mixture of the elements, and that too though it seems to be in a purer form in the first case ? (One might also ask why the soul in the air is purer and less mortal than the soul in living creatures.) Either way the conclusion is absurd and irrational ; for to describe fire or air as living creatures is highly irrational, and yet to refuse to call them living creatures, if there is a soul in them, is absurd. They appear to suppose that soul is found in these

elements, on the ground that a whole is homogeneous
with its parts ; so they are compelled to say that the
soul also is homogeneous with its parts, if living
creatures become possessed of soul because some
part of the surrounding air is cut off and enclosed in
them. But if the air detached is homogeneous, while
the soul has parts of different kinds, then evidently
although one part of the soul will be present in this
air, another will not. So that either the soul must be
of similar parts, or else it does not exist in any and
every part of the universe.

From what has been said it is obvious that the
faculty of knowing does not belong to the soul because it is composed of the elements, nor is it right
or true to say that it is moved. But since knowing,
perceiving, and the forming of opinions are opera-
tions of the soul, besides desiring, wishing, and the
appetites in general, and again since movement in
space is induced in living creatures by the soul,
besides growth, maturity, and decay, does each of
these belong to the soul as a whole ? Do we think,
perceive, and do or suffer everything else with the
whole soul, or do some functions belong to one part
and others to another ? Does life reside in one or
several or all of these parts or is something else the
cause of it ? Some say that the soul has parts, and
thinks with one part, and desires with another. In
this case what is it which holds the soul together,
if it naturally consists of parts ? Certainly not the
body : on the contrary the soul seems rather to hold
the body together ; at any rate when the soul is gone
the body dissolves into air and decays. If then some
other thing gives the soul unity, this would really be
the soul. But we shall have to inquire again, whether

Can the soul be divided into parts according to its functions ?

this is a unity or has many parts. If it is a unity, why should not the soul be directly described as a unit? And if it has parts, the progress of the argument will again demand to know what is its combining principle, and thus we shall proceed *ad infinitum*. There may also be some doubt about the parts of the soul, as to what is the function of each in the body. For if the soul as a whole holds together the whole body, it is natural that each of the parts should hold together some part of the body. But this seems impossible; for it is hard even to imagine what part the mind will hold together, or how it will do it. Moreover plants clearly live even when divided, and some of the insects also; which implies that the parts have a soul specifically if not numerically the same as that of the whole; at any rate each of the two parts has sensation and moves in space for some time. It is not at all surprising that they do not continue to do so; for they have not the organs necessary to maintain their natural state. But none the less all the parts of the soul are present in each of the two segments, and the two half-souls are homogeneous both with each other and with the whole; which implies that although the parts of the soul are inseparable from one another, the soul as a whole is divisible. The first principle in plants, too, seems to be a kind of soul; for this principle alone is common to both animals and plants. It can exist in separation from the sensitive principle, but nothing can have sensation without it.

B

I. Τὰ μὲν δὴ ὑπὸ τῶν πρότερον παραδεδομένα περὶ
ψυχῆς εἰρήσθω· πάλιν δ' ὥσπερ ἐξ ὑπαρχῆς ἐπ-
ανίωμεν, πειρώμενοι διορίσαι τί ἐστι ψυχὴ καὶ τίς
ἂν εἴη κοινότατος λόγος αὐτῆς. λέγομεν δὴ γένος
ἕν τι τῶν ὄντων τὴν οὐσίαν, ταύτης δὲ τὸ μὲν ὡς
ὕλην, ὃ καθ' αὑτὸ μὲν οὐκ ἔστι τόδε τι, ἕτερον δὲ
μορφὴν καὶ εἶδος, καθ' ἣν ἤδη λέγεται τόδε τι,
καὶ τρίτον τὸ ἐκ τούτων. ἔστι δ' ἡ μὲν ὕλη
δύναμις, τὸ δ' εἶδος ἐντελέχεια, καὶ τοῦτο διχῶς,
τὸ μὲν ὡς ἐπιστήμη, τὸ δ' ὡς τὸ θεωρεῖν. οὐσίαι
δὲ μάλιστ' εἶναι δοκοῦσι τὰ σώματα, καὶ τούτων
τὰ φυσικά· ταῦτα γὰρ τῶν ἄλλων ἀρχαί. τῶν δὲ
φυσικῶν τὰ μὲν ἔχει ζωήν, τὰ δ' οὐκ ἔχει· ζωὴν
δὲ λέγομεν τὴν δι' αὑτοῦ τροφήν τε ˙καὶ αὔξησιν
καὶ φθίσιν. ὥστε πᾶν σῶμα φυσικὸν μετέχον
ζωῆς οὐσία ἂν εἴη, οὐσία δ' οὕτως ὡς συνθέτη.
ἐπεὶ δ' ἐστὶ σῶμα τοιόνδε, ζωὴν γὰρ ἔχον, οὐκ
ἂν εἴη τὸ σῶμα ψυχή· οὐ γάρ ἐστι τῶν καθ'
ὑποκειμένου τὸ σῶμα, μᾶλλον δ' ὡς ὑποκείμενον

ᵃ If you have the capacity to acquire knowledge of a
subject, you may be said to have potential knowledge of it,
which will become actual by study. In another sense, if

66

BOOK II

I. The theories of the soul handed down by our predecessors have been sufficiently discussed ; now let us start afresh, as it were, and try to determine what the soul is, and what definition of it will be most comprehensive. We describe one class of existing things as substance ; and this we subdivide into three : (1) matter, which in itself is not an individual thing ; (2) shape or form, in virtue of which individuality is directly attributed, and (3) the compound of the two. Matter is potentiality, while form is realization or actuality, and the word actuality is used in two senses, illustrated by the possession of knowledge and the exercise of it.[a] Bodies seem to be pre-eminently substances, and most particularly those which are of natural origin ; for these are the sources from which the rest are derived. But of natural bodies some have life and some have not ; by life we mean the capacity for self-sustenance, growth, and decay. Every natural body, then, which possesses life must be substance, and substance of the compound type. But since it is a body of a definite kind, viz., having life, the body cannot be soul, for the body is not something predicated of a subject, but rather is itself to be regarded as a sub-

you possess knowledge which you are not using, it may be called potential, actual only when you are using it.

67

412 a

20 καὶ ὕλη. ἀναγκαῖον ἄρα τὴν¹ ψυχὴν οὐσίαν εἶναι
ὡς εἶδος σώματος φυσικοῦ δυνάμει ζωὴν ἔχοντος.
ἡ δ' οὐσία ἐντελέχεια. τοιούτου ἄρα σώματος
ἐντελέχεια. αὕτη δὲ λέγεται διχῶς, ἡ μὲν ὡς
ἐπιστήμη, ἡ δ' ὡς τὸ θεωρεῖν. φανερὸν οὖν ὅτι
ὡς ἐπιστήμη· ἐν γὰρ τῷ ὑπάρχειν τὴν ψυχὴν καὶ
25 ὕπνος καὶ ἐγρήγορσίς ἐστιν, ἀνάλογον δ' ἡ μὲν
ἐγρήγορσις τῷ θεωρεῖν, ὁ δ' ὕπνος τῷ ἔχειν καὶ
μὴ ἐνεργεῖν. προτέρα δὲ τῇ γενέσει ἐπὶ τοῦ αὐτοῦ
ἡ ἐπιστήμη. διὸ ψυχή ἐστιν ἐντελέχεια ἡ πρώτη
σώματος φυσικοῦ δυνάμει ζωὴν ἔχοντος. τοιοῦτο
412 b δέ, ὃ ἂν ᾖ ὀργανικόν. (ὄργανα δὲ καὶ τὰ τῶν
φυτῶν μέρη, ἀλλὰ παντελῶς ἁπλᾶ, οἷον τὸ φύλλον
περικαρπίου σκέπασμα, τὸ δὲ περικάρπιον καρποῦ.
αἱ δὲ ῥίζαι τῷ στόματι ἀνάλογον· ἄμφω γὰρ ἕλκει
τὴν τροφήν.) εἰ δή τι κοινὸν ἐπὶ πάσης ψυχῆς
5 δεῖ λέγειν, εἴη ἂν ἐντελέχεια ἡ πρώτη σώματος
φυσικοῦ ὀργανικοῦ. διὸ καὶ οὐ δεῖ ζητεῖν εἰ ἓν ἡ
ψυχὴ καὶ τὸ σῶμα, ὥσπερ οὐδὲ τὸν κηρὸν καὶ τὸ
σχῆμα, οὐδ' ὅλως τὴν ἑκάστου ὕλην καὶ τὸ οὗ
ὕλη· τὸ γὰρ ἓν καὶ τὸ εἶναι ἐπεὶ πλεοναχῶς
λέγεται, τὸ κυρίως ἡ ἐντελέχειά ἐστιν.

10   Καθόλου μὲν οὖν εἴρηται τί ἐστιν ἡ ψυχή· οὐσία
¹ τὸν (sic) Bekker.

ᵃ Every " substance " is composed of two factors—matter

ject, *i.e.*, as matter.[a] So the soul must be substance
in the sense of being the form of a natural body,
which potentially has life. And substance in this
sense is actuality. The soul, then, is the actuality
of the kind of body we have described. But actuality
has two senses, analogous to the possession of knowledge and the exercise of it. Clearly actuality in our
present sense is analogous to the possession of knowledge ; for both sleep and waking depend upon the
presence of soul, and waking is analogous to the
exercise of knowledge, sleep to its possession but
not its exercise. Now in one and the same person
the possession of knowledge comes first. The soul
may therefore be defined as the first actuality of a
natural body potentially possessing life ; and such
will be any body which possesses organs. (The parts
of plants are organs too, though very simple ones :
e.g., the leaf protects the pericarp, and the pericarp
protects the seed ; the roots are analogous to the
mouth, for both these absorb food.) If then one is to
find a definition which will apply to every soul, it
will be " the first actuality of a natural body possessed
of organs." So one need no more ask whether body
and soul are one than whether the wax and the impression it receives are one, or in general whether
the matter of each thing is the same as that of which
it is the matter ; for admitting that the terms unity
and being are used in many senses, the paramount
sense is that of actuality.

We have, then, given a general definition of what

The soul as form.

and form ; *e.g.* a billiard ball. Its matter is ivory, its form
spherical. An animate body, then, as it is a substance,
consists of matter and form. The body must be matter, for
it is not itself an attribute, but has attributes. Therefore
the soul is form.

412 b

γὰρ ἡ κατὰ τὸν λόγον. τοῦτο δὲ τὸ τί ἦν εἶναι
τῷ τοιῳδὶ σώματι, καθάπερ εἴ τι τῶν ὀργάνων
φυσικὸν ἦν σῶμα, οἷον πέλεκυς· ἦν γὰρ ἂν τὸ
πελέκει εἶναι ἡ οὐσία αὐτοῦ, καὶ ἡ ψυχὴ τοῦτο·
15 χωρισθείσης γὰρ ταύτης οὐκ ἂν ἔτι πέλεκυς ἦν,
ἀλλ᾽ ἢ ὁμωνύμως. νῦν δ᾽ ἐστὶ πέλεκυς· οὐ γὰρ
τοιούτου σώματος τὸ τί ἦν εἶναι καὶ ὁ λόγος ἡ
ψυχή, ἀλλὰ φυσικοῦ τοιουδὶ ἔχοντος ἀρχὴν κινήσεως καὶ στάσεως ἐν ἑαυτῷ. θεωρεῖν δὲ καὶ ἐπὶ
τῶν μερῶν δεῖ τὸ λεχθέν. εἰ γὰρ ἦν ὁ ὀφθαλμὸς
20 ζῷον, ψυχὴ ἂν ἦν αὐτοῦ ἡ ὄψις· αὕτη γὰρ οὐσία
ὀφθαλμοῦ ἡ κατὰ τὸν λόγον. ὁ δ᾽ ὀφθαλμὸς ὕλη
ὄψεως, ἧς ἀπολειπούσης οὐκ ἔστιν ὀφθαλμός, πλὴν
ὁμωνύμως, καθάπερ ὁ λίθινος καὶ ὁ γεγραμμένος.
δεῖ δὴ λαβεῖν τὸ ἐπὶ μέρους ἐφ᾽ ὅλου τοῦ ζῶντος
σώματος· ἀνάλογον γὰρ ἔχει ὡς τὸ μέρος πρὸς τὸ
25 μέρος, οὕτως ἡ ὅλη αἴσθησις πρὸς τὸ ὅλον σῶμα
τὸ αἰσθητικόν, ᾗ τοιοῦτο. ἔστι δὲ οὐ τὸ ἀπο-
βεβληκὸς τὴν ψυχὴν τὸ δυνάμει ὂν ὥστε ζῆν,
ἀλλὰ τὸ ἔχον. τὸ δὲ σπέρμα καὶ ὁ καρπὸς τὸ
δυνάμει τοιονδὶ σῶμα. ὡς μὲν οὖν ἡ τμῆσις καὶ
413 a ἡ ὅρασις, οὕτω καὶ ἡ ἐγρήγορσις ἐντελέχεια, ὡς
δ᾽ ἡ ὄψις καὶ ἡ δύναμις τοῦ ὀργάνου, ἡ ψυχή· τὸ
δὲ σῶμα τὸ δυνάμει ὄν· ἀλλ᾽ ὥσπερ ὁ ὀφθαλμὸς ἡ

---

[a] A.'s argument in the rest of this chapter is not quite
easy to follow. The introduction of the axe seems at first
irrelevant, because, as A. afterwards explains, being inanimate, it is not really parallel to the living creature. But
his point is clear, the axe consists of the matter (wood and
metal) of which it is composed, and its form (i.e. what makes
it an axe—cutting edge, weight, and so forth). If you take
away (e.g.) its edge, what remains? Still an axe, although
one that will not cut. But this is not true of the living
creature. It has a body which is its matter, and a soul

the soul is : it is substance in the sense of formula ;
i.e., the essence of such-and-such a body. Suppose [a]
that an implement, *e.g.* an axe, were a natural body ;
the substance of the axe would be that which makes
it an axe, and this would be its soul ; suppose this
removed, and it would no longer be an axe, except
equivocally. As it is, it remains an axe, because it is
not of this kind of body that the soul is the essence
or formula, but only of a certain kind of natural body
which has in itself a principle of movement and rest.
We must, however, investigate our definition in relation to the parts of the body. If the eye were a
living creature, its soul would be its vision ; for this
is the substance in the sense of formula of the eye.
But the eye is the matter of vision, and if vision fails
there is no eye, except in an equivocal sense, as for
instance a stone or painted eye. Now we must
apply what we have found true of the part to the
whole living body. For the same relation must hold
good of the whole of sensation to the whole sentient
body *qua* sentient as obtains between their respective parts. That which has the capacity to live
is not the body which has lost its soul, but that which
possesses its soul ; so seed and fruit are potentially
bodies of this kind. The waking state is actuality in
the same sense as the cutting of the axe or the seeing
of the eye, while the soul is actuality in the same
sense as the faculty of the eye for seeing, or of the
implement for doing its work. The body is that which
exists potentially ; but just as the pupil and the
faculty of seeing make an eye, so in the other case

which is its form. Take away the latter and the body
perishes, so that the whole is no longer a living creature ; by
removing the form of a living creature we destroy its identity.

the soul and body make a living creature. It is quite clear, then, that neither the soul nor certain parts of it, if it has parts, can be separated from the body; for in some cases the actuality belongs to the parts themselves. Not but what there is nothing to prevent some parts being separated, because they are not actualities of any body. It is also uncertain whether the soul as an actuality bears the same relation to the body as the sailor to the ship.[a] This must suffice as an attempt to determine in rough outline the nature of the soul.

II. But since the definite and logically more intelligible conception arises from the vague but more obvious data of sense, we must try to review the question of the soul in this light; for a definitive formula ought not merely to show the fact, as most definitions do, but to contain and exhibit the cause. But in practice the formulae of our definitions are like conclusions; for instance, what is squaring a rectangle? The construction of an equilateral rectangle equal to an oblong rectangle. Such a definition is merely a statement of the conclusion. But if a man says that squaring a rectangle is the finding of a mean proportional, he is giving the underlying cause of the thing to be defined.[b]

True definition.

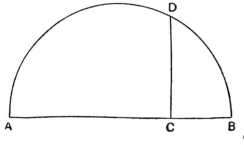

413 a

Λέγομεν οὖν ἀρχὴν λαβόντες τῆς σκέψεως,
διωρίσθαι τὸ ἔμψυχον τοῦ ἀψύχου τῷ ζῆν. πλεοναχῶς δὲ τοῦ ζῆν λεγομένου, κἂν ἕν τι τούτων
ἐνυπάρχῃ μόνον, ζῆν αὐτό φαμεν, οἷον νοῦς,
αἴσθησις, κίνησις καὶ στάσις ἡ κατὰ τόπον, ἔτι
25 κίνησις ἡ κατὰ τροφὴν καὶ φθίσις τε καὶ αὔξησις.
διὸ καὶ τὰ φυόμενα πάντα δοκεῖ ζῆν· φαίνεται γὰρ
ἐν αὐτοῖς ἔχοντα δύναμιν καὶ ἀρχὴν τοιαύτην, δι'
ἧς αὔξησίν τε καὶ φθίσιν λαμβάνουσι κατὰ τοὺς
ἐναντίους τόπους· οὐ γὰρ ἄνω μὲν αὔξεται, κάτω
δ' οὔ, ἀλλ' ὁμοίως ἐπ' ἄμφω καὶ πάντοσε καὶ
30 τρέφεται καὶ ζῇ διὰ τέλους, ἕως ἂν δύνηται λαμβάνειν τροφήν. χωρίζεσθαι δὲ τοῦτο μὲν τῶν
ἄλλων δυνατόν, τὰ δ' ἄλλα τούτου ἀδύνατον ἐν
τοῖς θνητοῖς. φανερὸν δ' ἐπὶ τῶν φυομένων·
413 b οὐδεμία γὰρ αὐτοῖς ὑπάρχει δύναμις ἄλλη ψυχῆς.

Τὸ μὲν οὖν ζῆν διὰ τὴν ἀρχὴν ταύτην ὑπάρχει
τοῖς ζῶσι, τὸ δὲ ζῷον διὰ τὴν αἴσθησιν πρώτως·
καὶ γὰρ τὰ μὴ κινούμενα μηδ' ἀλλάττοντα τόπον,
ἔχοντα δ' αἴσθησιν ζῷα λέγομεν καὶ οὐ ζῆν μόνον.
5 αἰσθήσεως δὲ πρῶτον ὑπάρχει πᾶσιν ἁφή. ὥσπερ
δὲ τὸ θρεπτικὸν δύναται χωρίζεσθαι τῆς ἁφῆς καὶ
πάσης αἰσθήσεως, οὕτως ἡ ἁφὴ τῶν ἄλλων
αἰσθήσεων. θρεπτικὸν δὲ λέγομεν τὸ τοιοῦτον
μόριον τῆς ψυχῆς οὗ καὶ τὰ φυτὰ μετέχει· τὰ δὲ
ζῷα πάντα φαίνεται τὴν ἁπτικὴν αἴσθησιν ἔχοντα·
10 δι' ἣν δ' αἰτίαν ἑκάτερον τούτων συμβέβηκεν,
ὕστερον ἐροῦμεν.

Νῦν δ' ἐπὶ τοσοῦτον εἰρήσθω μόνον, ὅτι ἐστὶν ἡ

We say then, assuming a fresh starting-point for our inquiry, that that which has soul is distinguished from that which has not by living. But the word living is used in many senses, and we say that a thing lives if any one of the following is present in it— mind, sensation, movement or rest in space, besides the movement implied in nutrition and decay or growth. Consequently all plants are considered to live, for they evidently have in themselves a capacity and first principle by means of which they exhibit both growth and decay in opposite directions ; for they do not grow up and not down, but equally in both directions, and in every direction, and they are nourished and continue to live, as long as they are able to absorb food. This capacity to absorb food may exist apart from all other powers, but the others cannot exist apart from this in mortal beings. This is evident in the case of plants ; for they have no other capacity of the soul.

This, then, is the principle through which all living things have life, but the first characteristic of an animal is sensation ; for even those which do not move or change their place, but have sensation, we call living creatures, and do not merely say that they live. The first essential factor of sensation, which we all share, is a sense of touch. Just as the merely nutritive faculty may exist apart from touch and from all sensation, so touch may exist apart from all other senses. (By " nutritive faculty " I mean that part of the soul which even the plants share ; all animals obviously possess the sense of touch.) Why each of these two facts is so, we shall explain later on.[a]

But for the moment let us be satisfied with saying

Sensation a first essential of life.

[a] In Book III. ch. xii.

413 b

ψυχὴ τῶν εἰρημένων τούτων ἀρχὴ καὶ τούτοις
ὥρισται, θρεπτικῷ, αἰσθητικῷ, διανοητικῷ, κινήσει.
πότερον δὲ τούτων ἕκαστόν ἐστι ψυχὴ ἢ μόριον
ψυχῆς, καὶ εἰ μόριον, πότερον οὕτως ὥστ' εἶναι
15 χωριστὸν λόγῳ μόνον ἢ καὶ τόπῳ, περὶ μὲν τινῶν
τούτων οὐ χαλεπὸν ἰδεῖν, ἔνια δὲ ἀπορίαν ἔχει.
ὥσπερ γὰρ ἐπὶ τῶν φυτῶν ἔνια διαιρούμενα φαίνεται ζῶντα καὶ χωριζόμενα ἀπ' ἀλλήλων, ὡς
οὔσης τῆς ἐν τούτοις ψυχῆς ἐντελεχείᾳ μὲν μιᾶς
ἐν ἑκάστῳ φυτῷ, δυνάμει δὲ πλειόνων, οὕτως
20 ὁρῶμεν καὶ περὶ ἑτέρας διαφορὰς τῆς ψυχῆς συμβαῖνον ἐπὶ τῶν ἐντόμων ἐν τοῖς διατεμνομένοις·
καὶ γὰρ αἴσθησιν ἑκάτερον τῶν μερῶν ἔχει καὶ
κίνησιν τὴν κατὰ τόπον, εἰ δ' αἴσθησιν, καὶ φαντασίαν καὶ ὄρεξιν· ὅπου μὲν γὰρ αἴσθησις, καὶ
λύπη τε καὶ ἡδονή, ὅπου δὲ ταῦτα, ἐξ ἀνάγκης
25 καὶ ἐπιθυμία. περὶ δὲ τοῦ νοῦ καὶ τῆς θεωρητικῆς
δυνάμεως οὐδέν πω φανερόν, ἀλλ' ἔοικε ψυχῆς
γένος ἕτερον εἶναι, καὶ τοῦτο μόνον ἐνδέχεται
χωρίζεσθαι, καθάπερ τὸ ἀΐδιον τοῦ φθαρτοῦ. τὰ
δὲ λοιπὰ μόρια τῆς ψυχῆς φανερὸν ἐκ τούτων ὅτι
οὐκ ἔστι χωριστά, καθάπερ τινές φασιν· τῷ δὲ
30 λόγῳ ὅτι ἕτερα, φανερόν· αἰσθητικῷ γὰρ εἶναι καὶ
δοξαστικῷ ἕτερον, εἴπερ καὶ τὸ αἰσθάνεσθαι τοῦ
δοξάζειν. ὁμοίως δὲ καὶ τῶν ἄλλων ἕκαστον τῶν
εἰρημένων. ἔτι δ' ἐνίοις μὲν τῶν ζῴων ἅπανθ'
ὑπάρχει ταῦτα, τισὶ δέ τινα τούτων, ἑτέροις δὲ ἓν
414 a μόνον. τοῦτο δὲ ποιεῖ διαφορὰν τῶν ζῴων· διὰ
τίνα δ' αἰτίαν, ὕστερον ἐπισκεπτέον. παραπλήσιον
δὲ καὶ περὶ τὰς αἰσθήσεις συμβέβηκεν· τὰ μὲν γὰρ

---

a Cf. Introduction.   *b* iii. 12, 13.

that the soul is the origin of the characteristics we have mentioned, and is defined by them, that is by the faculties of nutrition, sensation, thought and movement. The further questions, whether each of these faculties is a soul, or part of a soul, and, if a part, whether a part in the sense that it is only separable in thought or also in space, are in some cases easy of solution, but others involve difficulty. For just as in the case of plants some parts clearly live when divided and separated from each other, so that the soul in them appears to be one in actuality in each whole plant, but potentially more than one, so we can see that in other varieties of the soul the same thing happens, e.g., in insects which are divided; for each of the parts has sensation and movement in space; and, if it has sensation, it must also have imagination [a] and appetite; for, where sensation is, there is also pain and pleasure, and where these are there must also be desire. But in the case of the mind and the thinking faculty nothing is yet clear; it seems to be a distinct kind of soul, and it alone admits of being separated, as the immortal from the perishable. But it is quite clear from what we have said that the other parts of the soul are not separable, as some say; though it is obvious that they are theoretically different; for there is a difference between the abstract faculties of sensation and opinion, just as feeling is different from opining. The same is true of all the other faculties we have mentioned. Again, some animals have all these faculties, some only some of them, and others again only one. It is this which constitutes the differences between animals; the reason for it must be considered later.[b] It is much the same with the senses;

414 a

ἔχει πάσας, τὰ δὲ τινάς, τὰ δὲ μίαν τὴν ἀναγ-
καιοτάτην, ἁφήν.

5 Ἐπεὶ δὲ ᾧ ζῶμεν καὶ αἰσθανόμεθα διχῶς λέγεται,
καθάπερ ᾧ ἐπιστάμεθα (λέγομεν δὲ τὸ μὲν ἐπιστήμην τὸ δὲ ψυχήν· ἑκατέρῳ γὰρ τούτων φαμὲν
ἐπίστασθαι)· ὁμοίως δὲ καὶ [ᾧ]¹ ὑγιαίνομεν τὸ μὲν
ὑγιείᾳ,² τὸ δὲ μορίῳ τινὶ τοῦ σώματος ἢ καὶ ὅλῳ.
10 τούτων δ' ἡ μὲν ἐπιστήμη τε καὶ ὑγίεια μορφὴ
καὶ εἶδός τι καὶ λόγος καὶ οἷον ἐνέργεια τοῦ δεκτικοῦ, ἡ μὲν τοῦ ἐπιστημονικοῦ, ἡ δὲ τοῦ ὑγια-
στικοῦ (δοκεῖ γὰρ ἐν τῷ πάσχοντι καὶ διατιθεμένῳ
ἡ τῶν ποιητικῶν ὑπάρχειν ἐνέργεια), ἡ ψυχὴ δὲ
τοῦτο ᾧ ζῶμεν καὶ αἰσθανόμεθα καὶ διανοούμεθα
πρώτως, ὥστε λόγος τις ἂν εἴη καὶ εἶδος, ἀλλ'
15 οὐχ ὕλη καὶ τὸ ὑποκείμενον. τριχῶς γὰρ λεγομένης τῆς οὐσίας, καθάπερ εἴπομεν, ὧν τὸ μὲν
εἶδος, τὸ δὲ ὕλη, τὸ δὲ ἐξ ἀμφοῖν· τούτων δ' ἡ
μὲν ὕλη δύναμις, τὸ δὲ εἶδος ἐντελέχεια· ἐπεὶ δὲ
τὸ ἐξ ἀμφοῖν ἔμψυχον, οὐ τὸ σῶμά ἐστιν ἐντελέχεια
ψυχῆς, ἀλλ' αὕτη σώματός τινος. καὶ διὰ τοῦτο
20 καλῶς ὑπολαμβάνουσιν οἷς δοκεῖ μήτ' ἄνευ σώματος εἶναι μήτε σῶμά τι ἡ ψυχή· σῶμα μὲν γὰρ
οὐκ ἔστι, σώματος δέ τι, καὶ διὰ τοῦτο ἐν σώματι
ὑπάρχει, καὶ ἐν σώματι τοιούτῳ, καὶ οὐχ ὥσπερ
οἱ πρότερον εἰς σῶμα ἐνήρμοζον αὐτήν, οὐθὲν
προσδιορίζοντες ἐν τίνι καὶ ποίῳ, καίπερ οὐδὲ
25 φαινομένου τοῦ τυχόντος δέχεσθαι τὸ τυχόν. οὕτω

---

¹ Bywater.
² ὑγιείᾳ Sophonias, Torstrik, Hicks: ὑγίεια vel ὑγίειαν codd.

ᵃ 412 a 6.

for some animals have all, some only some, and some again one only—the most indispensable—touch.

The phrase " that whereby we live and perceive " has two senses, as has " that whereby we know " (in the one sense we mean knowledge and in the other the soul ; for we can say that we know by each of these) ; similarly we are healthy either by health or by part or the whole of the body. Now of these, knowledge and health are a kind of shape or form, or notion ; an actuality, as it were, of the recipient, *i.e.* of that which is capable of knowledge or health (for the actualization of active processes appears to reside in the patient upon which the effect is produced), and the soul is that whereby we live and perceive and think in the primary sense ; so that the soul would be the notion or form, and not the matter or substrate. As we have already said,[a] substance is used in three senses, form, matter, and a compound of the two. Of these matter is potentiality, and form actuality ; and since the compound is an animate thing, the body cannot be the actuality of a soul, but the soul is the actuality of some body. For this reason those are right in their view who maintain that the soul cannot exist without the body, but is not itself in any sense a body. It is not a body, it is associated with a body, and therefore resides in a body, and in a body of a particular kind ; not at all as our predecessors supposed, who fitted it to any body, without adding any limitations as to what body or what kind of body, although it is unknown for any chance thing to admit any other chance thing. But

δὲ γίνεται καὶ κατὰ λόγον· ἑκάστου γὰρ ἡ ἐντελέχεια ἐν τῷ δυνάμει ὑπάρχοντι καὶ τῇ οἰκείᾳ ὕλῃ
πέφυκεν ἐγγίνεσθαι. ὅτι μὲν οὖν ἐντελέχειά τίς
ἐστι καὶ λόγος τοῦ δύναμιν ἔχοντος εἶναι τοιούτου,
φανερὸν ἐκ τούτων.

III. Τῶν δὲ δυνάμεων τῆς ψυχῆς αἱ λεχθεῖσαι τοῖς
30 μὲν ὑπάρχουσι πᾶσαι, καθάπερ εἴπομεν, τοῖς δὲ
τινὲς αὐτῶν, ἐνίοις δὲ μία μόνη. δυνάμεις δ'
εἴπομεν θρεπτικόν, ὀρεκτικόν, αἰσθητικόν, κινητικὸν κατὰ τόπον, διανοητικόν. ὑπάρχει δὲ τοῖς
414 b μὲν φυτοῖς τὸ θρεπτικὸν μόνον, ἑτέροις δὲ τοῦτό
τε καὶ τὸ αἰσθητικόν. εἰ δὲ τὸ αἰσθητικόν, καὶ
τὸ ὀρεκτικόν· ὄρεξις μὲν γὰρ ἐπιθυμία καὶ θυμὸς
καὶ βούλησις, τὰ δὲ ζῷα πάντ' ἔχουσι μίαν γε
τῶν αἰσθήσεων, τὴν ἁφήν· ᾧ δ' αἴσθησις ὑπάρχει,
5 τούτῳ ἡδονή τε καὶ λύπη καὶ τὸ ἡδύ τε καὶ
λυπηρόν, οἷς δὲ ταῦτα, καὶ ἡ ἐπιθυμία· τοῦ γὰρ
ἡδέος ὄρεξις αὕτη. ἔτι δὲ τῆς τροφῆς αἴσθησιν
ἔχουσιν· ἡ γὰρ ἁφὴ τῆς τροφῆς αἴσθησις· ξηροῖς
γὰρ καὶ ὑγροῖς καὶ θερμοῖς καὶ ψυχροῖς τρέφεται
τὰ ζῷα πάντα, τούτων δ' αἴσθησις ἁφή· τῶν δ'
10 ἄλλων αἰσθητῶν κατὰ συμβεβηκός· οὐθὲν γὰρ εἰς
τροφὴν συμβάλλεται ψόφος οὐδὲ χρῶμα οὐδὲ
ὀσμή. ὁ δὲ χυμὸς ἕν τι τῶν ἁπτῶν ἐστιν. πεῖνα
δὲ καὶ δίψα ἐπιθυμία, καὶ ἡ μὲν πεῖνα ξηροῦ καὶ
θερμοῦ, ἡ δὲ δίψα ψυχροῦ καὶ ὑγροῦ· ὁ δὲ χυμὸς
15 οἷον ἥδυσμά τι τούτων ἐστίν. διασαφητέον δὲ
περὶ αὐτῶν ὕστερον, νῦν δ' ἐπὶ τοσοῦτον εἰρήσθω,
ὅτι τῶν ζῴων τοῖς ἔχουσιν ἁφὴν καὶ ὄρεξις ὑπάρχει. περὶ δὲ φαντασίας ἄδηλον, ὕστερον δ' ἐπι-

---

[a] *i.e.* on general grounds we should expect a particular
soul to belong to a particular body.

our view explains the facts quite reasonably [a]; for the actuality of each thing is naturally inherent in its potentiality, that is in its own proper matter. From all this it is clear that the soul is a kind of actuality or notion of that which has the capacity of having a soul.

III. Now of the faculties of the soul which we have mentioned, some living things, as we have said, have all, others only some, and others again only one. Those which we have mentioned are the faculties for nourishment, for appetite,[b] for sensation, for movement in space, and for thought. Plants have the nutritive faculty only, but other living things have the faculty for sensation too. But if for sensation then also for appetite; for appetite consists of desire, inclination, and wish, and all animals have at least one of the senses, that of touch; and that which has sensation knows pleasure and pain, the pleasant and the painful, and that which knows these has also desire; for desire is an appetite for what is pleasant. Again, they have a sense which perceives food; for touch is the sense which does this. All animals feed on what is dry or wet, hot or cold, and touch is the sense which apprehends these; the other objects of sense are only indirectly apprehended by touch. Neither sound, nor colour, nor smell contributes anything to nourishment; but flavour is one of the things apprehended by touch. Hunger and thirst are desire, the former for what is dry and hot, the latter for what is cold and wet; flavour is a kind of seasoning of these. We must be precise about these subjects later, but for the moment let it suffice to say that those animals which have a sense of touch have also appetite. The question of imagination is obscure,

All living creatures have not all faculties.

---

[b] Appetite is not included in the list in 413 b 13.

414 b

σκεπτέον. ἐνίοις δὲ πρὸς τούτοις ὑπάρχει καὶ τὸ
κατὰ τόπον κινητικόν, ἑτέροις δὲ καὶ τὸ διανοητικόν τε καὶ νοῦς, οἷον ἀνθρώποις καὶ εἴ τι τοιοῦτον
ἕτερόν ἐστιν ἢ καὶ τιμιώτερον.

20   Δῆλον οὖν ὅτι τὸν αὐτὸν τρόπον εἷς ἂν εἴη λόγος
ψυχῆς τε καὶ σχήματος· οὔτε γὰρ ἐκεῖ σχῆμα παρὰ
τὸ τρίγωνόν ἐστι καὶ τὰ ἐφεξῆς, οὔτ' ἐνταῦθα
ψυχὴ παρὰ τὰς εἰρημένας. γένοιτο δ' ἂν καὶ ἐπὶ
τῶν σχημάτων λόγος κοινός, ὃς ἐφαρμόσει μὲν
πᾶσιν, ἴδιος δ' οὐδενὸς ἔσται σχήματος· ὁμοίως
25 δὲ καὶ ἐπὶ ταῖς εἰρημέναις ψυχαῖς. διὸ γελοῖον
ζητεῖν τὸν κοινὸν λόγον καὶ ἐπὶ τούτων καὶ ἐφ'
ἑτέρων, ὃς οὐδενὸς ἔσται τῶν ὄντων ἴδιος λόγος,
οὐδὲ κατὰ τὸ οἰκεῖον καὶ ἄτομον εἶδος, ἀφέντας τὸν
τοιοῦτον.

Παραπλησίως δ' ἔχει τῷ περὶ τῶν σχημάτων
καὶ τὰ κατὰ ψυχήν· ἀεὶ γὰρ ἐν τῷ ἐφεξῆς
30 ὑπάρχει δυνάμει τὸ πρότερον ἐπί τε τῶν σχημάτων καὶ ἐπὶ τῶν ἐμψύχων, οἷον ἐν τετραγώνῳ μὲν
τρίγωνον, ἐν αἰσθητικῷ δὲ τὸ θρεπτικόν· ὥστε
καθ' ἕκαστον ζητητέον, τίς ἑκάστου ψυχή, οἷον
τίς φυτοῦ καὶ τίς ἀνθρώπου ἢ θηρίου. διὰ τίνα
415 a δ' αἰτίαν τῷ ἐφεξῆς οὕτως ἔχουσι, σκεπτέον.
ἄνευ μὲν γὰρ τοῦ θρεπτικοῦ τὸ αἰσθητικὸν οὐκ
ἔστιν· τοῦ δ' αἰσθητικοῦ χωρίζεται τὸ θρεπτικὸν
ἐν τοῖς φυτοῖς. πάλιν δ' ἄνευ μὲν τοῦ ἁπτικοῦ
τῶν ἄλλων αἰσθήσεων οὐδεμία ὑπάρχει, ἁφὴ δ'

---

[a] The argument of this paragraph is as follows : Just as
figure has a definition applicable to all varieties of figures,
so soul has a similar " common " definition ; but to understand individual types of soul, e.g., of man, animal, and plant,
we must study these types themselves.

and must be considered later. In addition to these senses some also possess the power of movement in space, and others again—*viz.*, man, and any other being similar or superior to him—have the power of thinking and intelligence.

Thus it is clear that there must be a single definition of soul, just as there is of rectilinear figure ; for as in the latter case there is no figure besides the triangle and those that follow it (*i.e.*, quadrilateral, pentagon, etc.), so there is no soul besides those we have mentioned. It would be possible in the case of figures to frame a common definition, which would fit them all, but would be descriptive of no particular figure ; and similarly in the case of the kinds of soul we have mentioned. Hence it would be ridiculous, in this case as in others, to look for the common definition, which is the particular definition of no existing thing, and has no reference to any special or individual species, while we neglect such a particular definition.[a]

The facts regarding the soul are much the same as those relating to figures ; for both in figures and in things which possess soul, the earlier type always exists potentially in that which follows ; *e.g.*, the triangle is implied by the quadrilateral, and the nutritive faculty by the sensitive. We must then inquire in each several case, what is the soul of each individual, for instance of the plant, the man, and the beast. But we must also consider why they are thus arranged in a series. For without the nutritive faculty the sensitive does not exist, but in plants the nutritive is divorced from the sensitive faculty. Again, without the sense of touch none of the other senses exists, but touch may exist without any

Definition of soul

similar to mathematical definition.

83

415 a

ἄνευ τῶν ἄλλων ὑπάρχει· πολλὰ γὰρ τῶν ζῴων
οὔτ' ὄψιν οὔτ' ἀκοὴν ἔχουσιν οὔτ' ὀσμῆς ὅλως
αἴσθησιν. καὶ τῶν αἰσθητικῶν δὲ τὰ μὲν ἔχει τὸ
κατὰ τόπον κινητικόν, τὰ δ' οὐκ ἔχει. τελευταῖον
δὲ καὶ ἐλάχιστα λογισμὸν καὶ διάνοιαν· οἷς μὲν
γὰρ ὑπάρχει λογισμὸς τῶν φθαρτῶν, τούτοις καὶ
10 τὰ λοιπὰ πάντα, οἷς δ' ἐκείνων ἕκαστον, οὐ πᾶσι
λογισμός, ἀλλὰ τοῖς μὲν οὐδὲ φαντασία, τὰ δὲ
ταύτῃ μόνῃ ζῶσιν. περὶ δὲ τοῦ θεωρητικοῦ νοῦ
ἕτερος λόγος. ὅτι μὲν οὖν ὁ περὶ τούτων ἑκάστου
λόγος οὗτος οἰκειότατος καὶ περὶ ψυχῆς, δῆλον.

IV. Ἀναγκαῖον δὲ τὸν μέλλοντα περὶ τούτων
15 σκέψιν ποιεῖσθαι λαβεῖν ἕκαστον αὐτῶν τί ἐστιν,
εἶθ' οὕτως περὶ τῶν ἐχομένων ἢ καὶ περὶ τῶν ἄλλων ἐπιζητεῖν. εἰ δὲ χρὴ λέγειν τί ἕκαστον αὐτῶν,
οἷον τί τὸ νοητικὸν ἢ τὸ αἰσθητικὸν ἢ τὸ θρεπτικόν, πρότερον ἔτι λεκτέον τί τὸ νοεῖν καὶ τί τὸ
20 αἰσθάνεσθαι· πρότερον γάρ εἰσι τῶν δυνάμεων αἱ
ἐνέργειαι καὶ αἱ πράξεις κατὰ τὸν λόγον. εἰ δ'
οὕτως, τούτων δ' ἔτι πρότερα τὰ ἀντικείμενα δεῖ
τεθεωρηκέναι, περὶ ἐκείνων πρῶτον ἂν δέοι διορίσαι διὰ τὴν αὐτὴν αἰτίαν, οἷον περὶ τροφῆς καὶ
αἰσθητοῦ καὶ νοητοῦ. ὥστε πρῶτον περὶ τροφῆς
καὶ γεννήσεως λεκτέον· ἡ γὰρ θρεπτικὴ ψυχὴ καὶ
τοῖς ἄλλοις ὑπάρχει, καὶ πρώτη καὶ κοινοτάτη
25 δύναμίς ἐστι ψυχῆς, καθ' ἣν ὑπάρχει τὸ ζῆν
ἅπασιν. ἧς ἐστιν ἔργα γεννῆσαι καὶ τροφῇ
χρῆσθαι· φυσικώτατον γὰρ τῶν ἔργων τοῖς ζῶσιν,

---

[a] We proceed from the exercise of the functions, which we
know and see, to assume the existence of the faculty ; nature
proceeds in the opposite order.

others ; for many of the animals have neither vision nor hearing nor any sense of smell at all. Again, of those which have sensation, some have the locomotive faculty, and some have not. And lastly, and most rarely, living creatures have power of reasoning and thought. For those perishable creatures which have reasoning power have all the other powers as well, but not all those which have any one of them have reasoning power ; some have not even imagination, while others live in virtue of this alone. The consideration of the speculative faculty is another question. It is clear, then, that the account of each of these faculties is also the most relevant account that can be given of the soul.

IV. If one intends to investigate these faculties, one must first grasp what each of them is, and then proceed to inquire into secondary matters, and so on. But if one is to state what each of them—*e.g.*, the thinking, sensitive, or nutritive faculty—is, one must again first explain what thinking and perceiving are ; for logically [a] the exercise of their functions comes before the faculties themselves. And if this is so, and if one should have examined, even before these functions, the objects corresponding to them, then for the same reason one must first of all determine the facts about those objects, *e.g.*, about food or the object of perception or thought. First, then, we must speak of food and reproduction ; for the nutritive soul belongs to all other living creatures besides man, and is the first and most widely shared faculty of the soul, in virtue of which they all have life. Its functions are reproduction and the assimilation of food. For this is the most natural of all func-

The method of inquiry.

415 a

ὅσα τέλεια καὶ μὴ πηρώματα, ἢ τὴν γένεσιν αὐτο-
μάτην ἔχει, τὸ ποιῆσαι ἕτερον οἷον αὐτό, ζῷον
μὲν ζῷον, φυτὸν δὲ φυτόν, ἵνα τοῦ ἀεὶ καὶ τοῦ
415 b θείου μετέχωσιν ᾗ δύνανται· πάντα γὰρ ἐκείνου
ὀρέγεται, κἀκείνου ἕνεκα πράττει ὅσα πράττει κατὰ
φύσιν. τὸ δ' οὗ ἕνεκα διττόν, τὸ μὲν οὗ, τὸ δὲ
ᾧ. ἐπεὶ οὖν κοινωνεῖν ἀδυνατεῖ τοῦ ἀεὶ καὶ τοῦ
θείου τῇ συνεχείᾳ, διὰ τὸ μηδὲν ἐνδέχεσθαι τῶν
5 φθαρτῶν ταὐτὸ καὶ ἓν ἀριθμῷ διαμένειν, ᾗ δύναται
μετέχειν ἕκαστον, κοινωνεῖ ταύτῃ, τὸ μὲν μᾶλλον
τὸ δ' ἧττον· καὶ διαμένει οὐκ αὐτὸ ἀλλ' οἷον αὐτό,
ἀριθμῷ μὲν οὐχ ἕν, εἴδει δ' ἕν.

Ἔστι δὲ ἡ ψυχὴ τοῦ ζῶντος σώματος αἰτία καὶ
10 ἀρχή. ταῦτα δὲ πολλαχῶς λέγεται. ὁμοίως δ' ἡ
ψυχὴ κατὰ τοὺς διωρισμένους τρόπους τρεῖς αἰτία·
καὶ γὰρ ὅθεν ἡ κίνησις αὐτή, καὶ οὗ ἕνεκα, καὶ ὡς
ἡ οὐσία τῶν ἐμψύχων σωμάτων ἡ ψυχὴ αἰτία.
ὅτι μὲν οὖν ὡς οὐσία, δῆλον· τὸ γὰρ αἴτιον τοῦ
εἶναι πᾶσιν ἡ οὐσία, τὸ δὲ ζῆν τοῖς ζῶσι τὸ εἶναί
ἐστιν, αἰτία δὲ καὶ ἀρχὴ τούτων ἡ ψυχή. ἔτι τοῦ
15 δυνάμει ὄντος λόγος ἡ ἐντελέχεια. φανερὸν δ' ὡς
καὶ οὗ ἕνεκεν ἡ ψυχὴ αἰτία· ὥσπερ γὰρ ὁ νοῦς

---

[a] Cf. Metaph. 983 a 26 " the word cause is used . . .
(1) as the essential nature, (2) as the source of movement,
(3) as the final cause."

[b] i.e., existence and life.

[c] e.g., a sphere exists potentially in any lump of matter,
but the sphere is not actualized until in that matter there is
induced the form or essential formula of sphericity.

tions among living creatures, provided that they are perfect and not maimed, and do not have spontaneous generation : *viz.*, to reproduce one's kind, an animal producing an animal, and a plant a plant, in order that they may have a share in the immortal and divine in the only way they can ; for every creature strives for this, and for the sake of this performs all its natural functions. " That for sake of which " has two meanings : (1) that for the purpose of which, and (2) that for the benefit of which. Since, then, they cannot share in the immortal and divine by continuity of existence, because no perishable thing can remain numerically one and the same, they share in these in the only way they can, some to a greater and some to a lesser extent ; what persists is not the individual itself, but something in its image, identical not numerically but specifically.

The soul is the cause and first principle of the living body. The words cause and first principle are used in several separate senses. But the soul is equally the cause in each of the three senses which we have distinguished [a] ; for it is the cause in the sense of being that from which motion is derived, in the sense of the purpose or final cause, and as being the substance of all bodies that have souls. (1) That the soul is the cause in the sense of substance is obvious ; for substance is the cause of existence in all things, and for living creatures existence is life, and of these [b] the soul is the cause and first principle. Also the actuality of that which exists potentially is its essential formula.[c] (2) Clearly the soul is also the cause in the final sense. For just as mind acts with

The nature of the soul.

87

415 b

ἕνεκά του ποιεῖ, τὸν αὐτὸν τρόπον καὶ ἡ φύσις,
καὶ τοῦτ' ἔστιν αὐτῇ τέλος. τοιοῦτον δ' ἐν τοῖς
ζῴοις ἡ ψυχὴ καὶ κατὰ φύσιν· πάντα γὰρ τὰ
φυσικὰ σώματα τῆς ψυχῆς ὄργανα, καὶ καθάπερ
20 τὰ τῶν ζῴων, οὕτω καὶ τὰ τῶν φυτῶν, ὡς ἕνεκα
τῆς ψυχῆς ὄντα. διττῶς δὲ τὸ οὗ ἕνεκα, τό τε
οὗ καὶ τὸ ᾧ. ἀλλὰ μὴν καὶ ὅθεν πρῶτον ἡ κατὰ
τόπον κίνησις, ψυχή· οὐ πᾶσι δ' ὑπάρχει τοῖς
ζῶσιν ἡ δύναμις αὕτη. ἔστι δὲ καὶ ἀλλοίωσις καὶ
αὔξησις κατὰ ψυχήν· ἡ μὲν γὰρ αἴσθησις ἀλλοίωσίς
25 τις εἶναι δοκεῖ, αἰσθάνεται δ' οὐθὲν ὃ μὴ ἔχει
ψυχήν. ὁμοίως δὲ καὶ περὶ αὐξήσεώς τε καὶ
φθίσεως ἔχει· οὐδὲν γὰρ φθίνει οὐδ' αὔξεται
φυσικῶς μὴ τρεφόμενον, τρέφεται δ' οὐθὲν ὃ μὴ
κοινωνεῖ ζωῆς.

Ἐμπεδοκλῆς δ' οὐ καλῶς εἴρηκε τοῦτο, προστιθεὶς τὴν αὔξησιν συμβαίνειν τοῖς φυτοῖς κάτω
416 a μὲν συρριζουμένοις διὰ τὸ τὴν γῆν οὕτω φέρεσθαι
κατὰ φύσιν, ἄνω δὲ διὰ τὸ πῦρ ὡσαύτως. οὔτε
γὰρ τὸ ἄνω καὶ κάτω καλῶς λαμβάνει· οὐ γὰρ
ταὐτὸ πᾶσι τὸ ἄνω καὶ κάτω καὶ τῷ παντί, ἀλλ'
ὡς ἡ κεφαλὴ τῶν ζῴων, οὕτως αἱ ῥίζαι τῶν φυτῶν,
5 εἰ χρὴ τὰ ὄργανα λέγειν ἕτερα καὶ ταὐτὰ τοῖς
ἔργοις. πρὸς δὲ τούτοις τί τὸ συνέχον εἰς τἀναντία
φερόμενα τὸ πῦρ καὶ τὴν γῆν; διασπασθήσεται
γάρ, εἰ μή τι ἔσται τὸ κωλῦσον· εἰ δ' ἔσται, τοῦτ'
ἐστὶν ἡ ψυχὴ καὶ τὸ αἴτιον τοῦ αὐξάνεσθαι καὶ
τρέφεσθαι.

some purpose in view, so too does nature, and this purpose is its end. In living creatures the soul supplies such a purpose, and this is in accordance with nature, for all natural bodies are instruments of the soul; and just as is the case with the bodies of animals, so with those of plants. This shows that they exist for the sake of the soul. ("That for the sake of which" has two meanings—"that for the purpose of which" and "that for the benefit of which.") (3) Lastly, the soul is the primary source of locomotion; but this capacity does not belong to all living creatures. Change of state and growth are also due to the soul; for sensation is held to be change of state, and nothing feels which has not a soul. The same is true about growth and decay; for nothing decays or grows in nature without nourishment, and nothing has nourishment which does not share in life.

Empedocles is mistaken in his account of this, when he adds that the growth in plants, when their roots spread downwards, is due to the fact that earth naturally tends in this direction, and that when they grow upwards, it is due to the natural movement of fire. His theory of "upwards" and "downwards" is wrong; for up and down are not the same for all individuals as for the universe, but the head in animals corresponds to the roots in plants, if we are to identify and distinguish organs by their functions. But in addition to this, what is it that holds fire and earth together when they tend to move in contrary directions? For they will be torn apart, unless there is something to prevent this; but if there is anything of the sort this will be the soul, and the cause of growth and nourishment.

10 Δοκεῖ δέ τισιν ἡ τοῦ πυρὸς φύσις ἁπλῶς αἰτία
τῆς τροφῆς καὶ τῆς αὐξήσεως εἶναι· καὶ γὰρ
αὐτὸ φαίνεται μόνον τῶν σωμάτων ἢ τῶν στοιχείων τρεφόμενον καὶ αὐξόμενον. διὸ καὶ ἐν τοῖς
φυτοῖς καὶ ἐν τοῖς ζῴοις ὑπολάβοι τις ἂν τοῦτο
εἶναι τὸ ἐργαζόμενον. τὸ δὲ συναίτιον μέν πώς
15 ἐστιν, οὐ μὴν ἁπλῶς γε αἴτιον, ἀλλὰ μᾶλλον ἡ
ψυχή· ἡ μὲν γὰρ τοῦ πυρὸς αὔξησις εἰς ἄπειρον,
ἕως ἂν ᾖ τὸ καυστόν, τῶν δὲ φύσει συνισταμένων
πάντων ἐστὶ πέρας καὶ λόγος μεγέθους τε καὶ αὐξή-
σεως· ταῦτα δὲ ψυχῆς, ἀλλ' οὐ πυρός, καὶ λόγου
μᾶλλον ἢ ὕλης.

Ἐπεὶ δ' ἡ αὐτὴ δύναμις τῆς ψυχῆς θρεπτικὴ
20 καὶ γεννητική, περὶ τροφῆς ἀναγκαῖον διωρίσθαι
πρῶτον· ἀφορίζεται γὰρ πρὸς τὰς ἄλλας δυνάμεις
τῷ ἔργῳ τούτῳ. δοκεῖ δ' εἶναι ἡ τροφὴ τὸ
ἐναντίον τῷ ἐναντίῳ, οὐ πᾶν δὲ παντί, ἀλλ' ὅσα
τῶν ἐναντίων μὴ μόνον γένεσιν ἐξ ἀλλήλων ἔχουσιν
ἀλλὰ καὶ αὔξησιν· γίνεται γὰρ πολλὰ ἐξ ἀλλήλων,
25 ἀλλ' οὐ πάντα ποσά, οἷον ὑγιὲς ἐκ κάμνοντος.
φαίνεται δ' οὐδ' ἐκεῖνα τὸν αὐτὸν τρόπον ἀλλήλοις
εἶναι τροφή, ἀλλὰ τὸ μὲν ὕδωρ τῷ πυρὶ τροφή, τὸ
δὲ πῦρ οὐ τρέφει τὸ ὕδωρ. ἐν μὲν οὖν τοῖς ἁπλοῖς
σώμασι ταῦτ' εἶναι δοκεῖ μάλιστα τὸ μὲν τροφὴ
τὸ δὲ τρεφόμενον. ἀπορίαν δ' ἔχει· φασὶ γὰρ οἱ
30 μὲν τὸ ὅμοιον τῷ ὁμοίῳ τρέφεσθαι, καθάπερ καὶ
αὐξάνεσθαι, τοῖς δ' ὥσπερ εἴπομεν τοὔμπαλιν δοκεῖ,
τὸ ἐναντίον τῷ ἐναντίῳ, ὡς ἀπαθοῦς ὄντος τοῦ
ὁμοίου ὑπὸ τοῦ ὁμοίου, τὴν δὲ τροφὴν μεταβάλλειν
καὶ πέττεσθαι· ἡ δὲ μεταβολὴ πᾶσιν εἰς τὸ ἀντι-

To some the nature of fire seems by itself to be the Fire in relation to growth. cause of nutrition and growth; for it alone of all bodies and elements seems to be nourished and grow of itself. Hence one might suppose that it is the operating principle in both plants and animals. It is in a sense a contributory cause, but not absolutely the cause, which is much more properly the soul; for the growth of fire is without limit, so long as there is something to be burned, but of all things naturally composed there is a limit or proportion of size and growth; this is due to the soul, not to fire, and to the essential formula rather than to matter.

Since the same faculty of the soul is at once nutri- The nature of food. tive and generative, we must first define nutriment carefully; for the nutritive faculty is distinguished from the others by its function of nutrition. There is a general opinion that contrary is nutriment to contrary; not of course in every case, but among such contraries as have not merely their birth from each other, but their growth as well; for many things arise from each other, but they are not all quantities; *e.g.*, a healthy from a diseased thing. But not even the things mentioned seem to be food for each other in the same way; water feeds fire,[a] but fire does not feed water. It seems, then, that in simple bodies especially the food and the thing fed are contraries. But this presents a difficulty; for some say that like is fed, as also it grows, by like, but others, as we have said, hold the opposite view, that contrary is fed by contrary, on the ground that like is unaffected by like, but that food changes and is digested. But all change is to the opposite, or to an

---

[a] It was supposed that wood, when perfectly dry, would not keep a fire alight.

35 κείμενον ἢ τὸ μεταξύ. ἔτι πάσχει τι ἡ τροφὴ ὑπὸ
416 b τοῦ τρεφομένου, ἀλλ' οὐ τοῦτο ὑπὸ τῆς τροφῆς,
ὥσπερ οὐδ' ὁ τέκτων ὑπὸ τῆς ὕλης, ἀλλ' ὑπ' ἐκείνου αὕτη· ὁ δὲ τέκτων μεταβάλλει μόνον εἰς ἐνέργειαν ἐξ ἀργίας.

Πότερον δ' ἐστὶν ἡ τροφὴ τὸ τελευταῖον προσ-
γινόμενον ἢ τὸ πρῶτον, ἔχει διαφοράν. εἰ δ'
5 ἄμφω, ἀλλ' ἡ μὲν ἄπεπτος ἡ δὲ πεπεμμένη,
ἀμφοτέρως ἂν ἐνδέχοιτο τὴν τροφὴν λέγειν· ᾗ μὲν
γὰρ ἄπεπτος, τὸ ἐναντίον τῷ ἐναντίῳ τρέφεται,
ᾗ δὲ πεπεμμένη, τὸ ὅμοιον τῷ ὁμοίῳ. ὥστε
φανερὸν ὅτι λέγουσί τινα τρόπον ἀμφότεροι καὶ
ὀρθῶς καὶ οὐκ ὀρθῶς. ἐπεὶ δ' οὐθὲν τρέφεται
10 μὴ μετέχον ζωῆς, τὸ ἔμψυχον ἂν εἴη σῶμα
τὸ τρεφόμενον, ᾗ ἔμψυχον, ὥστε καὶ ἡ τροφὴ
πρὸς ἔμψυχόν ἐστι καὶ οὐ κατὰ συμβεβηκός.
ἔστι δ' ἕτερον τροφῇ καὶ αὐξητικῷ εἶναι· ᾗ μὲν
γὰρ ποσόν τι τὸ ἔμψυχον, αὐξητικόν, ᾗ δὲ τόδε τι
καὶ οὐσία, τροφή· σώζει γὰρ τὴν οὐσίαν, καὶ
15 μέχρι τούτου ἐστὶν ἕως ἂν καὶ τρέφηται· καὶ
γενέσεως ποιητικὸν οὐ τοῦ τρεφομένου, ἀλλ' οἷον
τὸ τρεφόμενον· ἤδη γάρ ἐστιν αὐτὴ ἡ οὐσία, γεννᾷ
δ' οὐθὲν αὐτὸ ἑαυτό, ἀλλὰ σώζει. ὥσθ' ἡ μὲν
τοιαύτη τῆς ψυχῆς ἀρχὴ δύναμίς ἐστιν οἵα σώζειν
τὸ ἔχον αὐτὴν ᾗ τοιοῦτον, ἡ δὲ τροφὴ παρασκευάζει
20 ἐνεργεῖν. διὸ στερηθὲν τροφῆς οὐ δύναται εἶναι.
ἐπεὶ δ' ἐστὶ τρία, τὸ τρεφόμενον καὶ ᾧ τρέφεται
καὶ τὸ τρέφον, τὸ μὲν τρέφον ἐστὶν ἡ πρώτη
ψυχή, τὸ δὲ τρεφόμενον τὸ ἔχον αὐτὴν σῶμα, ᾧ
δὲ τρέφεται, ἡ τροφή. ἐπεὶ δὲ ἀπὸ τοῦ τέλους

intermediate state. Again, the food is affected by the thing fed, and not *vice versa*, just as the carpenter is not affected by his material, but the material by the carpenter ; the carpenter merely changes from idleness to activity.

Now it makes a difference whether " food " means the last or the first form of what is added. If both are food, the one being undigested and the other digested, we might speak of food in both the ways referred to above ; for when the food is undigested, contrary feeds on contrary, but when it is digested, like feeds on like. Thus clearly both views are, in a sense, both right and wrong. But since nothing is fed which does not share in life, that which is fed must be the body which has a soul, *qua* having a soul, so that food is related to that which has a soul and that not accidentally. But nutrivity and the promotion of growth are not the same ; for it is *qua* quantitative that that which has soul has its growth promoted by food, and *qua* individual and substance that it is nourished by it ; for it preserves its substance and continues to exist, so long as it is nourished, and it causes the generation not of that which is nourished, but of another like it ; for its actual substance already exists, and a thing cannot generate but only preserves itself. Thus the soul-principle in question is a power of preserving what possesses it as an individual, while food prepares it for work. For this reason it cannot continue to exist when deprived of food. Now there are three separate factors : the thing fed, the means by which it is fed, and the feeding agent. The feeding agent is soul in the primary sense ; the thing fed is the body which contains the soul, and the means by which it is fed is the food. But since everything

416 b

ἅπαντα προσαγορεύειν δίκαιον, τέλος δὲ τὸ γεννῆσαι
25 οἷον αὐτό, εἴη ἂν ἡ πρώτη ψυχὴ γεννητικὴ οἷον
αὐτό. ἔστι δὲ ᾧ τρέφεται διττόν, ὥσπερ καὶ ᾧ
κυβερνᾷ, ἡ χεὶρ καὶ τὸ πηδάλιον, τὸ μὲν κινοῦν
καὶ κινούμενον, τὸ δὲ κινοῦν μόνον. πᾶσαν δ'
ἀναγκαῖον τροφὴν δύνασθαι πέττεσθαι, ἐργάζεται
δὲ τὴν πέψιν τὸ θερμόν· διὸ πᾶν ἔμψυχον ἔχει θερ-
30 μότητα. τύπῳ μὲν οὖν ἡ τροφὴ τί ἐστιν εἴρηται·
διασαφητέον δ' ἐστὶν ὕστερον περὶ αὐτῆς ἐν τοῖς
οἰκείοις λόγοις.

V. Διωρισμένων δὲ τούτων λέγωμεν κοινῇ περὶ
πάσης αἰσθήσεως. ἡ δ' αἴσθησις ἐν τῷ κινεῖσθαί
τε καὶ πάσχειν συμβαίνει, καθάπερ εἴρηται· δοκεῖ
35 γὰρ ἀλλοίωσίς τις εἶναι. φασὶ δέ τινες καὶ τὸ
417 a ὅμοιον ὑπὸ τοῦ ὁμοίου πάσχειν. τοῦτο δὲ πῶς
δυνατὸν ἢ ἀδύνατον, εἰρήκαμεν ἐν τοῖς καθόλου
λόγοις περὶ τοῦ ποιεῖν καὶ πάσχειν. ἔχει δ'
ἀπορίαν διὰ τί καὶ τῶν αἰσθήσεων αὐτῶν οὐ
γίνεται αἴσθησις, καὶ διὰ τί ἄνευ τῶν ἔξω οὐ
5 ποιοῦσιν αἴσθησιν, ἐνόντος πυρὸς καὶ γῆς καὶ τῶν
ἄλλων στοιχείων, ὧν ἐστιν ἡ αἴσθησις καθ' αὑτὰ
ἢ τὰ συμβεβηκότα τούτοις. δῆλον οὖν ὅτι τὸ
αἰσθητικὸν οὐκ ἔστιν ἐνεργείᾳ, ἀλλὰ δυνάμει μόνον.
διὸ καθάπερ τὸ καυστὸν οὐ καίεται αὐτὸ καθ' αὑτὸ
ἄνευ τοῦ καυστικοῦ· ἔκαιε γὰρ ἂν ἑαυτό, καὶ οὐθὲν
10 ἐδεῖτο τοῦ ἐντελεχείᾳ πυρὸς ὄντος. ἐπειδὴ δὲ τὸ
αἰσθάνεσθαι λέγομεν διχῶς (τό τε γὰρ δυνάμει
ἀκοῦον καὶ ὁρῶν ἀκούειν καὶ ὁρᾶν λέγομεν, κἂν

---

* No such treatise has come down to us.

should be named in view of its end, and in this case
the end is the reproduction of the species, primary
soul will be that which reproduces another like itself.
But " the means by which it is nourished " is ambigu-
ous, just like the phrase " that by which the helms-
man steers," meaning either his hand or the rudder,
the latter both moving and being moved, and the
former only moving. Now all food requires digestion,
and that which produces digestion is heat ; therefore
everything which has a soul has heat. The nature of
food has now been described in outline ; later on we
must be more precise about it in a treatise of its own.[a]

V. Having established these points, let us dis- Sensation
in general.
cuss sensation in general. Sensation consists, as has
been said, in being moved and acted upon ; for it
is held to be sort of change of state. Now some say
that like is affected only by like. But the sense
in which this is possible or impossible we have already
stated in our general account of acting and being
acted upon.[b] The question arises as to why we have
no sensation of the senses [c] themselves ; that is, why
they give no sensation apart from external objects,
although they contain fire and earth and the other
elements which (either in themselves, or by their
attributes) excite sensation. It is clear from this that
the faculty of sensation has no actual but only
potential existence. So it is like the case of fuel,
which does not burn by itself without something
to set fire to it ; for otherwise it would burn itself,
and would not need any fire actually at work. But
since we speak of perceiving in two senses (for we
say that that which has the power of hearing and

[b] *De Gen. et Corr.* i. 7. 323 b 18 *sqq.*
[c] Or rather " sense-organs."

seeing hears and sees, even if it happens to be asleep, as well as when the faculty is actually operative), so the term sensation must be used in two senses, as potential and as actual. Similarly to perceive means both to possess the faculty and to exercise it.

To begin with, let us assume that being acted upon and moved is the same as exercising the function; for movement is a form of activity, though incomplete, as has been said elsewhere.[a] But everything is acted upon and moved by something which produces an effect and actually exists. Therefore, as we have said, a thing is acted upon in one sense by like, in another by unlike; for while it is being acted upon it is unlike, but when the action is complete, it is like.

But we must also distinguish certain senses of potentiality and actuality; for so far we have been using these terms quite generally. One sense of "instructed" is that in which we might call a man instructed because he is one of a class of instructed persons who have knowledge; but there is another sense in which we call instructed a person who knows (say) grammar. Each of these two has capacity, but in a different sense: the former, because the class to which he belongs, *i.e.*, his *matter*, is of a certain kind, the latter, because he is capable of exercising his knowledge whenever he likes, provided that external causes do not prevent him. But there is a third kind of instructed person—the man who is already exercising his knowledge; he is in actuality instructed and in the strict sense knows (*e.g.*) this particular A. The first two men are both only potentially instructed; but whereas the one becomes so in actuality through a qualitative alteration by means of learning, and after frequent changes from a contrary state, the other passes by a different process

Potential and actual.

417 b αἴσθησιν ἢ τὴν γραμματικήν, μὴ ἐνεργεῖν δ' εἰς
τὸ ἐνεργεῖν ἄλλον τρόπον. οὐκ ἔστι δ' ἁπλοῦν
οὐδὲ τὸ πάσχειν, ἀλλὰ τὸ μὲν φθορά τις ὑπὸ τοῦ
ἐναντίου, τὸ δὲ σωτηρία μᾶλλον τοῦ δυνάμει ὄντος
5 ὑπὸ τοῦ ἐντελεχείᾳ ὄντος καὶ ὁμοίου, οὕτως ὡς
δύναμις ἔχει πρὸς ἐντελέχειαν· θεωροῦν γὰρ
γίγνεται τὸ ἔχον τὴν ἐπιστήμην, ὅπερ ἢ οὐκ ἔστιν
ἀλλοιοῦσθαι (εἰς αὑτὸ γὰρ ἡ ἐπίδοσις καὶ εἰς
ἐντελέχειαν) ἢ ἕτερον γένος ἀλλοιώσεως. διὸ οὐ
καλῶς ἔχει λέγειν τὸ φρονοῦν, ὅταν φρονῇ, ἀλλοιοῦσθαι, ὥσπερ οὐδὲ τὸν οἰκοδόμον ὅταν οἰκοδομῇ.
10 τὸ μὲν οὖν εἰς ἐντελέχειαν ἄγον ἐκ δυνάμει ὄντος
κατὰ τὸ νοοῦν καὶ φρονοῦν οὐ διδασκαλίαν ἀλλ'
ἑτέραν ἐπωνυμίαν ἔχειν δίκαιον· τὸ δ' ἐκ δυνάμει
ὄντος μανθάνον καὶ λαμβάνον ἐπιστήμην ὑπὸ τοῦ
ἐντελεχείᾳ ὄντος καὶ διδασκαλικοῦ ἤτοι οὐδὲ
πάσχειν φατέον, ὥσπερ εἴρηται, ἢ δύο τρόπους
15 εἶναι ἀλλοιώσεως, τήν τε ἐπὶ τὰς στερητικὰς
διαθέσεις μεταβολὴν καὶ τὴν ἐπὶ τὰς ἕξεις καὶ τὴν
φύσιν. τοῦ δ' αἰσθητικοῦ ἡ μὲν πρώτη μεταβολὴ
γίνεται ὑπὸ τοῦ γεννῶντος, ὅταν δὲ γεννηθῇ, ἔχει
ἤδη ὥσπερ ἐπιστήμην καὶ τὸ αἰσθάνεσθαι. καὶ
τὸ κατ' ἐνέργειαν δὲ ὁμοίως λέγεται τῷ θεωρεῖν·
20 διαφέρει δέ, ὅτι τοῦ μὲν τὰ ποιητικὰ τῆς ἐνεργείας
ἔξωθεν, τὸ ὁρατὸν καὶ τὸ ἀκουστόν, ὁμοίως δὲ
καὶ τὰ λοιπὰ τῶν αἰσθητῶν. αἴτιον δ' ὅτι τῶν
καθ' ἕκαστον ἡ κατ' ἐνέργειαν αἴσθησις, ἡ δ'

---

^a When we see the colour red we see a particular object,
which is red, at a particular time. But knowledge of " red "
is not knowledge of any red object at any one time but of the
colour red detached from such objects, *i.e.* in A.'s phrase-
ology as " universal."

from the inactive possession of sensation or grammar to its active exercise. Even the term " being acted upon " is not used in a single sense, but sometimes it means a form of destruction of something by its contrary, and sometimes rather a preservation of that which is potential by something actual which is like it, in accordance with the relation of potentiality to actuality ; for that which merely possesses knowledge comes to exercise it by a process which either is not alteration at all (for the development is into its real self or actuality), or else is a unique kind of alteration. So it is not sound to describe that which thinks as being altered when it thinks, any more than it is true to say that the builder is altered when he builds. That which produces development from potential to actual in the matter of understanding and thought ought not to be called teaching, but needs some other name ; and that which, starting with a potentiality for knowledge, learns and acquires knowledge from what is actual and able to teach, either ought not to be described as " being acted upon," as has been said, or else there are two senses of alteration, one a change to a negative condition, and the other a change to a positive state, that is, a realization of its nature. In sentient creatures the first change is caused by the male parent, and at birth the subject has sensation in the sense in which we spoke of the mere possession of knowledge. Again, actual sensation corresponds to the exercise of knowledge ; with this difference, that the objects of sight and hearing (and similarly those of the other senses), which produce the actuality of sensation, are external.[a] This is because actual sensation is of particulars, whereas knowledge is of universals ;

these in a sense exist in the soul itself. So it lies in man's power to use his mind whenever he chooses, but it is not in his power to experience sensation; for the presence of the sensible object is essential. The same thing is true of our knowledge of sensible objects, and for the same reason, *viz.*, that sensible objects are particular and external.

But there will be a later opportunity to clear our impressions about these things. For the moment it will be enough to establish that the term " potential " is used with two meanings; first as we might say of a boy that he is a potential general, and secondly as we might say it of an adult. These two meanings apply also to the potentially sentient. But since there is no name corresponding to this difference in meaning, and we have now explained that the meanings differ, and how they differ, we must continue to use the phrases " to be acted upon " and " altered " as though they were precise terms. The sentient subject, as we have said, is potentially such as the object of sense is actually. Thus during the process of being acted upon it is unlike, but at the end of the process it has become like that object, and shares its quality.

VI. In discussing the several senses we must speak first of their respective objects. The term " object of sense " is used of three types; two of them we say that we perceive directly, and one indirectly. Of the first two, one is an object proper to a given sense, and the other is an object perceptible by all the senses. By proper object I mean that which cannot be perceived by any other sense, and concerning which error is impossible; *e.g.*, sight is concerned with

Potential as applied to sensation.

The objects of sense.

418 a

οἷον ὄψις χρώματος καὶ ἀκοὴ ψόφου καὶ γεῦσις
χυμοῦ. ἡ δ᾽ ἀφὴ πλείους μὲν ἔχει διαφοράς· ἀλλ᾽
15 ἑκάστη γε κρίνει περὶ τούτων, καὶ οὐκ ἀπατᾶται
ὅτι χρῶμα οὐδ᾽ ὅτι ψόφος, ἀλλὰ τί τὸ κεχρω-
σμένον ἢ ποῦ, ἢ τί τὸ ψοφοῦν ἢ ποῦ. τὰ μὲν οὖν
τοιαῦτα λέγεται ἴδια ἑκάστου, κοινὰ δὲ κίνησις,
ἠρεμία, ἀριθμός, σχῆμα, μέγεθος· τὰ γὰρ τοιαῦτα
20 οὐδεμιᾶς ἐστιν ἴδια, ἀλλὰ κοινὰ πάσαις· καὶ γὰρ
ἀφῇ κίνησίς τίς ἐστιν αἰσθητὴ καὶ ὄψει. κατὰ
συμβεβηκὸς δὲ λέγεται αἰσθητόν, οἷον εἰ τὸ λευκὸν
εἴη Διάρους υἱός· κατὰ συμβεβηκὸς γὰρ τούτου
αἰσθάνεται, ὅτι τῷ λευκῷ συμβέβηκε τοῦτο οὗ
αἰσθάνεται. διὸ καὶ οὐδὲν πάσχει ᾗ τοιοῦτον ὑπὸ
τοῦ αἰσθητοῦ. τῶν δὲ καθ᾽ αὑτὰ αἰσθητῶν τὰ
25 ἴδια κυρίως ἐστὶν αἰσθητά, καὶ πρὸς ἃ ἡ οὐσία
πέφυκεν ἑκάστης αἰσθήσεως.

VII. Οὗ μὲν οὖν ἐστιν ἡ ὄψις, τοῦτ᾽ ἐστὶν
ὁρατόν. ὁρατὸν δ᾽ ἔστιν χρῶμά τε, καὶ ὃ λόγῳ
μὲν ἔστιν εἰπεῖν, ἀνώνυμον δὲ τυγχάνει ὄν· δῆλον
δὲ ἔσται ὃ λέγομεν προελθοῦσι μάλιστα. τὸ γὰρ
ὁρατόν ἐστι χρῶμα. τοῦτο δ᾽ ἐστὶ τὸ ἐπὶ τοῦ
30 καθ᾽ αὑτὸ ὁρατοῦ· καθ᾽ αὑτὸ δὲ οὐ τῷ λόγῳ, ἀλλ᾽
ὅτι ἐν ἑαυτῷ ἔχει τὸ αἴτιον τοῦ εἶναι ὁρατόν. πᾶν
418 b δὲ χρῶμα κινητικόν ἐστι τοῦ κατ᾽ ἐνέργειαν
διαφανοῦς, καὶ τοῦτ᾽ ἔστιν αὐτοῦ ἡ φύσις. διόπερ
οὐχ ὁρατὸν ἄνευ φωτός, ἀλλὰ πᾶν τὸ ἑκάστου
χρῶμα ἐν φωτὶ ὁρατόν.

colour, hearing with sound, and taste with flavour.
Touch of course has many varieties of object. Each
sense has its proper sphere, nor is it deceived as to
the fact of colour or sound, but only as to the nature
and position of the coloured object or the thing which
makes the sound. Such objects we call proper to a
particular sense, but perception of movement, rest,
number, shape and size is shared by several senses.
For things of this kind are not proper to any one
sense, but are common to all ; for instance, some kinds
of movement are perceptible both by touch and by
sight. I call an object indirectly perceived if, for
instance, the white thing seen is the son of Diares ;
this is an indirect perception, because that which
is perceived (the son of Diares) only belongs incidentally to the whiteness. Hence the percipient is not
acted upon by the thing perceived as such. But of
per se perceptibles those are most strictly perceptible
which are proper to a given sense, and it is to these
that the special nature of the several senses is
adapted.

VII. The object of sight is the visible. This is Vision.
either colour, or something which can be described
in words, but has in fact no name ; what we mean
by this will become quite clear as we proceed. The
visible, then, is colour, *i.e.* that which overlies what
is in itself visible ; by " in itself " we mean not that
the object is by its definition visible but that it has
in itself the cause of its visibility. Every colour can
produce movement in that which is actually transparent, and it is its very nature to do so. This
is why it is not visible without light, but it is only
in light that the colour of each individual thing is
seen.

103

418 b

Διὸ περὶ φωτὸς πρῶτον λεκτέον τί ἐστιν. ἔστι
δή τι διαφανές. διαφανὲς δὲ λέγω ὃ ἔστι μὲν
5 ὁρατόν, οὐ καθ᾽ αὑτὸ δὲ ὁρατὸν ὡς ἁπλῶς εἰπεῖν,
ἀλλὰ δι᾽ ἀλλότριον χρῶμα. τοιοῦτον δέ ἐστιν ἀὴρ
καὶ ὕδωρ καὶ πολλὰ τῶν στερεῶν· οὐ γὰρ ᾗ ὕδωρ
οὐδ᾽ ᾗ ἀήρ, διαφανές, ἀλλ᾽ ὅτι ἐστὶ φύσις ἐνυπ-
άρχουσα ἡ αὐτὴ ἐν τούτοις ἀμφοτέροις καὶ ἐν τῷ
ἀιδίῳ τῷ ἄνω σώματι. φῶς δέ ἐστιν ἡ τούτου
10 ἐνέργεια τοῦ διαφανοῦς ᾗ διαφανές. δυνάμει δὲ
ἐν ᾧ τοῦτ᾽ ἐστί, καὶ τὸ σκότος. τὸ δὲ φῶς οἷον
χρῶμά ἐστι τοῦ διαφανοῦς, ὅταν ᾖ ἐντελεχείᾳ
διαφανὲς ὑπὸ πυρὸς ἢ τοιούτου οἷον τὸ ἄνω σῶμα·
καὶ γὰρ τούτῳ τι ὑπάρχει ἓν καὶ ταὐτόν. τί μὲν
15 οὖν τὸ διαφανὲς καὶ τί τὸ φῶς, εἴρηται, ὅτι οὔτε
πῦρ οὔθ᾽ ὅλως σῶμα οὐδ᾽ ἀπορροὴ σώματος
οὐδενός (εἴη γὰρ ἂν σῶμά τι καὶ οὕτως), ἀλλὰ
πυρὸς ἢ τοιούτου τινὸς παρουσία ἐν τῷ διαφανεῖ·
οὐδὲ γὰρ δύο σώματα ἅμα δυνατὸν ἐν τῷ αὐτῷ
εἶναι. δοκεῖ δὲ τὸ φῶς ἐναντίον εἶναι τῷ σκότει.
ἔστι δὲ τὸ σκότος στέρησις τῆς τοιαύτης ἕξεως ἐκ
20 διαφανοῦς, ὥστε δῆλον ὅτι καὶ ἡ τούτου παρουσία
τὸ φῶς ἐστιν. καὶ οὐκ ὀρθῶς Ἐμπεδοκλῆς, οὐδ᾽
εἴ τις ἄλλος οὕτως εἴρηκεν, ὡς φερομένου τοῦ
φωτὸς καὶ γιγνομένου ποτὲ μεταξὺ τῆς γῆς καὶ
τοῦ περιέχοντος, ἡμᾶς δὲ λανθάνοντος· τοῦτο γάρ
ἐστι καὶ παρὰ τὴν τοῦ λόγου ἐνάργειαν[1] καὶ παρὰ
25 τὰ φαινόμενα· ἐν μικρῷ μὲν γὰρ διαστήματι λάθοι
ἂν, ἀπ᾽ ἀνατολῆς δ᾽ ἐπὶ δυσμὰς τὸ λανθάνειν μέγα

---

[1] ἐνάργειαν TWy, Sophonias, Torstrik : ἐνέργειαν E : ἀλήθειαν vulgo.

Consequently we must explain in the first place what light is. Transparency evidently exists. By transparent I mean that which is visible, only not absolutely and in itself, but owing to the colour of something else. This character is shared by air, water, and many solid objects ; it is not *qua* water or air that water or air is transparent, but because the same nature belongs to these two as to the everlasting upper firmament. Now light is the activity of this transparent substance *qua* transparent ; and, wherever it is present, darkness also is potentially present. Light is then in a sense the colour of the transparent, owing to fire or any such agency as the upper firmament ; for one and the same quality belongs to this also. We have thus described what the transparent is, and what light is : it is neither fire, nor in general any body, nor an emanation from any body (for in that case too it would be a body of some kind), but the presence of fire, or something of the kind, in the transparent ; for there cannot be two bodies in the same place at the same time. Light is considered to be the contrary of darkness ; but darkness is a removal from the transparent of the active condition described above, so that obviously light is the presence of this. Empedocles, and anyone else who has argued on similar lines, is wrong in saying that light travels, and arrives at a certain time between the earth and its envelope, without our noticing it ; this is contrary both to the light of reason, and to observed facts ; it would be possible for it to escape our observation in a small intervening space, but that it does so all the way between east and west is too large a claim.

418 b

λίαν τὸ αἴτημα. ἔστι δὲ χρώματος μὲν δεκτικὸν
τὸ ἄχρουν, ψόφου δὲ τὸ ἄψοφον. ἄχρουν δ᾽ ἐστὶ
τὸ διαφανὲς καὶ τὸ ἀόρατον ἢ τὸ μόλις ὁρώμενον,
οἷον δοκεῖ τὸ σκοτεινόν. τοιοῦτον δὲ τὸ διαφανὲς
30 μέν, ἀλλ᾽ οὐχ ὅταν ᾖ ἐντελεχείᾳ διαφανές, ἀλλ᾽
ὅταν δυνάμει· ἡ γὰρ αὐτὴ φύσις ὁτὲ μὲν σκότος
419 a ὁτὲ δὲ φῶς ἐστιν. οὐ πάντα δὲ ὁρατὰ ἐν φωτὶ
ἐστίν, ἀλλὰ μόνον ἑκάστου τὸ οἰκεῖον χρῶμα·
ἔνια γὰρ ἐν μὲν τῷ φωτὶ οὐχ ὁρᾶται, ἐν δὲ τῷ
σκότει ποιεῖ αἴσθησιν, οἷον τὰ πυρώδη φαινόμενα
καὶ λάμποντα (ἀνώνυμα δ᾽ ἐστὶ ταῦτα ἑνὶ ὀνόματι),
5 οἷον μύκης, κρέας,[1] κεφαλαὶ ἰχθύων καὶ λεπίδες καὶ
ὀφθαλμοί· ἀλλ᾽ οὐδενὸς ὁρᾶται τούτων τὸ οἰκεῖον
χρῶμα. δι᾽ ἣν μὲν οὖν αἰτίαν ταῦτα ὁρᾶται, ἄλλος
λόγος· νῦν δ᾽ ἐπὶ τοσοῦτον φανερόν ἐστιν, ὅτι τὸ
μὲν ἐν φωτὶ ὁρώμενον χρῶμα. διὸ καὶ οὐχ ὁρᾶται
10 ἄνευ φωτός· τοῦτο γὰρ ἦν αὐτῷ τὸ χρώματι εἶναι
τὸ κινητικῷ εἶναι τοῦ κατ᾽ ἐνέργειαν διαφανοῦς·
ἡ δ᾽ ἐντελέχεια τοῦ διαφανοῦς φῶς ἐστίν. σημεῖον
δὲ τούτου φανερόν· ἐὰν γάρ τις θῇ τὸ ἔχον χρῶμα
ἐπ᾽ αὐτὴν τὴν ὄψιν, οὐκ ὄψεται· ἀλλὰ τὸ μὲν
χρῶμα κινεῖ τὸ διαφανές, οἷον τὸν ἀέρα, ὑπὸ
15 τούτου δὲ συνεχοῦς ὄντος κινεῖται τὸ αἰσθητήριον.
οὐ γὰρ καλῶς τοῦτο λέγει Δημόκριτος οἰόμενος,
εἰ γένοιτο κενὸν τὸ μεταξύ, ὁρᾶσθαι ἂν ἀκριβῶς
καὶ εἰ μύρμηξ ἐν τῷ οὐρανῷ εἴη· τοῦτο γὰρ
ἀδύνατόν ἐστιν. πάσχοντος γάρ τι τοῦ αἰσθητικοῦ
γίνεται τὸ ὁρᾶν· ὑπ᾽ αὐτοῦ μὲν οὖν τοῦ ὁρωμένου
20 χρώματος ἀδύνατον, λείπεται δὲ ὑπὸ τοῦ μεταξύ,
ὥστ᾽ ἀναγκαῖόν τι εἶναι μεταξύ· κενοῦ δὲ γενομένου οὐχ ὅτι ἀκριβῶς, ἀλλ᾽ ὅλως οὐθὲν ὀφθήσεται.

---

[1] κρέας Chandler : κέρας.

It is the colourless which is receptive of colour, as the soundless is of sound. The transparent is colourless, and so is the invisible or barely visible, such as the dark is held to be. This, then, is the nature of the transparent, when it is not actually, but potentially transparent ; the same underlying nature is sometimes darkness and sometimes light. But not everything is visible in the light, but only the proper colour of each individual thing ; for some things are not seen in the light, but are only perceptible in the dark, such as those which appear fiery or luminous (there is no single name for these), like fungi, flesh, the heads, scales, and eyes of fishes ; but in none of these is the proper colour seen. Why such things are visible is another question ; but so much is now clear, that what is visible in light is colour. Hence too it is not seen without light ; for, as we saw, it is the essence of colour to produce movement in the actually transparent ; and the actuality of the transparent is light. The evidence for this is clear ; for if one puts that which has colour right up to the eye, it will not be visible. Colour moves the transparent medium, *e.g.*, the air, and this, being continuous, acts upon the sense organ. Democritus is mistaken in thinking that if the intervening space were empty, even an ant in the sky would be clearly visible ; for this is impossible. For vision occurs when the sensitive faculty is acted upom ; as it cannot be acted upon by the actual colour which is seen, there only remains the medium to act on it, so that some medium must exist ; in fact, if the intervening space were void, not merely would accurate vision be impossible, but nothing would be seen at all. We

The medium of vision.

419 a

δι' ἣν μὲν οὖν αἰτίαν τὸ χρῶμα ἀναγκαῖον ἐν φωτὶ
ὁρᾶσθαι, εἴρηται. πῦρ δὲ ἐν ἀμφοῖν ὁρᾶται, καὶ
ἐν σκότει καὶ ἐν φωτί, καὶ τοῦτο ἐξ ἀνάγκης· τὸ
25 γὰρ διαφανὲς ὑπὸ τούτου γίνεται διαφανές.

Ὁ δ' αὐτὸς λόγος καὶ περὶ ψόφου καὶ ὀσμῆς
ἐστίν· οὐθὲν γὰρ αὐτῶν ἁπτόμενον τοῦ αἰσθητηρίου
ποιεῖ τὴν αἴσθησιν, ἀλλ' ὑπὸ μὲν ὀσμῆς καὶ ψόφου
τὸ μεταξὺ κινεῖται, ὑπὸ δὲ τούτου τῶν αἰσθητηρίων
30 ἑκάτερον· ὅταν δ' ἐπ' αὐτὸ ἐπιτεθῇ τὸ αἰσθητήριον τὸ ψοφοῦν ἢ τὸ ὄζον, οὐδεμίαν αἴσθησιν
ποιήσει. περὶ δὲ ἁφῆς καὶ γεύσεως ἔχει μὲν
ὁμοίως, οὐ φαίνεται δέ· δι' ἣν δ' αἰτίαν, ὕστερον
ἔσται δῆλον. τὸ δὲ μεταξὺ ψόφων μὲν ἀήρ, ὀσμῆς
δ' ἀνώνυμον· κοινὸν γὰρ δή τι πάθος ἐπ' ἀέρος
καὶ ὕδατός ἐστιν, ὥσπερ τὸ διαφανὲς χρώματι,
35 οὕτω τῷ ἔχοντι ὀσμὴν ὃ ἐν ἀμφοτέροις ὑπάρχει
τούτοις· φαίνεται γὰρ καὶ τὰ ἔνυδρα τῶν ζῴων
419 b ἔχειν αἴσθησιν ὀσμῆς. ἀλλ' ὁ μὲν ἄνθρωπος καὶ
τῶν πεζῶν ὅσα ἀναπνεῖ, ἀδυνατεῖ ὀσμᾶσθαι μὴ
ἀναπνέοντα. ἡ δ' αἰτία καὶ περὶ τούτων ὕστερον
λεχθήσεται.

VIII. Νῦν δὲ πρῶτον περὶ ψόφου καὶ ἀκοῆς
5 διορίσωμεν. ἔστι δὲ διττὸς ὁ ψόφος· ὁ μὲν γὰρ
ἐνέργειά τις, ὁ δὲ δύναμις· τὰ μὲν γὰρ οὔ φαμεν
ἔχειν ψόφον, οἷον σπόγγον, ἔρια, τὰ δ' ἔχειν, οἷον
χαλκὸν καὶ ὅσα στερεὰ καὶ λεῖα, ὅτι δύναται
ψοφῆσαι· τοῦτο δ' ἐστὶν αὐτοῦ μεταξὺ καὶ τῆς

---

[a] 422 b 34 *sqq.*

[b] Fishes have certain olfactory apparatus, but it is very
doubtful whether they have a sense of smell.

[c] 421 b 13 to 422 a 6.

[d] There seems to be little point in this distinction, nor does

have then explained why colour can only be seen in the light. Now fire is visible in both darkness and light, and this is necessarily so ; for it is because of the fire that the transparent becomes transparent.

The same theory applies also to sound and smell ; no sound or smell provokes sensation because it touches the sense organ, but movement is produced in the medium by smell and sound, and in the appropriate sense organ by the medium ; but, when one puts the sounding or smelling object in contact with the sense organ, no sensation is produced. The same thing is true of touch and taste, although it is not apparent ; why this is so will become clear later on.[a] The medium in the case of sound is air, but in the case of smell has no name ; for air and water have certainly a common characteristic, which is present in both of them, and bears the same relation to that which emits smell as the transparent does to colour ; for even animals which live under water seem to have the sense of smell,[b] whereas man, and all the land animals which breathe, cannot smell except when they are breathing. The reason for this will be discussed later.[c]

VIII. Let us now first clear up certain points about sound and hearing. There are two kinds of sound, one actual, the other potential[d] ; for we say that some things have no sound, such as a sponge or wool, but that others have, such as bronze, and all things which are both solid and smooth, because they can give forth sound. That is to say, they can actually produce sound between the object itself and the organ

A medium is also necessary for sound and scent.

The conditions in which sound is heard.

A. make any use of it. What he means is that bronze has " potential " sound because sound can be produced from it, whereas from wool no sound can be produced.

419 b

ἀκοῆς ἐμποιῆσαι ψόφον ἐνεργείᾳ. γίνεται δ᾽ ὁ
10 κατ᾽ ἐνέργειαν ψόφος ἀεί τινος πρός τι καὶ ἔν
τινι· πληγὴ γάρ ἐστιν ἡ ποιοῦσα. διὸ καὶ ἀδύνατον
ἑνὸς ὄντος γενέσθαι ψόφον· ἕτερον γὰρ τὸ τύπτον
καὶ τὸ τυπτόμενον· ὥστε τὸ ψοφοῦν πρός τι
ψοφεῖ. πληγὴ δ᾽ οὐ γίνεται ἄνευ φορᾶς. ὥσπερ
δ᾽ εἴπομεν, οὐ τῶν τυχόντων πληγὴ ὁ ψόφος·
15 οὐθένα γὰρ ποιεῖ ψόφον ἔρια ἂν πληγῇ, ἀλλὰ
χαλκὸς καὶ ὅσα λεῖα καὶ κοῖλα, ὁ μὲν χαλκός, ὅτι
λεῖος· τὰ δὲ κοῖλα τῇ ἀνακλάσει πολλὰς ποιεῖ
πληγὰς μετὰ τὴν πρώτην, ἀδυνατοῦντος ἐξελθεῖν
τοῦ κινηθέντος. ἔτι ἀκούεται ἐν ἀέρι καὶ ὕδατι,
ἀλλ᾽ ἧττον. οὐκ ἔστι δὲ ψόφου κύριος ὁ ἀὴρ οὐδὲ
20 τὸ ὕδωρ· ἀλλὰ δεῖ στερεῶν πληγὴν γενέσθαι πρὸς
ἄλληλα καὶ πρὸς τὸν ἀέρα. τοῦτο δὲ γίνεται,
ὅταν ὑπομένῃ πληγεὶς ὁ ἀὴρ καὶ μὴ διαχυθῇ.
διὸ ἐὰν ταχέως καὶ σφοδρῶς πληγῇ, ψοφεῖ· δεῖ
γὰρ φθάσαι τὴν κίνησιν τοῦ ῥαπίζοντος τὴν
θρύψιν τοῦ ἀέρος, ὥσπερ ἂν εἰ σωρὸν ἢ ὁρμαθὸν
25 ψάμμου τύπτοι τις φερόμενον ταχύ.

Ἠχὼ δὲ γίνεται, ὅταν ἀπὸ τοῦ ἀέρος ἑνὸς
γενομένου διὰ τὸ ἀγγεῖον τὸ διορίσαν καὶ κωλῦσαν
θρυφθῆναι πάλιν ὁ ἀὴρ ἀπωσθῇ, ὥσπερ σφαῖρα.
ἔοικε δ᾽ ἀεὶ γίνεσθαι ἠχώ, ἀλλ᾽ οὐ σαφής, ἐπεὶ
συμβαίνει γε ἐπὶ τοῦ ψόφου καθάπερ καὶ ἐπὶ τοῦ
30 φωτός· καὶ γὰρ τὸ φῶς ἀεὶ ἀνακλᾶται (οὐδὲ γὰρ
ἂν ἐγίνετο πάντῃ φῶς, ἀλλὰ σκότος ἔξω τοῦ ἡλιου-

ᵃ Sc., when sound occurs at all.

of hearing. But the sound actually produced is of something striking against something else in a medium ; for that which produces the sound is a blow. So if there is only one condition present there can be no sound ; for the striker and the thing struck are two different things ; so that what produces the sound sounds against something else. And no blow occurs without movement. But, as we have said, sound is not caused by the collision of any two things ; for wool produces no sound when it is struck, but bronze and things which are hollow and smooth do ; bronze because it is smooth, and hollow things after the original blow produce a number of other sounds by reverberation, because that which is moved (*i.e.*, the medium) cannot escape. Sound is heard also in water as well as in air, but less loudly. But neither the air nor the water is responsible for the sound ; but there must be a striking of solid objects against each other, and against the air. This occurs when the air remains in its place when struck, and is not dispersed. Hence it can only sound under a sudden and violent blow ; for it is necessary that the movement of the striker should forestall the escape of air, just as it would be if one were to strike at a heap or revolving column of sand in rapid motion.

Echo occurs when air rebounds, like a bouncing ball, Echo. from another body of air unified by the vessel which confines it, and prevents it from escaping. It seems likely that there is always an echo,[a] but it is not always noticeable, since the same thing happens with sound as with light ; for light is always reflected (otherwise there would not be light everywhere, but there would be darkness in every region outside that

111

419 b

μένου), ἀλλ' οὐχ οὕτως ἀνακλᾶται ὥσπερ ἀφ'
ὕδατος ἢ χαλκοῦ ἢ καί τινος ἄλλου τῶν λείων,
ὥστε σκιὰν ποιεῖν, ᾗ τὸ φῶς ὁρίζομεν. τὸ δὲ
κενὸν ὀρθῶς λέγεται κύριον τοῦ ἀκούειν. δοκεῖ
γὰρ εἶναι κενὸν ὁ ἀήρ, οὗτος δ' ἐστὶν ὁ ποιῶν
35 ἀκούειν, ὅταν κινηθῇ συνεχὴς καὶ εἷς. ἀλλὰ διὰ
420 a τὸ ψαθυρὸς εἶναι οὐ γεγωνεῖ, ἂν μὴ λεῖον ᾖ τὸ
πληγέν. τότε δὲ εἷς γίνεται ἅμα διὰ τὸ ἐπίπεδον·
ἓν γὰρ τὸ τοῦ λείου ἐπίπεδον.

Ψοφητικὸν μὲν οὖν τὸ κινητικὸν ἑνὸς ἀέρος
5 συνεχείᾳ μέχρις ἀκοῆς, ἀκοὴ δὲ συμφυὴς ἀέρι.
διὰ δὲ τὸ ἐν ἀέρι εἶναι, κινουμένου τοῦ ἔξω τὸ
εἴσω κινεῖται.[1] διόπερ οὐ πάντῃ τὸ ζῷον ἀκούει,
οὐδὲ πάντῃ διέρχεται ὁ ἀήρ· οὐ γὰρ πάντῃ ἔχει
ἀέρα τὸ κινησόμενον μέρος καὶ ἔμψοφον.[2] αὐτὸς[3]
μὲν δὴ ἄψοφον ὁ ἀὴρ διὰ τὸ εὔθρυπτον· ὅταν δὲ
κωλυθῇ θρύπτεσθαι, ἡ τούτου κίνησις ψόφος. ὁ
10 δ' ἐν τοῖς ὠσὶν ἐγκατῳκοδόμηται πρὸς τὸ ἀκίνητος
εἶναι, ὅπως ἀκριβῶς αἰσθάνηται πάσας τὰς διαφορὰς τῆς κινήσεως. διὰ ταῦτα δὲ καὶ ἐν ὕδατι
ἀκούομεν, ὅτι οὐκ εἰσέρχεται πρὸς αὐτὸν τὸν
συμφυῆ ἀέρα· ἀλλ' οὐδ' εἰς τὸ οὖς διὰ τὰς ἕλικας.
15 ὅταν δὲ τοῦτο συμβῇ, οὐκ ἀκούει· οὐδ' ἂν ἡ
μῆνιγξ κάμῃ, ὥσπερ τὸ ἐπὶ τῇ κόρῃ δέρμα ὅταν
κάμῃ. ἀλλὰ καὶ σημεῖον τοῦ ἀκούειν ἢ μὴ τὸ

---

[1] κινεῖ STVW, Bekker.   [2] ἔμψοφον Torstrik : ἔμψυχον.
[3] αὐτὸς Torstrik : αὐτό.

directly illuminated by the sun), but it is not always reflected as it is by water or bronze or any other smooth surface, in such a way as to cause a shadow, which is our test of light. It is correct to say that " void " is essential to hearing ; for the air is commonly thought to be void, and it is air, when moved as one continuous whole, which causes hearing. But owing to its fragility the air produces no sound unless the object which it strikes is smooth ; when this is so, the air forms a single continuous mass, because the surface of the smooth object is a continuous unity.

That, then, is sound-producing, which can produce movement in a body of air, which is single and continuous as far as the organ of hearing, and the air is physically one with the organ of hearing ; and since this organ of hearing is in the air, when the air outside is moved, the air inside is moved too. So that the animal does not hear with every part of it, nor does the air penetrate everywhere ; for the part which will be affected and produce sound has not air everywhere in it. The air itself is incapable of sound, because it is easily dissipated ; only when there is something to prevent its dissipation does its movement result in sound. The air in the ears is lodged deep, so as to be unmoved, in order that it may accurately perceive all differences of motion. That is why we can hear even in water, for the water does nor enter as far as the air which forms part of the ear ; nor even into the ear itself, because of the convolutions. When this does occur, there is no hearing ; nor again if the membrane is damaged, just as when the membrane over the eye is damaged. A test of our hearing or not is the continual ringing in the ear

The process of hearing.

420 a

ἠχεῖν αἰεὶ τὸ οὖς ὥσπερ τὸ κέρας· ἀεὶ γὰρ οἰκείαν
τινὰ κίνησιν ὁ ἀὴρ κινεῖται ὁ ἐν τοῖς ὠσίν· ἀλλ' ὁ
ψόφος ἀλλότριος καὶ οὐκ ἴδιος. καὶ διὰ τοῦτό
φασιν ἀκούειν τῷ κενῷ καὶ ἠχοῦντι, ὅτι ἀκούομεν
20 τῷ ἔχοντι ὡρισμένον τὸν ἀέρα. πότερον δὲ ψοφεῖ τὸ τυπτόμενον ἢ τὸ τύπτον; ἢ καὶ ἄμφω,
τρόπον δ' ἕτερον· ἔστι γὰρ ὁ ψόφος κίνησις τοῦ
δυναμένου κινεῖσθαι τὸν τρόπον τοῦτον ὅνπερ τὰ
ἀφαλλόμενα ἀπὸ τῶν λείων, ὅταν τις κρούσῃ. οὐ
δὴ πᾶν, ὥσπερ εἴρηται, ψοφεῖ τυπτόμενον καὶ
25 τύπτον, οἷον ἐὰν πατάξῃ βελόνη βελόνην· ἀλλὰ
δεῖ τὸ τυπτόμενον ὁμαλὸν εἶναι, ὥστε τὸν ἀέρα
ἀθροῦν ἀφάλλεσθαι καὶ σείεσθαι. αἱ δὲ διαφοραὶ
τῶν ψοφούντων ἐν τῷ κατ' ἐνέργειαν ψόφῳ
δηλοῦνται· ὥσπερ γὰρ ἄνευ φωτὸς οὐχ ὁρᾶται τὰ
χρώματα, οὕτως οὐδ' ἄνευ ψόφου τὸ ὀξὺ καὶ τὸ
βαρύ. ταῦτα δὲ λέγεται κατὰ μεταφορὰν ἀπὸ
30 τῶν ἁπτῶν· τὸ μὲν γὰρ ὀξὺ κινεῖ τὴν αἴσθησιν ἐν
ὀλίγῳ χρόνῳ ἐπὶ πολύ, τὸ δὲ βαρὺ ἐν πολλῷ ἐπ'
ὀλίγον. οὐ δὴ ταχὺ τὸ ὀξύ, τὸ δὲ βαρὺ βραδύ,
ἀλλὰ γίνεται τοῦ μὲν διὰ τὸ τάχος ἡ κίνησις
420 b τοιαύτη, τοῦ δὲ διὰ βραδυτῆτα. καὶ ἔοικεν ἀνάλογον ἔχειν τῷ περὶ τὴν ἁφὴν ὀξεῖ καὶ ἀμβλεῖ·
τὸ μὲν γὰρ ὀξὺ οἷον κεντεῖ, τὸ δ' ἀμβλὺ οἷον
ὠθεῖ διὰ τὸ κινεῖν, τὸ μὲν ἐν ὀλίγῳ, τὸ δὲ ἐν
5 πολλῷ, ὥστε συμβαίνει τὸ μὲν ταχὺ τὸ δὲ βραδὺ
εἶναι. περὶ μὲν οὖν ψόφου ταύτῃ διωρίσθω.

Ἡ δὲ φωνὴ ψόφος τίς ἐστιν ἐμψύχου· τῶν γὰρ
ἀψύχων οὐθὲν φωνεῖ, ἀλλὰ καθ' ὁμοιότητα λέγεται

114

like a horn ; for the air in the ear always moves with a special movement of its own ; but sound is from an outside source, and not a property of the ear. This is why they say that we hear by something which is empty and resonant, because we hear by that which has the air enclosed in it. Now which makes the sound—the thing struck or the striker ? Surely both, but in different senses ; for sound is the movement of what can be moved, in the way that things rebound from a smooth surface when struck against it. But, as has been said,[a] not everything produces a sound, when it strikes or is struck, for instance, if one needle strikes another ; but that which is struck must be flat, so that the air may rebound and vibrate as one mass. But the differences in things which sound are shown in sound actually realized ; for just as colours cannot be seen without light, so sharp and heavy noises cannot be distinguished without sound. These terms are used by analogy from the sense of touch. The sharp sensation excites to a great extent in a short time, the heavy to a slight extent in a long time. It is not that the sharp is itself quick, while the heavy is slow, but that their respective movements differ in quality because of their speed and slowness. There seems to be an analogy to the sharp and blunt in the sphere of touch ; for the sharp stabs, so to speak, but the blunt pushes its way in, because the former produces its effect in a short, the latter in a long time, so that the one is swift and the other slow. So much for our analysis of the properties of sound.

Voice is the sound produced by a creature posses- Voice. sing a soul ; for inanimate things never have a voice ; they can only metaphorically be said to give voice, *e.g.*,

[a] 419 b 6.

420 b

φωνεῖν, οἷον αὐλὸς καὶ λύρα καὶ ὅσα ἄλλα τῶν
ἀψύχων ἀπότασιν ἔχει καὶ μέλος καὶ διάλεκτον·
ἔοικε γὰρ ὅτι καὶ ἡ φωνὴ ταῦτ' ἔχει, πολλὰ δὲ
10 τῶν ζῴων οὐκ ἔχουσι φωνήν, οἷον τά τε ἄναιμα
καὶ τῶν ἐναίμων ἰχθύες. καὶ τοῦτ' εὐλόγως, εἴπερ
ἀέρος κίνησίς τίς ἐστιν ὁ ψόφος. ἀλλ' οἱ λεγό-
μενοι φωνεῖν, οἷον ἐν τῷ Ἀχελῴῳ, ψοφοῦσι τοῖς
βραγχίοις ἤ τινι ἑτέρῳ τοιούτῳ. φωνὴ δ' ἐστὶ
ζῴου ψόφος, καὶ οὐ τῷ τυχόντι μορίῳ. ἀλλ' ἐπεὶ
15 πᾶν ψοφεῖ τύπτοντός τινος καί τι καὶ ἔν τινι,
τοῦτο δ' ἐστὶν ἀήρ, εὐλόγως ἂν φωνοίη ταῦτα
μόνα ὅσα δέχεται τὸν ἀέρα. ἤδη γὰρ τῷ ἀναπνεο-
μένῳ καταχρῆται ἡ φύσις ἐπὶ δύο ἔργα, καθάπερ
τῇ γλώττῃ ἐπί τε τὴν γεῦσιν καὶ τὴν διάλεκτον,
ὧν ἡ μὲν γεῦσις ἀναγκαῖον (διὸ καὶ πλείοσιν
20 ὑπάρχει), ἡ δ' ἑρμηνεία ἕνεκα τοῦ εὖ, οὕτω καὶ
τῷ πνεύματι πρός τε τὴν θερμότητα τὴν ἐντὸς
ὡς ἀναγκαῖον (τὸ δ' αἴτιον ἐν ἑτέροις εἰρήσεται)
καὶ πρὸς τὴν φωνήν, ὅπως ὑπάρχῃ τὸ εὖ. ὄργανον δὲ τῇ ἀναπνοῇ ὁ φάρυγξ· οὗ δ' ἕνεκα καὶ τὸ
μόριόν ἐστι τοῦτο, πλεύμων· τούτῳ γὰρ τῷ μορίῳ
25 πλεῖστον ἔχει τὸ θερμὸν τὰ πεζὰ τῶν ἄλλων.
δεῖται δὲ τῆς ἀναπνοῆς καὶ ὁ περὶ τὴν καρδίαν
τόπος πρῶτος. διὸ ἀναγκαῖον εἴσω ἀναπνεομένου
εἰσιέναι τὸν ἀέρα. ὥστε ἡ πληγὴ τοῦ ἀναπνεομένου
ἀέρος ὑπὸ τῆς ἐν τούτοις τοῖς μορίοις ψυχῆς πρὸς

---

[a] *Hist. Anim.* iv. 9, 535 b 14 " the boar in the Achelous."
A cat-fish has been found in the Achelous of which this is
true.

[b] *Cf. De Resp.* ch. viii.

a flute or a lyre, and all the other inanimate things which have a musical compass, and tune, and modulation. The metaphor is due to the fact that the voice also has these, but many animals—*e.g.*, those which are bloodless, and of animals which have blood, fish —have no voice. And this is quite reasonable, since sound is a kind of movement of the air. The fish, such as those in the Achelous,[a] which are said to have a voice, only make a sound with their gills, or with some other such part. Voice, then, is a sound made by a living animal, and that not with any part of it indiscriminately. But, since sound only occurs when something strikes something else in a certain medium, and this medium is the air, it is natural that only those things should have voice which admit the air. As air is breathed in Nature makes use of it for two functions : just as she uses the tongue both for taste and for articulation, of which taste is an essential to life (and consequently belongs to more species), and articulate speech is an aid to living well ; so in the same way she employs breath both to conserve internal heat, as something essential (why it is so will be explained in another treatise),[b] and also for the voice, that life may be of good standard. The organ of respiration is the throat, and the part which this is designed to serve is the lung ; it is because of this part that the land animals have more heat than the rest. But the region about the heart also has a primary need of respiration. Hence it is necessary that in respiration the air should enter the body. Hence voice consists in the impact of the inspired air upon what is called the windpipe under the agency of the soul in those parts. For, as we have said, not every sound made by a living creature is a voice (for one can make a sound even with the tongue, or as in coughing), but that which even causes the impact, must have a soul, and use some imagination ; for the voice is a sound which means something, and is not merely indicative of air inhaled, as a cough is ; in uttering voice the agent uses the respired air to strike the air in the windpipe against the windpipe itself. Proof of this lies in the fact that it is impossible to speak either when inhaling or exhaling, but only when holding the breath ; for it is only in holding the breath that one can make this movement. It is clear also why fish are dumb ; it is because they have no throat. They have not this organ because they do not take in air or breathe. The reason for this is another question.[a]

IX. Concerning sense of smell and objects smelt, Smell. it is less easy to give a precise account than in the subjects we have already discussed, for the character of smell is not so obvious as that of sound and colour. The reason is that this sense with us is not highly discriminating, far less so indeed than with many animals ; for man's sense of smell is inferior, and it is also incapable of apprehending the object smelt without a consciousness of either pleasure or pain, which shows that the sense organ is not discriminating. It is probable that the hard-eyed animals perceive colours in a similar way, and that they are incapable of distinguishing colours except as to be feared or the reverse. The human race is in the same position towards smells ; it would seem that there is an analogy between smell and taste, and that the species of tastes correspond to those of smells, but that taste is with us more discriminating because it is

421 a

20 ἀφήν τινα, ταύτην δ' ἔχειν τὴν αἴσθησιν τὸν
ἄνθρωπον ἀκριβεστάτην· ἐν μὲν γὰρ ταῖς ἄλλαις
λείπεται πολλῶν¹ τῶν ζῴων, κατὰ δὲ τὴν ἀφὴν
πολλῷ² τῶν ἄλλων διαφερόντως ἀκριβοῖ. διὸ καὶ
φρονιμώτατόν ἐστι τῶν ζῴων. σημεῖον δὲ τὸ καὶ
ἐν τῷ γένει τῶν ἀνθρώπων παρὰ τὸ αἰσθητήριον
25 τοῦτο εἶναι εὐφυεῖς καὶ ἀφυεῖς, παρ' ἄλλο δὲ μηδέν·
οἱ μὲν γὰρ σκληρόσαρκοι ἀφυεῖς τὴν διάνοιαν, οἱ δὲ
μαλακόσαρκοι εὐφυεῖς.

Ἔστι δ', ὥσπερ χυμὸς ὁ μὲν γλυκὺς ὁ δὲ πικρός,
οὕτω καὶ ὀσμαί. ἀλλὰ τὰ μὲν ἔχουσι τὴν ἀνάλογον
ὀσμὴν καὶ χυμόν (λέγω δὲ οἷον γλυκεῖαν ὀσμὴν
30 καὶ γλυκὺν χυμόν), τὰ δὲ τοὐναντίον. ὁμοίως δὲ
καὶ δριμεῖα καὶ αὐστηρὰ καὶ ὀξεῖα καὶ λιπαρά ἐστιν
ὀσμή. ἀλλ' ὥσπερ εἴπομεν, διὰ τὸ μὴ σφόδρα δια-
δήλους εἶναι τὰς ὀσμὰς ὥσπερ τοὺς χυμούς, ἀπὸ
421 b τούτων εἴληφε τὰ ὀνόματα καθ' ὁμοιότητα τῶν
πραγμάτων· ἡ μὲν γὰρ γλυκεῖα ἀπὸ τοῦ κρόκου
καὶ τοῦ μέλιτος, ἡ δὲ δριμεῖα θύμου καὶ τῶν
τοιούτων· τὸν αὐτὸν δὲ τρόπον καὶ ἐπὶ τῶν ἄλλων.
ἔστι δ' ὥσπερ ἡ ἀκοὴ καὶ ἑκάστη τῶν αἰσθήσεων,
5 ἡ μὲν τοῦ ἀκουστοῦ καὶ ἀνηκούστου, ἡ δὲ τοῦ
ὁρατοῦ καὶ ἀοράτου, καὶ ἡ ὄσφρησις τοῦ ὀσφραντοῦ
καὶ ἀνοσφράντου. ἀνόσφραντον δὲ τὸ μὲν παρὰ
τὸ ὅλως ἀδύνατον ἔχειν ὀσμήν, τὸ δὲ μικρὰν ἔχον
καὶ φαύλην. ὁμοίως δὲ καὶ τὸ ἄγευστον λέγεται.
ἔστι δὲ καὶ ἡ ὄσφρησις διὰ τοῦ μεταξύ, οἷον ἀέρος
10 ἢ ὕδατος· καὶ γὰρ τὰ ἔνυδρα δοκοῦσιν ὀσμῆς
αἰσθάνεσθαι. ὁμοίως δὲ καὶ τὰ ἔναιμα καὶ τὰ
ἄναιμα, ὥσπερ καὶ τὰ ἐν τῷ ἀέρι· καὶ γὰρ τούτων

---

¹ πολλῶν comm. vet. : πολλῷ ESUV, Bekker.
² πολλῶν Bekker.

120

itself a form of touch, and this sense in man is highly discriminating ; in the other senses he is behind many kinds of animal, but in touch he is much more discriminating than the other animals. This is why he is of all living creatures the most intelligent. Proof of this lies in the fact that among the human race men are well or poorly endowed with intelligence in proportion to their sense of touch, and no other sense ; for men of hard skin and flesh are poorly, and men of soft flesh well endowed with intelligence.

Just as flavours are sweet and bitter, so are smells. In some things the smell corresponds to the taste, *e.g.*, both smell and taste are sweet ; in others they are contrasted. Smell, like flavour, may be pungent, rough, acid or oily. But, as we have said, smells being, unlike flavours, not easily differentiated, they have taken their names from flavours on the ground of a correspondence between them ; the smell of saffron or honey is called sweet, while that of thyme and similar herbs is called pungent ; and similarly in other cases. Just as hearing or any other given sense has for its object both the audible and the inaudible, or both the visible and the invisible, so smell has both the odorous and the odourless. " Odourless " means both that which has no smell at all and that which has a very small and slight smell. The sense of smell also operates through a medium, such as air or water ; for the water animals too, whether they have blood or not, seem to have a sense of smell, just like those which live in the air; for some of them, guided by the scent, come from a great distance to find their food.

Hence a problem presents itself. All animals smell in the same way, while man only smells during inhalation; when not inhaling, but either exhaling or holding the breath, he cannot smell either at a distance or at close range, not even if the object of smell is placed inside and in contact with the nostril. That what is placed on the sense organ should be imperceptible is common to all senses; but to perceive no smell without inhaling seems to be peculiar to man. The fact, when tested, is obvious. So that bloodless animals, since they do not inhale, would seem to have another sense beyond the usually accepted ones. But this is impossible, if what they perceive is smell; for perception of the odorous, and of sweet or foul smell, is an act of smelling. Again, we can observe that they are destroyed by the same strong smells, such as bitumen, sulphur and the like, which destroy man. So they must smell, but without inhaling.

Probably this sense organ in man differs from its counterpart in other living creatures, just as human eyes are different from those of hard-eyed animals; for human eyes have lids as a covering, and a sheath as it were, without moving and raising which they cannot see; but hard-eyed animals have nothing of the kind, but see directly what appears in the transparency. Similarly in some animals the organ of smell is uncovered like the eye, but others which admit the air have a veil which is lifted when they inhale, the veins and passages dilating. This is why animals which inhale do not smell in water; for to

How the sense of smell operates.

Man differs from the animal.

123

422 a
5 ἐν τῷ ὑγρῷ· ἀναγκαῖον γὰρ ὀσφρανθῆναι ἀναπνεύσαντα, τοῦτο δὲ ποιεῖν ἐν τῷ ὑγρῷ ἀδύνατον. ἔστι
δ' ἡ ὀσμὴ τοῦ ξηροῦ, ὥσπερ ὁ χυμὸς τοῦ ὑγροῦ·
τὸ δὲ ὀσφραντικὸν αἰσθητήριον δυνάμει τοιοῦτον.

X. Τὸ δὲ γευστόν ἐστιν ἁπτόν τι· καὶ τοῦτ'
αἴτιον τοῦ μὴ εἶναι αἰσθητὸν διὰ τοῦ μεταξὺ
10 ἀλλοτρίου ὄντος σώματος· οὐδὲ γὰρ ἡ ἁφή. καὶ
τὸ σῶμα δὲ ἐν ᾧ ὁ χυμός, τὸ γευστόν, ἐν ὑγρῷ
ὡς ὕλῃ· τοῦτο δ' ἁπτόν τι. διὸ κἂν εἰ ἐν ὕδατι
εἶμεν, αἰσθανοίμεθ' ἂν ἐμβληθέντος τοῦ γλυκέος,
οὐκ ἦν δ' ἂν ἡ αἴσθησις ἡμῖν διὰ τοῦ μεταξύ,
ἀλλὰ τῷ μιχθῆναι τῷ ὑγρῷ, καθάπερ ἐπὶ τοῦ
15 ποτοῦ. τὸ δὲ χρῶμα οὐχ οὕτως ὁρᾶται τῷ
μίγνυσθαι, οὐδὲ ταῖς ἀπορροίαις. ὡς μὲν οὖν τὸ
μεταξὺ οὐθέν ἐστιν· ὡς δὲ χρῶμα τὸ ὁρατόν, οὕτω
τὸ γευστὸν ὁ χυμός. οὐθὲν δὲ ποιεῖ χυμοῦ αἴσθησιν ἄνευ ὑγρότητος, ἀλλ' ἔχει ἐνεργείᾳ ἢ δυνάμει
ὑγρότητα, οἷον τὸ ἁλμυρόν· εὔτηκτόν τε γὰρ αὐτὸ
καὶ συντηκτικὸν γλώττης.

20 Ὥσπερ δὲ καὶ ἡ ὄψις ἐστὶ τοῦ τε ὁρατοῦ καὶ
τοῦ ἀοράτου (τὸ γὰρ σκότος ἀόρατον, κρίνει δὲ
καὶ τοῦτο ἡ ὄψις), ἔτι τοῦ λίαν λαμπροῦ (καὶ γὰρ
τοῦτο ἀόρατον, ἄλλον δὲ τρόπον τοῦ σκότους),
ὁμοίως δὲ καὶ ἡ ἀκοὴ ψόφου τε καὶ σιγῆς, ὧν
τὸ μὲν ἀκουστὸν τὸ δ' οὐκ ἀκουστόν, καὶ μεγάλου
25 ψόφου, καθάπερ ἡ ὄψις τοῦ λαμπροῦ· ὥσπερ γὰρ
ὁ μικρὸς ψόφος ἀνήκουστος τρόπον τινά, καὶ ὁ
μέγας τε καὶ ὁ βίαιος· ἀόρατον δὲ τὸ μὲν ὅλως
λέγεται, ὥσπερ καὶ ἐπ' ἄλλων τὸ ἀδύνατον, τὸ δ',
124

smell they must first inhale, and it is impossible to do this in water. Smell belongs to the dry, as flavour does to the wet ; and the organ of smell is potentially dry.

X. The tasteable is a kind of tangible ; and this is the reason why it is not perceptible through the medium of any foreign body ; for the same thing is true of touch. Further, the tasteable body in which flavour resides is in a liquid material medium ; and this is tangible. Hence if we lived in water, we should perceive sweetness injected into it, but our perception would not come through any medium, but would be due to the mixing of the sweet stuff with the water, just as in a drink. But colour is not seen by being mixed, nor by an emanation. Taste has nothing, then, to act as a medium ; although as the object of sight is colour, so the object of taste is flavour. But nothing can produce a perception of flavour without liquid ; it must possess wetness actually or potentially, like salt, which is both soluble itself and solvent of the tongue.

Taste like touch requires no medium,

but liquid must be present.

Now sight is concerned both with what can and with what cannot be seen (for darkness cannot be seen, and the power of sight distinguishes darkness), and also with that which is too bright (for this also cannot be seen, though in a different sense from darkness) ; and in the same way the power of hearing is concerned with both sound and silence, the former being that which is heard, and the latter that which is not heard, and also with very loud noise, just as sight is concerned with the too bright ; for as a slight sound is in a sense inaudible, so also is a loud and violent one. The word invisible is used in one sense quite generally, like other terms which deny a

125

422 a

ἐὰν πεφυκὸς μὴ ἔχῃ ἢ φαύλως, ὥσπερ τὸ ἄπουν
καὶ τὸ ἀπύρηνον· οὕτω δὴ καὶ ἡ γεῦσις τοῦ
30 γευστοῦ τε καὶ ἀγεύστου· τοῦτο δὲ τὸ μικρὸν ἢ
φαῦλον ἔχον χυμὸν ἢ φθαρτικὸν τῆς γεύσεως.
δοκεῖ δ' εἶναι ἀρχὴ τὸ ποτὸν καὶ ἄποτον· γεῦσις
γάρ τις ἀμφότερα· ἀλλὰ τὸ μὲν φαύλη καὶ φθαρτικὴ τῆς γεύσεως, τὸ δὲ κατὰ φύσιν. ἔστι δὲ
κοινὸν ἁφῆς καὶ γεύσεως τὸ ποτόν.

422 b Ἐπεὶ δ' ὑγρὸν τὸ γευστόν, ἀνάγκη καὶ τὸ
αἰσθητήριον αὐτοῦ μήτε ὑγρὸν εἶναι ἐντελεχείᾳ
μήτε ἀδύνατον ὑγραίνεσθαι· πάσχει γάρ τι ἡ γεῦσις ὑπὸ τοῦ γευστοῦ, ᾗ γευστόν. ἀναγκαῖον ἄρα
ὑγρανθῆναι τὸ δυνάμενον μὲν ὑγραίνεσθαι σωζό-
5 μενον, μὴ ὑγρὸν δέ, τὸ γευστικὸν αἰσθητήριον.
σημεῖον δὲ τὸ μήτε κατάξηρον οὖσαν τὴν γλῶτταν
αἰσθάνεσθαι μήτε λίαν ὑγράν· αὕτη γὰρ ἁφὴ γίνεται τοῦ πρώτου ὑγροῦ, ὥσπερ ὅταν προγευμα-
τίσας τις ἰσχυροῦ χυμοῦ γεύηται ἑτέρου· καὶ
οἷον τοῖς κάμνουσι πικρὰ πάντα φαίνεται διὰ τὸ
10 τῇ γλώττῃ πλήρει τοιαύτης ὑγρότητος αἰσθάνεσθαι. τὰ δ' εἴδη τῶν χυμῶν, ὥσπερ καὶ ἐπὶ τῶν
χρωμάτων, ἁπλᾶ μὲν τἀναντία, τὸ γλυκὺ καὶ τὸ
πικρόν, ἐχόμενα δὲ τοῦ μὲν τὸ λιπαρόν, τοῦ δὲ τὸ
ἁλμυρόν· μεταξὺ δὲ τούτων τό τε δριμὺ καὶ τὸ
αὐστηρὸν καὶ στρυφνὸν καὶ ὀξύ· σχεδὸν γὰρ αὗται
15 δοκοῦσιν εἶναι διαφοραὶ χυμῶν. ὥστε τὸ γευ-
στικόν ἐστι τὸ δυνάμει τοιοῦτον, γευστὸν δὲ τὸ
ποιητικὸν ἐντελεχείᾳ αὐτοῦ.

XI. Περὶ δὲ τοῦ ἁπτοῦ καὶ ἁφῆς ὁ αὐτὸς λόγος.
126

capacity, and also in the sense of not having the quality or having it to a very small extent, though by nature qualified to possess it, like the words footless or stoneless. In just the same way the sense of taste is concerned with both that which is tasted and that which is not ; the latter being that which has little or poor taste, or which is destructive of taste. The ultimate distinction seems to lie in the drinkable and undrinkable ; for each implies a tasting, but the latter is bad and destructive of taste, while the former is natural. The drinkable is an object common to both touch and taste.

Since what is tasted is wet, the organ which perceives it must be neither actually liquid nor incapable of liquefaction ; for taste is affected by the object of taste, in so far as it is tasted. Hence there must be liquefaction of the organ of taste, which must be liquefiable without loss of identity, but not liquid. This is proved by the fact that the tongue is as insensitive when too wet as when quite dry ; what happens is a contact with the moisture already in the tongue, as when a man having first tasted a strong flavour then tastes another ; or as when a man is ill everything tastes bitter, because he perceives it with a tongue filled with bitter fluid. The types of flavours, just as in the case of colours, in their simplest form are contraries, sweet and bitter ; next to these respectively are oily and saline ; between these latter come pungent, rough, astringent and acid. These seem to be nearly all the differences in flavours. Hence what is capable of tasting is that which potentially has these qualities ; and the tasteable is that which actualizes this potentiality.

XI. The same account applies to the tangible and Touch.

127

422 b

εἰ γὰρ ἡ ἀφὴ μὴ μία ἐστὶν αἴσθησις ἀλλὰ πλείους,
ἀναγκαῖον καὶ τὰ ἁπτὰ αἰσθητὰ πλείω εἶναι.
20 ἔχει δ' ἀπορίαν πότερον πλείους εἰσὶν ἢ μία, καὶ
τί τὸ αἰσθητήριον τὸ τοῦ ἁπτοῦ ἁπτικόν, πότερον
ἡ σὰρξ καὶ ἐν τοῖς ἄλλοις τὸ ἀνάλογον, ἢ οὔ, ἀλλὰ
τοῦτο μέν ἐστι τὸ μεταξύ, τὸ δὲ πρῶτον αἰσθητήριον ἄλλο τί ἐστιν ἐντός. πᾶσά τε γὰρ αἴσθησις
μιᾶς ἐναντιώσεως εἶναι δοκεῖ, οἷον ὄψις λευκοῦ
25 καὶ μέλανος καὶ ἀκοὴ ὀξέος καὶ βαρέος καὶ γεῦσις
πικροῦ καὶ γλυκέος· ἐν δὲ τῷ ἁπτῷ πολλαὶ ἔνεισιν
ἐναντιώσεις, θερμὸν ψυχρόν, ξηρὸν ὑγρόν, σκληρὸν
μαλακόν, καὶ τῶν ἄλλων ὅσα τοιαῦτα. ἔχει δέ
τινα λύσιν πρός γε ταύτην τὴν ἀπορίαν, ὅτι καὶ
ἐπὶ τῶν ἄλλων αἰσθήσεών εἰσιν ἐναντιώσεις
30 πλείους, οἷον ἐν φωνῇ οὐ μόνον ὀξύτης καὶ
βαρύτης, ἀλλὰ καὶ μέγεθος καὶ μικρότης καὶ
λειότης καὶ τραχύτης φωνῆς καὶ τοιαῦθ' ἕτερα.
εἰσὶ δὲ καὶ περὶ χρῶμα διαφοραὶ τοιαῦται ἕτεραι.
ἀλλὰ τί τὸ ἓν τὸ ὑποκείμενον, ὥσπερ ἀκοῇ ψόφος,
οὕτω τῇ ἀφῇ, οὐκ ἔστιν ἔνδηλον.

Πότερον δ' ἐστὶ τὸ αἰσθητήριον ἐντός, ἢ οὔ,
423 a ἀλλ' εὐθέως ἡ σάρξ; οὐδὲν δοκεῖ σημεῖον εἶναι
τὸ γίνεσθαι τὴν αἴσθησιν ἅμα θιγγανομένων. καὶ
γὰρ νῦν εἴ τις περὶ τὴν σάρκα περιτείνειεν οἷον
ὑμένα ποιήσας, ὁμοίως τὴν αἴσθησιν εὐθέως ἁψά-
μενος ἐνσημαίνει· καίτοι δῆλον ὡς οὐκ ἔστιν ἐν
5 τούτῳ τὸ αἰσθητήριον· εἰ δὲ καὶ συμφυὲς γένοιτο,
θᾶττον ἔτι διϊκνοῖτ' ἂν ἡ αἴσθησις. διὸ τὸ τοιοῦτο
μόριον τοῦ σώματος ἔοικεν οὕτως ἔχειν ὥσπερ ἂν
εἰ κύκλῳ ἡμῖν περιεπεφύκει ὁ ἀήρ· ἐδοκοῦμεν

the sense of touch. For if touch is not one sense, but several, there must be several kinds of tangibles. It is difficult to say whether touch is one sense or more than one, and also what the organ is which is perceptive of the object of touch ; whether it is flesh, and whatever is analogous to this in creatures without flesh, or whether this is only the medium, and the primary sense organ is something distinct and internal. For every sensation appears to be concerned with one pair of contraries, e.g., vision is of white and black, hearing of high and low pitch, and taste of bitter and sweet ; but in the tangible there are many pairs of contraries, hot and cold, dry and wet, hard and soft, and all other like qualities. Some solution may be found to this difficulty in the fact that the other senses too are conscious of more than one pair of contraries : so in sound there is not merely high and low pitch, but also loud and soft, smooth and rough, and so on. There are similarly other differences in colour. But what in the case of touch is the single substrate corresponding to sound in hearing is not obvious.

Whether the sense organ is within, or whether the flesh feels directly, is not decided by the fact that sensation occurs instantly upon contact. For even as it is, if the flesh is surrounded with a closely fitting fabric, as soon as this is touched sensation is registered as before ; yet it is quite clear that the sense organ is not in the fabric. And if the fabric actually grew on the flesh, the sensation would traverse it even more quickly. So this part of the body seems to have much the same effect as that of a natural envelope of air ; for in that case we should suppose that our

What is the sense organ of touch ?

129

γὰρ ἂν ἑνί τινι αἰσθάνεσθαι καὶ ψόφου καὶ χρώματος καὶ ὀσμῆς, καὶ μία τις αἴσθησις εἶναι ὄψις
10 ἀκοὴ ὄσφρησις. νῦν δὲ διὰ τὸ διωρίσθαι δι' οὗ
γίνονται αἱ κινήσεις, φανερὰ τὰ εἰρημένα αἰσθητήρια ἕτερα ὄντα. ἐπὶ δὲ τῆς ἁφῆς τοῦτο νῦν
ἄδηλον· ἐξ ἀέρος μὲν γὰρ ἢ ὕδατος ἀδύνατον συ-
στῆναι τὸ ἔμψυχον σῶμα· δεῖ γάρ τι στερεὸν
εἶναι. λείπεται δὴ μικτὸν ἐκ γῆς καὶ τούτων
15 εἶναι, οἷον βούλεται ἡ σὰρξ καὶ τὸ ἀνάλογον· ὥστε
ἀναγκαῖον καὶ τὸ σῶμα εἶναι μεταξὺ τοῦ ἁπτικοῦ
προσπεφυκός, δι' οὗ γίνονται αἱ αἰσθήσεις πλείους
οὖσαι. δηλοῖ δ' ὅτι πλείους ἡ ἐπὶ τῆς γλώττης
ἁφή· ἁπάντων γὰρ τῶν ἁπτῶν αἰσθάνεται κατὰ
τὸ αὐτὸ μόριον καὶ χυμοῦ. εἰ μὲν οὖν καὶ ἡ
20 ἄλλη σὰρξ ᾐσθάνετο τοῦ χυμοῦ, ἐδόκει ἂν ἡ αὐτὴ
καὶ μία εἶναι αἴσθησις ἡ γεῦσις καὶ ἡ ἁφή· νῦν
δὲ δύο διὰ τὸ μὴ ἀντιστρέφειν.

Ἀπορήσειε δ' ἄν τις, εἰ πᾶν σῶμα βάθος ἔχει,
τοῦτο δ' ἐστὶ τὸ τρίτον μέγεθος· ὧν δ' ἐστὶ δύο
σωμάτων μεταξὺ σῶμά τι, οὐκ ἐνδέχεται ταῦτα
ἀλλήλων ἅπτεσθαι. τὸ δ' ὑγρὸν οὐκ ἔστιν ἄνευ
25 σώματος, οὐδὲ τὸ διερόν, ἀλλ' ἀναγκαῖον ὕδωρ
εἶναι ἢ ἔχειν ὕδωρ. τὰ δὲ ἁπτόμενα ἀλλήλων ἐν
τῷ ὕδατι, μὴ ξηρῶν τῶν ἄκρων ὄντων, ἀναγκαῖον
ὕδωρ ἔχειν μεταξύ, οὗ ἀνάπλεα τὰ ἔσχατα. εἰ
δὲ τοῦτ' ἀληθές, ἀδύνατον ἅψασθαι ἄλλο ἄλλου
ἐν ὕδατι. τὸν αὐτὸν δὲ τρόπον καὶ ἐν τῷ ἀέρι·
30 ὁμοίως γὰρ ἔχει ὁ ἀὴρ πρὸς τὰ ἐν αὐτῷ καὶ τὸ
ὕδωρ πρὸς τὰ ἐν τῷ ὕδατι. λανθάνει δὲ μᾶλλον
423 b ἡμᾶς, ὥσπερ καὶ τὰ ἐν τῷ ὕδατι ζῷα, εἰ διερὸν

perception of sound, colour, and smell were all due to the one thing, and that vision, hearing, and smell were all one and the same sense. But, as it is, since the medium through which the movements occur is detached, it is obvious that the sense organs in question are different. But in the case of touch the fact is still obscure. The animate body cannot be made of water or air ; it must be something solid. The alternative is that it is a mixture of these elements with earth, as flesh and its equivalent tend to be ; so that the tactual medium through which the several sensations are felt must be an organically attached body. That they are several is clear from a consideration of touch in the case of the tongue ; for the tongue perceives all tangible objects with the same part with which it perceives flavour. If then the rest of the flesh also could perceive flavour, taste and touch would seem to be one and the same sense. But, as it is, they are proved two, because they are not convertible.

But here a difficulty arises. Every body has depth, *i.e.*, the third dimension ; and when two bodies have a third between them, they cannot touch each other. But the liquid or wet cannot exist without a body, and must either be, or contain, water. Those things, then, which touch each other in water, as their extremities are not dry, must have water between them, of which their extremities are full. If this is true, then it is impossible for one thing to touch another in water. The same thing will be true of air, for air has the same relation to things in it, as water has to things in water. But we tend to overlook this point, just as animals living in water do not notice if the surfaces of things which touch are wet.

The medium in the case of touch.

131

423 b

διεροῦ ἅπτεται. πότερον οὖν πάντων ὁμοίως ἐστὶν
ἡ αἴσθησις, ἢ ἄλλων ἄλλως, καθάπερ νῦν δοκεῖ
ἡ μὲν γεῦσις καὶ ἡ ἁφὴ τῷ ἅπτεσθαι, αἱ δ' ἄλλαι
ἄποθεν; τὸ δ' οὐκ ἔστιν, ἀλλὰ καὶ τὸ σκληρὸν
5 καὶ τὸ μαλακὸν δι' ἑτέρων αἰσθανόμεθα, ὥσπερ
καὶ τὸ ψοφητικὸν καὶ τὸ ὁρατὸν καὶ τὸ ὀσφραντόν·
ἀλλὰ τὰ μὲν πόρρωθεν, τὰ δ' ἐγγύθεν. διὸ λανθάνει, ἐπεὶ αἰσθανόμεθά γε πάντων διὰ τοῦ μέσου·
ἀλλ' ἐπὶ τούτων λανθάνει. καίτοι καθάπερ εἴπαμεν
καὶ πρότερον, κἂν εἰ δι' ὑμένος αἰσθανοίμεθα τῶν
10 ἁπτῶν ἁπάντων λανθάνοντος ὅτι διείργει, ὁμοίως
ἂν ἔχοιμεν ὥσπερ καὶ νῦν ἐν τῷ ὕδατι καὶ ἐν τῷ
ἀέρι· δοκοῦμεν γὰρ αὐτῶν ἅπτεσθαι καὶ οὐδὲν
εἶναι διὰ μέσου. ἀλλὰ διαφέρει τὸ ἁπτὸν τῶν
ὁρατῶν καὶ τῶν ψοφητικῶν, ὅτι ἐκείνων μὲν
αἰσθανόμεθα τῷ τὸ μεταξὺ ποιεῖν τι ἡμᾶς, τῶν
15 δὲ ἁπτῶν οὐχ ὑπὸ τοῦ μεταξὺ ἀλλ' ἅμα τῷ μεταξύ,
ὥσπερ ὁ δι' ἀσπίδος πληγείς· οὐ γὰρ ἡ ἀσπὶς
πληγεῖσα ἐπάταξεν, ἀλλ' ἅμ' ἄμφω[1] συνέβη πλη-
γῆναι. ὅλως δ' ἔοικεν ἡ σὰρξ καὶ ἡ γλῶττα, ὡς
ὁ ἀὴρ καὶ τὸ ὕδωρ πρὸς τὴν ὄψιν καὶ τὴν ἀκοὴν
καὶ τὴν ὄσφρησιν ἔχουσιν, οὕτως ἔχειν πρὸς τὸ
20 αἰσθητήριον ὥσπερ ἐκείνων ἕκαστον. αὐτοῦ δὲ
τοῦ αἰσθητηρίου ἁπτομένου οὔτ' ἐκεῖ οὔτ' ἐνταῦθα
γένοιτ' ἂν αἴσθησις, οἷον εἴ τις σῶμα τὸ λευκὸν
ἐπὶ τοῦ ὄμματος θείη τὸ ἔσχατον. ᾗ καὶ δῆλον
ὅτι ἐντὸς τὸ τοῦ ἁπτοῦ αἰσθητικόν. οὕτω γὰρ
ἂν συμβαίνοι ὅπερ καὶ ἐπὶ τῶν ἄλλων· ἐπι-
25 τιθεμένων γὰρ ἐπὶ τὸ αἰσθητήριον οὐκ αἰσθάνεται,

[1] ἀλλ' ἅμ' ἄμφω E, Themistius, Sophonias, vet. trans.,
Torstrik : ἀλλ' ἄμφω vulgo.

Is then the perception of all things one only, or is it different of different things, just as it is now generally supposed that taste and touch both act by contact, but that the other senses act at a distance? This is not the truth; we perceive hard and soft through a medium, just as we apprehend what sounds, or is seen, or smelt; but since we perceive the latter from a distance, and the former only from near by, the facts escape us. We perceive all things through a medium; but in this case the medium is not obvious. Still, as we have said before, if we were to perceive all tangible things through a fabric, without noticing the separation caused by it, we should react exactly in the same way as we do now in water and in air; for we seem to touch them directly without the intervention of any medium. But there is a difference between tangible things, and visible or audible things. We perceive the latter because some medium acts on us, but we perceive tangible things not by a medium, but at the same time as the medium, like a man wounded through his shield; for it is not the stricken shield that struck him, but both he and the shield were struck simultaneously. In a general sense we may say that as air and water are related to vision, hearing and smell, so is the relation of the flesh and the tongue to the sense organ in the case of touch. In neither class of case mentioned would sensation result from touching the sense organ; for instance, if one were to put a white body on the surface of the eye. From this it is clear that that which is perceptive of what is touched is within. Thus would occur what is true in the other cases; for when objects are placed on the other sense organs no sensation occurs, but when

133

they are placed on the flesh it does; hence the medium of the tangible is flesh.

The distinguishing characteristics of the body, *qua* body, are tangible; by distinguishing character- istics I mean those which differentiate the elements hot and cold, dry and wet, about which we have spoken before in our discussion of the elements.[a] The tactual organ which perceives them, *i.e.*, that in which the sense of touch, as it is called, primarily resides, is a part which has potentially the qualities of the objects touched. For perception is a form of being acted upon. Hence that which an object makes actually like itself is potentially such already. This is why we have no sensation of what is as hot, cold, hard, or soft as we are, but only of what is more so, which implies that the sense is a sort of mean between the relevant sensible extremes. That is how it can discern sensible objects. It is the mean that has the power of discernment; for it becomes an extreme in relation to each of the extremes in turn; and just as that which is to perceive white and black must be actually neither, but potentially both (and similarly with the other senses), so in the case of touch it must be neither hot nor cold. And just as we saw [b] that sight is in a sense concerned with both visible and invisible, and the other senses similarly with opposite objects, so touch is concerned with both tangible and intangible; by intangible we mean what has the quality of the tangible to an extremely small extent, as is the case with air, and also those tangibles which show excess, such as those which are destructive. Now we have described in outline each of the senses.

The sense is a mean.

XII. Καθόλου δὲ περὶ πάσης αἰσθήσεως δεῖ
λαβεῖν ὅτι ἡ μὲν αἴσθησίς ἐστι τὸ δεκτικὸν τῶν
αἰσθητῶν εἰδῶν ἄνευ τῆς ὕλης, οἷον ὁ κηρὸς τοῦ
20 δακτυλίου ἄνευ τοῦ σιδήρου καὶ τοῦ χρυσοῦ δέχεται τὸ σημεῖον, λαμβάνει δὲ τὸ χρυσοῦν ἢ τὸ
χαλκοῦν σημεῖον, ἀλλ' οὐχ ᾗ χρυσὸς ἢ χαλκός,
ὁμοίως δὲ καὶ ἡ αἴσθησις ἑκάστου ὑπὸ τοῦ ἔχοντος
χρῶμα ἢ χυμὸν ἢ ψόφον πάσχει, ἀλλ' οὐχ ᾗ
ἕκαστον ἐκείνων λέγεται, ἀλλ' ᾗ τοιονδί, καὶ κατὰ
25 τὸν λόγον. αἰσθητήριον δὲ πρῶτον ἐν ᾧ ἡ τοιαύτη
δύναμις. ἔστι μὲν οὖν ταὐτόν, τὸ δ' εἶναι ἕτερον·
μέγεθος μὲν γὰρ ἄν τι εἴη τὸ αἰσθανόμενον· οὐ
μὴν τό γε αἰσθητικῷ εἶναι, οὐδ' ἡ αἴσθησις
μέγεθός ἐστιν, ἀλλὰ λόγος τις καὶ δύναμις ἐκείνου.
φανερὸν δ' ἐκ τούτων καὶ διὰ τί ποτε τῶν
30 αἰσθητῶν αἱ ὑπερβολαὶ φθείρουσι τὰ αἰσθητήρια·
ἐὰν γὰρ ᾖ ἰσχυροτέρα τοῦ αἰσθητηρίου ἡ κίνησις,
λύεται ὁ λόγος· (τοῦτο δ' ἦν ἡ αἴσθησις), ὥσπερ
καὶ ἡ συμφωνία καὶ ὁ τόνος κρουομένων σφόδρα
τῶν χορδῶν. καὶ διὰ τί ποτε τὰ φυτὰ οὐκ αἰσθάνεται, ἔχοντά τι μόριον ψυχικὸν καὶ πάσχοντά
424 b τι ὑπὸ τῶν ἁπτῶν· καὶ γὰρ ψύχεται καὶ θερ-
μαίνεται· αἴτιον γὰρ τὸ μὴ ἔχειν μεσότητα, μηδὲ
τοιαύτην ἀρχὴν οἵαν τὰ εἴδη δέχεσθαι τῶν αἰσθητῶν, ἀλλὰ πάσχειν μετὰ τῆς ὕλης. ἀπορήσειε
5 δ' ἄν τις εἰ πάθοι ἄν τι ὑπ' ὀσμῆς τὸ ἀδύνατον
ὀσφρανθῆναι, ἢ ὑπὸ χρώματος τὸ μὴ δυνάμενον
ἰδεῖν· ὁμοίως δὲ καὶ ἐπὶ τῶν ἄλλων. εἰ δὲ τὸ
ὀσφραντὸν ὀσμή, εἴ τι ποιεῖ τὴν ὄσφρησιν, ἡ

ᵃ Sc., for receiving the impression.

XII. We must understand as true generally of every sense (1) that sense is that which is receptive of the form of sensible objects without the matter, just as the wax receives the impression of the signet-ring without the iron or the gold, and receives the impression of the gold or bronze, but not as gold or bronze ; so in every case sense is affected by that which has colour, or flavour, or sound, but by it, not *qua* having a particular identity, but *qua* having a certain quality, and in virtue of its formula ; (2) the sense organ in its primary meaning is that in which this potentiality [a] lies. The organ and the potentiality are identified, but their essential nature is not the same. The sentient subject must be extended, but sensitivity and sense cannot be extended ; they are a kind of ratio and potentiality of the said subject. From this it is also clear why excess in the perceptibility of objects destroys the sense organs ; for if the excitement of the sense organ is too strong, the ratio of its adjustment (which, as we saw, constitutes the sense) is destroyed ; just as the adjustment and pitch of a lyre is destroyed when the strings are struck hard. It is also clear why plants do not feel, though they have one part of the soul, and are affected to some extent by objects touched, for they show both cold and heat ; the reason is that they have no mean, *i.e.*, no first principle such as to receive the form of sensible objects, but are affected by the matter at the same time as the form. One might wonder whether anything that cannot smell is affected at all by smell, or that which cannot see by colour ; and in the same way with all other sensible objects. But if the object of smell is smell, if anything affects the sense of smell, it must be smell, so that it is

Definition of sensation.

137

impossible for anything which cannot smell to be affected by a smell ; and the same argument applies to the other senses ; nor can any of those things which can be acted upon be affected, except in so far as each has the sense in question. And the point is equally clear from the following argument. For neither light and darkness, nor sound, nor smell affects bodies at all : it is the objects in which they reside that produce the effect, just as it is the air with the thunderbolt that splits the timber. But it may be said that tangible objects and flavours do affect bodies; otherwise, by what could inanimate objects be affected and altered ? Will then the objects of other senses affect things ? Perhaps it is not every body that is affected by smell and sound : the things affected are indefinable and impermanent, such as air ; for it smells as though affected somehow. What, then, is smelling apart from being affected in some way ? Probably the act of smelling is an act of perception, whereas the air, being only temporarily affected, merely becomes perceptible.

Γ

I. Ὅτι δ' οὐκ ἔστιν αἴσθησις ἑτέρα παρὰ τὰς
πέντε (λέγω δὲ ταύτας ὄψιν, ἀκοήν, ὄσφρησιν,
γεῦσιν, ἀφήν), ἐκ τῶνδε πιστεύσειεν ἄν τις. εἰ
25 γὰρ παντὸς οὗ ἐστιν αἴσθησις ἁφή, καὶ νῦν
αἴσθησιν ἔχομεν (πάντα γὰρ τὰ τοῦ ἁπτοῦ ᾗ
ἁπτὸν πάθη τῇ ἁφῇ ἡμῖν αἰσθητά ἐστιν), ἀνάγκη
τ', εἴπερ ἐκλείπει τις αἴσθησις, καὶ αἰσθητήριόν
τι ἡμῖν ἐκλείπειν· καὶ ὅσων μὲν αὐτῶν ἁπτόμενοι
αἰσθανόμεθα, τῇ ἁφῇ αἰσθητά ἐστιν, ἣν τυγχάνομεν ἔχοντες· ὅσα δὲ διὰ τῶν μεταξύ, καὶ μὴ
30 αὐτῶν ἁπτόμενοι, τοῖς ἁπλοῖς, λέγω δ' οἷον ἀέρι
καὶ ὕδατι· ἔχει δ' οὕτως, ὥστ' εἰ μὲν δι' ἑνὸς
πλείω αἰσθητὰ ἕτερα ὄντα ἀλλήλων τῷ γένει,
ἀνάγκη τὸν ἔχοντα τὸ τοιοῦτον αἰσθητήριον ἀμφοῖν αἰσθητικὸν εἶναι (οἷον εἰ ἐξ ἀέρος ἐστὶ τὸ
αἰσθητήριον, καὶ ἔστιν ὁ ἀὴρ καὶ ψόφου καὶ
425 a χρόας), εἰ δὲ πλείω τοῦ αὐτοῦ, οἷον χρόας καὶ
ἀὴρ καὶ ὕδωρ (ἄμφω γὰρ διαφανῆ), καὶ ὁ τὸ
ἕτερον αὐτῶν ἔχων μόνον αἰσθήσεται ἀμφοῖν·
τῶν δὲ ἁπλῶν ἐκ δύο τούτων αἰσθητήρια μόνον
ἐστίν, ἐξ ἀέρος καὶ ὕδατος (ἡ μὲν γὰρ κόρη

BOOK III

I. One may be satisfied that there are no senses apart from the five (I mean vision, hearing, smell, taste and touch) from the following arguments. We may assume that we actually have perception of everything which is apprehended by touch (for by touch we perceive all those things which are qualities of the tangible object, *qua* tangible). Again, if we lack some sense, we must lack some sense organ ; and, again, all the things which we perceive by direct contact are perceptible by touch, a sense which we in fact possess ; but all those things which are perceived through media, and not by direct contact, are perceptible by means of the elements, *viz.*, air and water. Again, the facts are such that, if objects of more than one kind are perceived through one medium, the possessor of the appropriate sense organ will apprehend both (for instance, if the sense organ is composed of air, and air is the medium both of sound and of colour), but if there is more than one medium of the same thing, as for instance both air and water are media of colour (for both are transparent), then he that has either of these will perceive what is perceptible through both. But sense organs are composed of only two of these elements, air and water (for the pupil of the eye is composed of water,

5 ὕδατος, ἡ δ' ἀκοὴ ἀέρος, ἡ δ' ὄσφρησις θατέρου
τούτων), τὸ δὲ πῦρ ἢ οὐθενὸς ἢ κοινὸν πάντων
(οὐθὲν γὰρ ἄνευ θερμότητος αἰσθητικόν), γῆ δὲ
ἢ οὐθενός, ἢ ἐν τῇ ἁφῇ μάλιστα μέμικται ἰδίως·
διὸ λείποιτ' ἂν μηθὲν εἶναι αἰσθητήριον ἔξω ὕδατος
καὶ ἀέρος· ταῦτα δὲ καὶ νῦν ἔχουσιν ἔνια ζῷα·
10 πᾶσαι ἄρα αἱ αἰσθήσεις ἔχονται ὑπὸ τῶν μὴ
ἀτελῶν μηδὲ πεπηρωμένων· φαίνεται γὰρ καὶ ἡ
σπάλαξ ὑπὸ τὸ δέρμα ἔχουσα ὀφθαλμούς. ὥστ'
εἰ μή τι ἕτερόν ἐστι σῶμα, καὶ πάθος ὃ μηθενός
ἐστι τῶν ἐνταῦθα σωμάτων, οὐδεμία ἂν ἐκλίποι
αἴσθησις.

Ἀλλὰ μὴν οὐδὲ τῶν κοινῶν οἷόν τ' εἶναι αἰ-
15 σθητήριόν τι ἴδιον, ὧν ἑκάστῃ αἰσθήσει αἰσθανόμεθα κατὰ συμβεβηκός, οἷον κινήσεως, στάσεως,
σχήματος, μεγέθους, ἀριθμοῦ, ἑνός· ταῦτα γὰρ
πάντα κινήσει αἰσθανόμεθα, οἷον μέγεθος κινήσει
ὥστε καὶ σχῆμα· μέγεθος γάρ τι τὸ σχῆμα. τὸ
δ' ἠρεμοῦν τῷ μὴ κινεῖσθαι· ὁ δ' ἀριθμὸς τῇ ἀπο-
20 φάσει τοῦ συνεχοῦς καὶ τοῖς ἰδίοις· ἑκάστη γὰρ
ἓν αἰσθάνεται αἴσθησις. ὥστε δῆλον ὅτι ἀδύνατον
ὁτουοῦν ἰδίαν αἴσθησιν εἶναι τούτων, οἷον κινή-
σεως· οὕτω γὰρ ἔσται ὥσπερ νῦν τῇ ὄψει τὸ
γλυκὺ αἰσθανόμεθα. τοῦτο δ' ὅτι ἀμφοῖν ἔχοντες
τυγχάνομεν αἴσθησιν, ᾗ καὶ ὅταν συμπέσωσιν ἀνα-
γνωρίζομεν· εἰ δὲ μή, οὐδαμῶς ἂν ἀλλ' ἢ κατὰ

---

[a] This is the conclusion to which all the preceding arguments lead.

[b] *i.e.*, when and because they move.

[c] When a thing is " continuous " it is " one," which in the
Greek mind was not a number.

and the hearing organ of air, while the organ of smell is composed of one or other of these). But fire is the medium of no perception, or else is common to them all (for there is no possibility of perception without heat), and earth is the medium of no sense perception, or else is connected in a special way with the sense of touch. So we are left to suppose that there is no sense organ apart from water and air ; and some animals actually have organs composed of these.[a] The conclusion is that all the senses are possessed by all such animals as are neither undeveloped nor maimed ; even the mole, we find, has eyes under the skin. If then there is no other body, and no property other than those which belong to the bodies of this world, there can be no sense perception omitted from our list.

But, again, it is impossible that there should be a special sense organ to perceive common sensibles, which we perceive incidentally by each sense, such, I mean, as motion, rest, shape, magnitude, number and unity ; for we perceive all these things by movement [b] ; for instance we perceive magnitude by movement, and shape also ; for shape is a form of magnitude. What is at rest is perceived by absence of movement ; number by the negation of continuity,[c] and by the special sensibles ; for each sense perceives one kind of object. Thus it is clearly impossible for there to be a special sense of any of these common sensibles, e.g., movement ; if there were, we should perceive them in the same way as we now perceive what is sweet by sight. But we do this because we happen to have a sense for each of these qualities, and so recognize them when they occur together ; otherwise we should never perceive them except

The perception of common sensibles.

143

425 a

25 συμβεβηκὸς ᾐσθανόμεθα, οἷον τὸν Κλέωνος υἱὸν
οὐχ ὅτι Κλέωνος υἱός, ἀλλ' ὅτι λευκός· τούτῳ δὲ
συμβέβηκεν υἱῷ Κλέωνος εἶναι. τῶν δὲ κοινῶν ἤδη
ἔχομεν αἴσθησιν κοινήν, οὐ κατὰ συμβεβηκός· οὐκ
ἄρ' ἐστὶν ἰδία· οὐδαμῶς γὰρ ἂν ᾐσθανόμεθα ἀλλ'
30 ἢ οὕτως ὥσπερ εἴρηται τὸν Κλέωνος υἱὸν ἡμᾶς
ὁρᾶν. τὰ δ' ἀλλήλων ἴδια κατὰ συμβεβηκὸς
αἰσθάνονται αἱ αἰσθήσεις, οὐχ ᾗ αὐταί,[1] ἀλλ' ᾗ
425 b μία, ὅταν ἅμα γένηται ἡ αἴσθησις ἐπὶ τοῦ αὐτοῦ,
οἷον χολὴν ὅτι πικρὰ καὶ ξανθή· οὐ γὰρ δὴ ἑτέρας
γε τὸ εἰπεῖν ὅτι ἄμφω ἕν· διὸ καὶ ἀπατᾶται, καὶ
ἐὰν ᾖ ξανθόν, χολὴν οἴεται εἶναι. ζητήσειε δ' ἄν
τις τίνος ἕνεκα πλείους ἔχομεν αἰσθήσεις, ἀλλ' οὐ
μίαν μόνην. ἢ ὅπως ἧττον λανθάνῃ τὰ ἀκολου-
θοῦντα καὶ κοινά, οἷον κίνησις καὶ μέγεθος καὶ
ἀριθμός· εἰ γὰρ ἦν ἡ ὄψις μόνη, καὶ αὕτη[2] λευκοῦ,
ἐλάνθανεν ἂν μᾶλλον καὶ ἐδόκει ταὐτὸ εἶναι πάντα
διὰ τὸ ἀκολουθεῖν ἀλλήλοις ἅμα χρῶμα καὶ
10 μέγεθος. νῦν δ' ἐπεὶ καὶ ἐν ἑτέρῳ αἰσθητῷ τὰ
κοινὰ ὑπάρχει, δῆλον ποιεῖ ὅτι ἄλλο τι ἕκαστον
αὐτῶν.

II. Ἐπεὶ δ' αἰσθανόμεθα ὅτι ὁρῶμεν καὶ ἀκούομεν, ἀνάγκη ἢ τῇ ὄψει αἰσθάνεσθαι ὅτι ὁρᾷ, ἢ
ἑτέρᾳ. ἀλλ' ἡ αὐτὴ ἔσται τῆς ὄψεως καὶ τοῦ
15 ὑποκειμένου χρώματος. ὥστε ἢ δύο τοῦ αὐτοῦ
ἔσονται ἢ αὐτὴ αὑτῆς. ἔτι δ' εἰ καὶ ἑτέρα εἴη ἡ[3]
τῆς ὄψεως αἴσθησις, ἢ εἰς ἄπειρον εἶσιν ἢ αὐτή

---

[1] αἱ αὐταί Bekker. [2] αὕτη Jackson : αὐτή.
[3] ἡ om. Bekker.

incidentally, as, *e.g.*, we perceive of Cleon's son, not that he is Cleon's son, but that he is white; and this white object is incidentally Cleon's son. But we have already a common faculty which apprehends common sensibles directly. Therefore there is no special sense for them. If there were, we should have no perception of them, except as we said that we saw Cleon's son. The senses perceive each other's proper objects incidentally, not in their own identity, but acting together as one, when sensation occurs simultaneously in the case of the same object, as for instance of bile, that it is bitter and yellow; for it is not the part of any single sense to state that both objects are one. Thus sense may be deceived, and, if an object is yellow, may think that it is bile. One might ask why we have several senses and not one only. It may be in order that the accompanying common sensibles, such as movement, size and number, may escape us less; for if vision were our only sense, and it perceived mere whiteness, they would be less apparent; indeed all sensibles would be indistinguishable, because of the concomitance of, *e.g.*, colour and size. As it is, the fact that common sensibles inhere in the objects of more than one sense shows that each of them is something distinct.

II. Since we can perceive that we see and hear, it must be either by sight itself, or by some other sense. But then the same sense must perceive both sight and colour, the object of sight. So that either two senses perceive the same object, or sight perceives itself. Again, if there is a separate sense perceiving sight, either the process will go on *ad infinitum*,

How do we know that we see?

425 b

τις ἔσται αὐτῆς. ὥστ᾽ ἐπὶ τῆς πρώτης τοῦτο
ποιητέον. ἔχει δ᾽ ἀπορίαν· εἰ γὰρ τὸ τῇ ὄψει
αἰσθάνεσθαί ἐστιν ὁρᾶν, ὁρᾶται δὲ χρῶμα ἢ τὸ
ἔχον, εἰ ὄψεταί τις τὸ ὁρῶν, καὶ χρῶμα ἕξει τὸ
20 ὁρῶν πρῶτον. φανερὸν τοίνυν ὅτι οὐχ ἓν τὸ τῇ
ὄψει αἰσθάνεσθαι· καὶ γὰρ ὅταν μὴ ὁρῶμεν, τῇ
ὄψει κρίνομεν καὶ τὸ σκότος καὶ τὸ φῶς, ἀλλ᾽
οὐχ ὡσαύτως. ἔτι δὲ καὶ τὸ ὁρῶν ἔστιν ὡς κε-
χρωμάτισται· τὸ γὰρ αἰσθητήριον δεκτικὸν τοῦ
αἰσθητοῦ ἄνευ τῆς ὕλης ἕκαστον. διὸ καὶ ἀπ-
25 ελθόντων τῶν αἰσθητῶν ἔνεισιν αἱ αἰσθήσεις καὶ
φαντασίαι ἐν τοῖς αἰσθητηρίοις.

Ἡ δὲ τοῦ αἰσθητοῦ ἐνέργεια καὶ τῆς αἰσθήσεως
ἡ αὐτὴ μέν ἐστι καὶ μία, τὸ δ᾽ εἶναι οὐ ταὐτὸν
αὐταῖς· λέγω δ᾽ οἷον ψόφος ὁ κατ᾽ ἐνέργειαν καὶ
ἀκοὴ ἡ κατ᾽ ἐνέργειαν· ἔστι γὰρ ἀκοὴν ἔχοντα
μὴ ἀκούειν, καὶ τὸ ἔχον ψόφον οὐκ ἀεὶ ψοφεῖ.
30 ὅταν δ᾽ ἐνεργῇ τὸ δυνάμενον ἀκούειν καὶ ψοφῇ τὸ
δυνάμενον ψοφεῖν, τότε ἡ κατ᾽ ἐνέργειαν ἀκοὴ
426 a ἅμα γίνεται καὶ ὁ κατ᾽ ἐνέργειαν ψόφος, ὧν
εἴπειεν ἄν τις τὸ μὲν εἶναι ἄκουσιν τὸ δὲ ψόφησιν.

Εἰ δ᾽ ἔστιν ἡ κίνησις καὶ ἡ ποίησις καὶ τὸ πάθος
ἐν τῷ ποιουμένῳ, ἀνάγκη καὶ τὸν ψόφον καὶ τὴν
ἀκοὴν τὴν κατ᾽ ἐνέργειαν ἐν τῇ κατὰ δύναμιν
5 εἶναι· ἡ γὰρ τοῦ ποιητικοῦ καὶ κινητικοῦ ἐνέργεια
ἐν τῷ πάσχοντι ἐγγίνεται. διὸ οὐκ ἀνάγκη τὸ
κινοῦν κινεῖσθαι. ἡ μὲν οὖν τοῦ ψοφητικοῦ ἐνέργειά ἐστι ψόφος ἢ ψόφησις, ἡ δὲ τοῦ ἀκουστικοῦ

---

[a] If we suppose a special sense to apprehend that we see,
we must suppose another to apprehend this and so on.

[b] Sc., as we discern colours.   [c] Cf. ch. viii.

146

or a sense must perceive itself.[a] So we may assume that this occurs with the first sense. But here is a difficulty ; for if perception by vision is seeing, and that which is seen either is colour or has colour, then if one is to see that which sees, it follows that what primarily sees will possess colour. It is therefore obvious that the phrase " perceiving by vision " has not merely one meaning ; for, even when we do not see, we discern darkness and light by vision, but not in the same way.[b] Moreover that which sees does in a sense possess colour ; for each sense organ is receptive of the perceived object, but without its matter. This is why, even when the objects of perception are gone, sensations and mental images are still present in the sense organ.

The activity of the sensible object and of the sensation is one and the same,[c] though their essence is not the same ; in saying that they are the same, I mean the actual sound and the actual hearing ; for it is possible for one who possesses hearing not to hear, and that which has sound is not always sounding. But when that which has the power of hearing is exercising its power, and that which can sound is sounding, then the active hearing and the active sound occur together ; we may call them respectively audition and sonance.

What is sensation ?

If then the movement, that is, the acting and being acted upon, takes place in that which is acted upon, then the sound and the hearing in a state of activity must reside in the potential hearing ; for the activity of what is moving and active takes place in what is being acted upon. Hence that which causes motion need not be moved. The activity, then, of the object producing sound is sound, or sonance, and of that

426 a

ἀκοὴ ἢ ἄκουσις· διττὸν γὰρ ἡ ἀκοή, καὶ διττὸν ὁ
ψόφος. ὁ δ' αὐτὸς λόγος καὶ ἐπὶ τῶν ἄλλων
αἰσθήσεων καὶ αἰσθητῶν. ὥσπερ γὰρ ἡ ποίησις
10 καὶ ἡ πάθησις ἐν τῷ πάσχοντι ἀλλ' οὐκ ἐν τῷ
ποιοῦντι, οὕτω καὶ ἡ τοῦ αἰσθητοῦ ἐνέργεια καὶ
ἡ τοῦ αἰσθητικοῦ ἐν τῷ αἰσθητικῷ. ἀλλ' ἐπ'
ἐνίων μὲν ὠνόμασται, οἷον ἡ ψόφησις καὶ ἡ
ἄκουσις, ἐπὶ δ' ἐνίων ἀνώνυμον θάτερον· ὅρασις
15 γὰρ λέγεται ἡ τῆς ὄψεως ἐνέργεια, ἡ δὲ τοῦ
χρώματος ἀνώνυμος, καὶ γεῦσις ἡ τοῦ γευστικοῦ,
ἡ δὲ τοῦ χυμοῦ ἀνώνυμος. ἐπεὶ δὲ μία μέν ἐστιν
ἐνέργεια ἡ τοῦ αἰσθητοῦ καὶ ἡ τοῦ αἰσθητικοῦ,
τὸ δ' εἶναι ἕτερον, ἀνάγκη ἅμα φθείρεσθαι καὶ
σώζεσθαι τὴν οὕτω λεγομένην ἀκοὴν καὶ ψόφον,
καὶ χυμὸν δὴ καὶ γεῦσιν καὶ τὰ ἄλλα ὁμοίως·
20 τὰ δὲ κατὰ δύναμιν λεγόμενα οὐκ ἀνάγκη, ἀλλ'
οἱ πρότερον φυσιολόγοι τοῦτο οὐ καλῶς ἔλεγον,
οὐθὲν οἰόμενοι οὔτε λευκὸν οὔτε μέλαν εἶναι ἄνευ
ὄψεως, οὐδὲ χυμὸν ἄνευ γεύσεως. τῇ μὲν γὰρ
ἔλεγον ὀρθῶς, τῇ δ' οὐκ ὀρθῶς· διχῶς γὰρ λεγο-
25 μένης τῆς αἰσθήσεως καὶ τοῦ αἰσθητοῦ, τῶν μὲν
κατὰ δύναμιν τῶν δὲ κατ' ἐνέργειαν, ἐπὶ τούτων
μὲν συμβαίνει τὸ λεχθέν, ἐπὶ δὲ τῶν ἑτέρων οὐ
συμβαίνει. ἀλλ' ἐκεῖνοι ἁπλῶς ἔλεγον περὶ τῶν
λεγομένων οὐχ ἁπλῶς.

Εἰ δ' ἡ συμφωνία φωνή τίς ἐστιν, ἡ δὲ φωνὴ
καὶ ἡ ἀκοή ἔστιν ὡς ἕν ἐστι [καὶ ἔστιν ὡς οὐχ ἕν
τὸ αὐτό],[1] λόγος δ' ἡ συμφωνία, ἀνάγκη καὶ τὴν
30 ἀκοὴν λόγον τινὰ εἶναι. καὶ διὰ τοῦτο καὶ φθείρει

[1] incl. Torstrik.

---

[a] If a red object is in the dark it does not appear red.
Some philosophers maintained that it is not red, and that
148

producing hearing is hearing or audition, for hearing is used in two senses, and so is sound. The same argument applies to all other senses and sensible objects. For just as acting and being acted upon reside in that which is acted upon, and not in the agent, so also the activity of the sensible object and that of the sensitive subject lie in the latter. In some cases we have names for both, such as sonance and audition, but in others one of the terms has no name ; for the activity of vision is called seeing, but that of colour has no name ; the activity of taste is called tasting, but that of flavour has no name. But since the activity of the sensible and of the sensitive is the same, though their essence is different, it follows that hearing in the active sense must cease or continue simultaneously with the sound, and so with flavour and taste and the rest ; but this does not apply to their potentialities. The earlier natural philosophers were at fault in this, supposing that white and black have no existence without vision, nor flavour without taste.[a] In one sense they were right, but in another wrong ; for the terms sensation and sensible being used in two senses, that is potentially and actually, their statements apply to the latter class, but not to the former. These thinkers did not distinguish the meanings of terms which have more than one meaning.

If harmony is a species of voice, and voice and hearing are in one sense one and the same, and if harmony is a ratio, then it follows that hearing must be in some sense a ratio. That is why both high and

Sensation as a harmony.

[a] " red " has no existence except when we see it. A. argues that such an object is " potentially " red, because, given the right conditions, it will appear red.

426 a

ἕκαστον ὑπερβάλλον, καὶ τὸ ὀξὺ καὶ τὸ βαρύ, τὴν
426 b ἀκοήν· ὁμοίως δὲ καὶ ἐν χυμοῖς τὴν γεῦσιν, καὶ
ἐν χρώμασι τὴν ὄψιν τὸ σφόδρα λαμπρὸν ἢ
ζοφερόν, καὶ ἐν ὀσφρήσει ἡ ἰσχυρὰ ὀσμὴ καὶ
γλυκεῖα καὶ πικρά, ὡς λόγου τινὸς ὄντος τῆς
αἰσθήσεως. διὸ καὶ ἡδέα μέν, ὅταν εἰλικρινῆ καὶ
5 ἀμιγῆ ἄγηται εἰς τὸν λόγον, οἷον τὸ ὀξὺ ἢ γλυκὺ
ἢ ἁλμυρόν· ἡδέα γὰρ τότε. ὅλως δὲ μᾶλλον τὸ
μικτὸν συμφωνία ἢ τὸ ὀξὺ ἢ βαρύ, ἁφῇ δὲ τὸ
θερμαντὸν ἢ ψυκτόν· ἡ δ' αἴσθησις ὁ λόγος·
ὑπερβάλλοντα δὲ λυπεῖ ἢ φθείρει.

Ἑκάστη μὲν οὖν αἴσθησις τοῦ ὑποκειμένου
αἰσθητοῦ ἐστίν, ὑπάρχουσα ἐν τῷ αἰσθητηρίῳ ᾗ
10 αἰσθητήριον, καὶ κρίνει τὰς τοῦ ὑποκειμένου αἰσθητοῦ διαφοράς, οἷον λευκὸν μὲν καὶ μέλαν ὄψις,
γλυκὺ δὲ καὶ πικρὸν γεῦσις. ὁμοίως δ' ἔχει τοῦτο
καὶ ἐπὶ τῶν ἄλλων. ἐπεὶ δὲ καὶ τὸ λευκὸν καὶ
τὸ γλυκὺ καὶ ἕκαστον τῶν αἰσθητῶν πρὸς ἕκαστον
κρίνομεν, τίνι¹ καὶ αἰσθανόμεθα ὅτι διαφέρει;
15 ἀνάγκη δὴ αἰσθήσει· αἰσθητὰ γάρ ἐστιν. ᾗ καὶ
δῆλον ὅτι ἡ σὰρξ οὐκ ἔστι τὸ ἔσχατον αἰσθητήριον·
ἀνάγκη γὰρ ἦν ἁπτόμενον αὐτοῦ κρίνειν τὸ κρῖνον.
οὔτε δὴ κεχωρισμένοις ἐνδέχεται κρίνειν ὅτι ἕτερον
τὸ γλυκὺ τοῦ λευκοῦ, ἀλλὰ δεῖ ἑνί τινι ἄμφω δῆλα
εἶναι. οὕτω μὲν γὰρ κἂν εἰ τοῦ μὲν ἐγὼ τοῦ δὲ
20 σὺ αἴσθοιο, δῆλον ἂν εἴη ὅτι ἕτερα ἀλλήλων. δεῖ
δὲ τὸ ἓν λέγειν ὅτι ἕτερον· ἕτερον γὰρ τὸ γλυκὺ
τοῦ λευκοῦ. λέγει ἄρα τὸ αὐτό, ὥστε, ὡς λέγει,

¹ τίνι . . . διαφέρει; Trendelenburg: τινὶ . . . διαφέρει.

low pitch, if excessive, destroy hearing ; in the same way in flavours excess destroys taste, and in colours the over-brilliant or over-dark destroys vision, and in smelling the strong scent, whether sweet or bitter, destroys smell ; which implies that sense is some kind of ratio. That is also why things are pleasant when they enter pure and unmixed into the ratio,[a] *e.g.*, acid, sweet or salt ; for in that case they are pleasant. But generally speaking a mixed constitution produces a better harmony than the high or low pitch, and to the touch that is more pleasant which can be warmed or cooled ; the sense is the ratio, and excess hurts or destroys.

Each sense then relates to its sensible subject-matter ; it resides in the sense organ as such, and discerns differences in the said subject-matter ; *e.g.*, vision discriminates between white and black, and taste between sweet and bitter ; and similarly in all other cases. But, since we also distinguish white and sweet, and compare all objects perceived with each other, by what sense do we perceive that they differ ? It must evidently be by some sense that we perceive the difference ; for they are objects of sense. Incidentally it becomes clear that flesh is not the ultimate sense organ ; for, if it were, judgement would depend on being in contact. Nor, again, is it possible to judge that sweet and white are different by separate senses, but both must be clearly presented to a single sense. For, in the other case, if you perceived one thing and I another, it would be obvious that they differed from each other. That which asserts the difference must be one ; for sweet differs from white. It is the same faculty, then, that asserts this ; hence as it asserts, so it thinks and

How do we apprehend the difference between the objects of different senses ?

---

[a] *i.e.*, the ratio which constitutes the sense-organ.

426 b

οὕτω καὶ νοεῖ καὶ αἰσθάνεται. ὅτι μὲν οὖν οὐχ
οἷόν τε κεχωρισμένοις κρίνειν τὰ κεχωρισμένα,
δῆλον· ὅτι δ᾽ οὐδ᾽ ἐν κεχωρισμένῳ χρόνῳ, ἐντεῦθεν.
25 ὥσπερ γὰρ τὸ αὐτὸ λέγει ὅτι ἕτερον τὸ ἀγαθὸν
καὶ τὸ κακόν, οὕτω καὶ ὅτε θάτερον λέγει ὅτι
ἕτερον, καὶ θάτερον οὐ κατὰ συμβεβηκὸς τὸ ὅτε
(λέγω δ᾽, οἷον νῦν λέγω ὅτι ἕτερον, οὐ μέντοι ὅτι
νῦν ἕτερον). ἀλλ᾽ οὕτω λέγει, καὶ νῦν, καὶ ὅτι
νῦν· ἅμα ἄρα. ὥστε ἀχώριστον καὶ ἐν ἀχωρίστῳ
30 χρόνῳ. ἀλλὰ μὴν ἀδύνατον ἅμα τὰς ἐναντίας
κινήσεις κινεῖσθαι τὸ αὐτὸ ᾗ ἀδιαίρετον καὶ ἐν
ἀδιαιρέτῳ χρόνῳ. εἰ γὰρ τὸ γλυκὺ ὡδὶ κινεῖ τὴν
427 a αἴσθησιν ἢ τὴν νόησιν, τὸ δὲ πικρὸν ἐναντίως,
καὶ τὸ λευκὸν ἑτέρως. ἆρ᾽ οὖν ἅμα μὲν καὶ
ἀριθμῷ ἀδιαίρετον καὶ ἀχώριστον τὸ κρῖνον, τῷ
εἶναι δὲ κεχωρισμένον; ἔστι δή πως ὡς τὸ δι-
αιρετὸν τῶν διῃρημένων αἰσθάνεται, ἔστι δ᾽ ὡς ᾗ
5 ἀδιαίρετον· τῷ εἶναι μὲν γὰρ διαιρετόν, τόπῳ δὲ
καὶ ἀριθμῷ ἀδιαίρετον. ἢ οὐχ οἷόν τε; δυνάμει
μὲν γὰρ τὸ αὐτὸ καὶ ἀδιαίρετον τἀναντία, τῷ δ᾽
εἶναι οὔ, ἀλλὰ τῷ ἐνεργεῖσθαι διαιρετόν, καὶ οὐχ
οἷόν τε ἅμα λευκὸν καὶ μέλαν εἶναι· ὥστ᾽ οὐδὲ
τὰ εἴδη πάσχειν αὐτῶν, εἰ τοιοῦτον ἡ αἴσθησις
10 καὶ ἡ νόησις, ἀλλ᾽ ὥσπερ ἦν καλοῦσί τινες στιγμήν,
ᾗ μία καὶ ᾗ δύο, ταύτῃ καὶ διαιρετή. ᾗ μὲν οὖν
ἀδιαίρετον, ἓν τὸ κρῖνόν ἐστι καὶ ἅμα, ᾗ δὲ

perceives. Evidently, therefore, it is impossible to pass judgement on separate objects by separate faculties; and it is also obvious from the following considerations that they are not judged at separate times. For just as the same faculty declares that good and evil are different, so also when it declares that one is different and the other different, the " time when " is not merely incidental (as when, e.g., I *now* say that there is a difference, but do not say that there is *now* a difference). The faculty says now, and also that the difference is now; hence both are different at once. So the judging sense must be undivided, and also must judge without an interval. But, again, it is impossible that the same faculty should be moved at the same time with contrary movements, in so far as it is indivisible, and in indivisible time. For if the object is sweet it excites sensation or thought in one way, but if bitter, in the contrary way, and if white, in a different way altogether. Are we, then, to suppose that the judging faculty is numerically indivisible and inseparable, but is divided in essence? Then in one sense it is what is divided that perceives divided things, but in sense it does this *qua* indivisible. For it is divisible in essence, but indivisible spatially and numerically. Or is this impossible? For although the same indivisible thing may be both contraries potentially, it is not so in essence, but it becomes divisible in actualization; the same thing cannot be at once white and black, and so the same thing cannot be acted upon by the forms of these, if this is what happens in perception and thought. The fact is that just as what some thinkers describe as a point is, as being both one and two, in this sense divisible, so too in so far as the judging faculty is indivisible, it is

one and instantaneous in action; but in so far as it is divisible, it uses the same symbol twice at the same time. In so far, then, as it treats the limit as two, it passes judgement on two distinct things, as being itself in a sense distinct; but in so far as it judges of it as only one, it judges by one faculty and at one time.

Concerning the principle in virtue of which we call a living creature sentient, let this account suffice.

III. Now there are two special characteristics which distinguish soul, *viz.*, (1) movement in space, and (2) thinking, judging and perceiving. Thinking, both speculative and practical, is regarded as a form of perceiving; for in both cases the soul judges and has cognizance of something which is. Indeed the older philosophers assert that thinking and perceiving are identical. For instance Empedocles has said " Understanding grows with a man according to what appears to him," and in another passage " whence it befalls them ever to think different thoughts." Homer's phrase, again, " Such is the nature of man's mind " [a] implies the same thing. For all these authors suppose the process of thinking to be a bodily function like perceiving, and that men both perceive and recognize like by like, as we have explained at the beginning of this treatise.[b] And yet they ought to have made some mention of error at the same time; for error seems to be more natural to living creatures, and the soul spends more time in it. From this belief it must follow either that, as some say, all appearances are true, or that error is contact with the unlike; for this is the opposite to recognizing like by like. But it appears that in the case of contraries error, like

Relation between sensation and thinking.

427 b

τῶν ἐναντίων ἡ αὐτὴ εἶναι. ὅτι μὲν οὖν οὐ ταὐτόν ἐστι τὸ αἰσθάνεσθαι καὶ τὸ φρονεῖν, φανερόν·
τοῦ μὲν γὰρ πᾶσι μέτεστι, τοῦ δὲ ὀλίγοις τῶν
ζῴων. ἀλλ' οὐδὲ τὸ νοεῖν, ἐν ᾧ ἐστι τὸ ὀρθῶς
10 καὶ τὸ μὴ ὀρθῶς, τὸ μὲν ὀρθῶς φρόνησις καὶ
ἐπιστήμη καὶ δόξα ἀληθής, τὸ δὲ μὴ ὀρθῶς
τἀναντία τούτων· οὐδὲ τοῦτο[1] ἐστὶ ταὐτὸ τῷ
αἰσθάνεσθαι· ἡ μὲν γὰρ αἴσθησις τῶν ἰδίων ἀεὶ
ἀληθής, καὶ πᾶσιν ὑπάρχει τοῖς ζῴοις, διανοεῖσθαι
δ' ἐνδέχεται καὶ ψευδῶς, καὶ οὐδενὶ ὑπάρχει ᾧ
15 μὴ καὶ λόγος· φαντασία γὰρ ἕτερον καὶ αἰσθήσεως
καὶ διανοίας· αὐτή τε οὐ γίγνεται ἄνευ αἰσθήσεως, καὶ ἄνευ ταύτης οὐκ ἔστιν ὑπόληψις. ὅτι
δ' οὐκ ἔστιν ἡ αὐτὴ νόησις καὶ ὑπόληψις, φανερόν.
τοῦτο μὲν γὰρ τὸ πάθος ἐφ' ἡμῖν ἐστιν, ὅταν
20 βουλώμεθα (πρὸ ὀμμάτων γὰρ ἔστι ποιήσασθαι,
ὥσπερ οἱ ἐν τοῖς μνημονικοῖς τιθέμενοι καὶ εἰδωλο-
ποιοῦντες), δοξάζειν δ' οὐκ ἐφ' ἡμῖν· ἀνάγκη γὰρ
ἢ ψεύδεσθαι ἢ ἀληθεύειν. ἔτι δὲ ὅταν μὲν δοξά-
σωμεν δεινόν τι ἢ φοβερόν, εὐθὺς συμπάσχομεν,
ὁμοίως δὲ κἂν θαρραλέον· κατὰ δὲ τὴν φαντασίαν
25 ὡσαύτως ἔχομεν ὥσπερ ἂν οἱ θεώμενοι ἐν γραφῇ
τὰ δεινὰ ἢ θαρραλέα. εἰσὶ δὲ καὶ αὐτῆς τῆς
ὑπολήψεως διαφοραί, ἐπιστήμη καὶ δόξα καὶ
φρόνησις καὶ τἀναντία τούτων, περὶ ὧν τῆς διαφορᾶς ἕτερος ἔστω λόγος.

Περὶ δὲ τοῦ νοεῖν, ἐπεὶ ἕτερον τοῦ αἰσθάνεσθαι,
τούτου δὲ τὸ μὲν φαντασία δοκεῖ εἶναι τὸ δὲ

---

[1] δ' post τοῦτο in plerisque codd. repertum delendum
censuit Vahlen.

    [a] In normal cases if a man sees a red object, it is red.
    [b] ? Eth. Nic. 1139 b 15.

156

knowledge, is one and the same. Now it is quite clear that perceiving and practical thinking are not the same; for all living creatures have a share in the former, but only a few in the latter. Nor again is speculative thinking, which involves being right or wrong—" being right " corresponding to intelligence and knowledge and true opinion, and " being wrong " to their contraries—the same thing as perceiving; for the perception of proper objects is always true,[a] and is a characteristic of all living creatures, but it is possible to think falsely, and thought belongs to no animal which has not reasoning power; for imagination is different from both perception and thought; imagination always implies perception, and is itself implied by judgement. But clearly imagination and judgement are different modes of thought. For the former is an affection which lies in our power whenever we choose (for it is possible to call up mental pictures, as those do who employ images in arranging their ideas under a mnemonic system), but it is not in our power to form opinions as we will; for we must either hold a false opinion or a true one. Again, when we form an opinion that something is threatening or frightening, we are immediately affected by it, and the same is true of our opinion of something that inspires courage; but in imagination we are like spectators looking at something dreadful or encouraging in a picture. Judgement itself, too, has various forms—knowledge, opinion, prudence, and their opposites, but their differences must be the subject of another discussion.[b]

As for thought, since it is distinct from perception, and is held to comprise imagination and judgement,

Imagination.

157

427 b

428 a

ὑπόληψις, περὶ φαντασίας διορίσαντας οὕτω περὶ
θατέρου λεκτέον. εἰ δή ἐστιν ἡ φαντασία καθ᾽
ἣν λέγομεν φάντασμά τι ἡμῖν γίγνεσθαι καὶ μὴ
εἴ τι κατὰ μεταφορὰν λέγομεν, μία τίς ἐστι τούτων δύναμις ἢ ἕξις, καθ᾽ ἣν κρίνομεν καὶ ἀλη-
θεύομεν ἢ ψευδόμεθα. τοιαῦται δ᾽ εἰσὶν αἴσθησις,
δόξα, ἐπιστήμη, νοῦς. ὅτι μὲν οὖν οὐκ ἔστιν
αἴσθησις, δῆλον ἐκ τῶνδε. αἴσθησις μὲν γὰρ ἤτοι
δύναμις ἢ ἐνέργεια, οἷον ὄψις καὶ ὅρασις, φαίνεται
δέ τι καὶ μηδετέρου ὑπάρχοντος τούτων, οἷον τὰ
ἐν τοῖς ὕπνοις. εἶτα αἴσθησις μὲν ἀεὶ πάρεστι,
φαντασία δ᾽ οὔ. εἰ δὲ τῇ ἐνεργείᾳ τὸ αὐτό, πᾶσιν
ἂν ἐνδέχοιτο τοῖς θηρίοις φαντασίαν ὑπάρχειν·
δοκεῖ δ᾽ οὔ, οἷον μύρμηκι ἢ μελίττῃ ἢ σκώληκι.
εἶτα αἱ μὲν ἀληθεῖς αἰεί, αἱ δὲ φαντασίαι γίνονται
αἱ πλείους ψευδεῖς. ἔπειτ᾽ οὐδὲ λέγομεν, ὅταν
ἐνεργῶμεν ἀκριβῶς περὶ τὸ αἰσθητόν, ὅτι φαίνεται
τοῦτο ἡμῖν ἄνθρωπος· ἀλλὰ μᾶλλον ὅταν μὴ ἐν-
αργῶς αἰσθανώμεθα.[1] καὶ ὅπερ δὲ ἐλέγομεν πρότερον, φαίνεται καὶ μύουσιν ὁράματα. ἀλλὰ μὴν
οὐδὲ τῶν ἀεὶ ἀληθευόντων οὐδεμία ἔσται, οἷον
ἐπιστήμη ἢ νοῦς· ἔστι γὰρ φαντασία καὶ ψευδής.
λείπεται ἄρα ἰδεῖν εἰ δόξα· γίνεται γὰρ δόξα καὶ
ἀληθὴς καὶ ψευδής. ἀλλὰ δόξῃ μὲν ἕπεται πίστις
(οὐκ ἐνδέχεται γὰρ δοξάζοντα οἷς δοκεῖ μὴ
πιστεύειν), τῶν δὲ θηρίων οὐθενὶ ὑπάρχει πίστις,
φαντασία δὲ πολλοῖς. ἔτι πάσῃ μὲν δόξῃ ἀκολουθεῖ πίστις, πίστει δὲ τὸ πεπεῖσθαι, πειθοῖ δὲ
λόγος· τῶν δὲ θηρίων ἐνίοις φαντασία μὲν ὑπάρχει,

[1] quae hic vulgo sequuntur τότε ἢ ἀληθὴς ἢ ψευδής unc.
inclusit Torstrik.

it will be best to discuss it after having completed our analysis of imagination. If imagination is (apart from any metaphorical sense of the word) the process by which we say that an image is presented to us, it is one of those faculties or states of mind by which we judge and are either right or wrong. Such are sensation, opinion, knowledge and intelligence. It is clear from the following considerations that imagination is not sensation. Sensation is either potential or actual, *e.g.*, either sight or seeing, but imagination occurs when neither of these is present, as when objects are seen in dreams. Secondly, sensation is always present but imagination is not. If sensation and imagination were identical in actuality, then imagination would be possible for all creatures ; but this appears not to be the case ; for instance it is not true of the ant, the bee, or the grub. Again, all sensations are true, but most imaginations are false. Nor do we say " I imagine that it is a man " when our sense is functioning accurately with regard to its object, but only when we do not perceive distinctly. And, as we have said before, visions are seen by men even with their eyes shut. Nor is imagination any one of the faculties which are always right, such as knowledge or intelligence ; for imagination may be false. It remains, then, to consider whether it is opinion ; for opinion may be either true or false. But opinion implies belief (for one cannot hold opinions in which one does not believe) ; and no animal has belief, but many have imagination. Again, every opinion is accompanied by belief, belief by conviction, and conviction by rational discourse ; but although some creatures have imagination, they

428 a
25 λόγος δ' οὔ. φανερὸν τοίνυν ὅτι οὐδὲ δόξα μετ'
αἰσθήσεως, οὐδὲ δι' αἰσθήσεως, οὐδὲ συμπλοκὴ
δόξης καὶ αἰσθήσεως φαντασία ἂν εἴη, διά τε
ταῦτα καὶ [δῆλον]¹ ὅτι οὐκ ἄλλου τινός ἐστιν ἡ
δόξα, ἀλλ' ἐκείνου ἐστὶν οὗ καὶ αἴσθησις· λέγω
δ', ἐκ τῆς τοῦ λευκοῦ δόξης καὶ αἰσθήσεως ἡ
30 συμπλοκὴ φαντασία ἐστίν· οὐ γὰρ δὴ ἐκ τῆς δόξης
428 b μὲν τῆς τοῦ ἀγαθοῦ, αἰσθήσεως δὲ τῆς τοῦ λευκοῦ.
τὸ οὖν φαίνεσθαί ἐστι τὸ δοξάζειν ὅπερ αἰσθάνεται
μὴ κατὰ συμβεβηκός. φαίνεται δὲ καὶ ψευδῆ,
περὶ ὧν ἅμα ὑπόληψιν ἀληθῆ ἔχει, οἷον φαίνεται
μὲν ὁ ἥλιος ποδιαῖος, πεπίστευται δ' εἶναι μείζων
5 τῆς οἰκουμένης· συμβαίνει οὖν ἤτοι ἀποβεβλη-
κέναι τὴν ἑαυτοῦ ἀληθῆ δόξαν, ἣν εἶχε, σωζομένου
τοῦ πράγματος, μὴ ἐπιλαθόμενον μηδὲ μεταπει-
σθέντα, ἢ εἰ ἔτι ἔχει, ἀνάγκη τὴν αὐτὴν ἀληθῆ
εἶναι καὶ ψευδῆ. ἀλλὰ ψευδὴς ἐγένετο, ὅτε λάθοι
μεταπεσὸν τὸ πρᾶγμα. οὔτ' ἄρα ἕν τι τούτων
10 ἐστὶν οὔτ' ἐκ τούτων ἡ φαντασία.

'Αλλ' ἐπειδὴ ἔστι κινηθέντος τουδὶ κινεῖσθαι
ἕτερον ὑπὸ τούτου, ἡ δὲ φαντασία κίνησίς τις
δοκεῖ εἶναι καὶ οὐκ ἄνευ αἰσθήσεως γίγνεσθαι ἀλλ'
αἰσθανομένοις καὶ ὧν αἴσθησις ἐστιν, ἔστι δὲ
γίγνεσθαι κίνησιν ὑπὸ τῆς ἐνεργείας τῆς αἰσθήσεως, καὶ ταύτην ὁμοίαν ἀνάγκη εἶναι τῇ αἰσθήσει,
15 εἴη ἂν αὕτη ἡ κίνησις οὔτε ἄνευ αἰσθήσεως ἐνδε-
χομένη οὔτε μὴ αἰσθανομένοις ὑπάρχειν, καὶ πολλὰ
κατ' αὐτὴν καὶ ποιεῖν καὶ πάσχειν τὸ ἔχον, καὶ
εἶναι καὶ ἀληθῆ καὶ ψευδῆ. τοῦτο δὲ συμβαίνει

¹ Shorey.

have no reasoning power. It is clear, then, that imagination cannot be either opinion in conjunction with sensation, or opinion based on sensation, or a blend of opinion and sensation, both for the reasons given, and because the opinion relates to nothing else but the object of sensation : I mean that imagination is the blend of the perception of white with the opinion that it is white—not, surely, of the perception of white with the opinion that it is good. To imagine, then, is to form an opinion exactly corresponding to a direct perception. But things about which we have at the same time a true belief may have a false appearance ; for instance the sun appears to measure a foot across, but we are convinced that it is greater than the inhabited globe ; it follows, then, that either the percipient, without any alteration in the thing itself, and without forgetting or changing his mind, has rejected the true opinion which he had, or, if he still holds that opinion, it must be at once true and false. But a true opinion only becomes false when the fact changes unnoticed. Imagination, then, is not one of these things, nor a compound of them.

But since when a particular thing is moved another thing may be moved by it, and since imagination seems to be some kind of movement, and not to occur apart from sensation, but only to men when perceiving, and in connexion with what is perceptible, and since movement may be caused by actual sensation, and this movement must be similar to the sensation, this movement cannot exist without sensation, or when we are not perceiving ; in virtue of it the possessor may act and be acted upon in various ways ; and the movement may be true or false. The reason

The region of error.

161

428 b

διὰ τάδε. ἡ αἴσθησις τῶν μὲν ἰδίων ἀληθής ἐστιν
ἢ ὅτι ὀλίγιστον ἔχουσα τὸ ψεῦδος. δεύτερον δὲ
20 τοῦ συμβεβηκέναι ταῦτα· καὶ ἐνταῦθα ἤδη ἐνδέχεται διαψεύδεσθαι· ὅτι μὲν γὰρ λευκόν, οὐ ψεύδεται, εἰ δὲ τοῦτο τὸ λευκὸν ἢ ἄλλο τι, ψεύδεται.
τρίτον δὲ τῶν κοινῶν καὶ ἑπομένων τοῖς συμ-
βεβηκόσιν, οἷς ὑπάρχει τὰ ἴδια· λέγω δ' οἷον
κίνησις καὶ μέγεθος, ἃ συμβέβηκε τοῖς αἰσθητοῖς,
25 περὶ ἃ μάλιστα ἤδη ἔστιν ἀπατηθῆναι κατὰ τὴν
αἴσθησιν. ἡ δὲ κίνησις ἡ ὑπὸ τῆς ἐνεργείας γινομένη διοίσει τῆς αἰσθήσεως τῆς ἀπὸ τούτων τῶν
τριῶν αἰσθήσεων. καὶ ἡ μὲν πρώτη παρούσης
τῆς αἰσθήσεως ἀληθής, αἱ δ' ἕτεραι καὶ παρούσης
καὶ ἀπούσης εἶεν ἂν ψευδεῖς, καὶ μάλιστα ὅταν
30 πόρρω τὸ αἰσθητὸν ᾖ. εἰ οὖν μηθὲν μὲν ἄλλο
429 a ἔχοι ἢ τὰ εἰρημένα ἡ φαντασία,[1] τοῦτο δ' ἐστὶ
τὸ λεχθέν, ἡ φαντασία ἂν εἴη κίνησις ὑπὸ τῆς
αἰσθήσεως τῆς κατ' ἐνέργειαν γιγνομένης. ἐπεὶ
δ' ἡ ὄψις μάλιστα αἴσθησίς ἐστι, καὶ τὸ ὄνομα
5 ἀπὸ τοῦ φάους εἴληφεν, ὅτι ἄνευ φωτὸς οὐκ ἔστιν
ἰδεῖν. καὶ διὰ τὸ ἐμμένειν καὶ ὁμοίως εἶναι ταῖς
αἰσθήσεσι, πολλὰ κατ' αὐτὰς πράττει τὰ ζῷα, τὰ
μὲν διὰ τὸ μὴ ἔχειν νοῦν, οἷον τὰ θηρία, τὰ δὲ
διὰ τὸ ἐπικαλύπτεσθαι τὸν νοῦν ἐνίοτε πάθει ἢ
νόσοις ἢ ὕπνῳ, οἷον οἱ ἄνθρωποι. περὶ μὲν οὖν
φαντασίας, τί ἐστι καὶ διὰ τί ἐστιν, εἰρήσθω ἐπὶ
τοσοῦτον.

10 IV. Περὶ δὲ τοῦ μορίου τοῦ τῆς ψυχῆς ᾧ γινώσκει τε ἡ ψυχὴ καὶ φρονεῖ, εἴτε χωριστοῦ ὄντος

---

[1] ἔχοι ἢ . . . ἡ φαντασία E: ἔχει . . . ἢ μὴ φαντασίαν
Bekker: alii aliter.

for this last fact is as follows. The perception of proper objects is true, or is only capable of error to the least possible degree. Next comes perception that they are attributes, and here a possibility of error at once arises ; for perception does not err in perceiving that an object is white, but only as to whether the white object is one thing or another. Thirdly comes perception of the common attributes which accompany the concomitants to which the proper sensibles belong (I mean, *e.g.*, motion and magnitude) ; it is about these that error is most likely to occur. But the movement produced by the sense-activity will differ from the actual sensation in each of these three modes of perception. The first is true whenever the sensation is present, but the others may be false both when it is present and when it is absent, and especially when the sensible object is at a distance. If, then, imagination involves nothing else than we have stated, and is as we have described it, then imagination must be a movement produced by sensation actively operating. Since sight is the chief sense, the name φαντασία (imagination) is derived from φάος (light), because without light it is impossible to see. Again, because imaginations persist in us and resemble sensations, living creatures frequently act in accordance with them, some, *viz.*, the brutes, because they have no mind, and some, *viz.*, men, because the mind is temporarily clouded over by emotion, or disease, or sleep. Let this suffice about the nature and cause of imagination.

IV. Concerning that part of the soul (whether it is separable in extended space, or only in thought) *Feeling and thinking are not analogous.*

429 a

εἴτε καὶ μὴ χωριστοῦ κατὰ μέγεθος ἀλλὰ κατὰ
λόγον, σκεπτέον τίν' ἔχει διαφοράν, καὶ πῶς ποτὲ
γίνεται τὸ νοεῖν. εἰ δή ἐστι τὸ νοεῖν ὥσπερ τὸ
αἰσθάνεσθαι, ἢ πάσχειν τι ἂν εἴη ὑπὸ τοῦ νοητοῦ
15 ἤ τι τοιοῦτον ἕτερον. ἀπαθὲς ἄρα δεῖ εἶναι, δεκτικὸν δὲ τοῦ εἴδους καὶ δυνάμει τοιοῦτον ἀλλὰ μὴ
τοῦτο, καὶ ὁμοίως ἔχειν, ὥσπερ τὸ αἰσθητικὸν
πρὸς τὰ αἰσθητά, οὕτω τὸν νοῦν πρὸς τὰ νοητά.
ἀνάγκη ἄρα, ἐπεὶ πάντα νοεῖ, ἀμιγῆ εἶναι, ὥσπερ
φησὶν Ἀναξαγόρας, ἵνα κρατῇ, τοῦτο δ' ἐστὶν ἵνα
20 γνωρίζῃ· παρεμφαινόμενον γὰρ κωλύει τὸ ἀλλότριον καὶ ἀντιφράττει, ὥστε μηδ' αὐτοῦ εἶναι
φύσιν μηδεμίαν ἀλλ' ἢ ταύτην, ὅτι δυνατόν. ὁ
ἄρα καλούμενος τῆς ψυχῆς νοῦς (λέγω δὲ νοῦν ᾧ
διανοεῖται καὶ ὑπολαμβάνει ἡ ψυχή) οὐθέν ἐστιν
ἐνεργείᾳ τῶν ὄντων πρὶν νοεῖν. διὸ οὐδὲ μεμῖχθαι
25 εὔλογον αὐτὸν τῷ σώματι· ποιός τις γὰρ ἂν
γίγνοιτο, ψυχρὸς ἢ θερμός, ἢ κἂν ὄργανόν τι εἴη,
ὥσπερ τῷ αἰσθητικῷ· νῦν δ' οὐθέν ἐστιν. καὶ εὖ
δὴ οἱ λέγοντες τὴν ψυχὴν εἶναι τόπον εἰδῶν, πλὴν
ὅτι οὔτε ὅλη ἀλλ' ἡ νοητική, οὔτε ἐντελεχείᾳ ἀλλὰ
30 δυνάμει τὰ εἴδη. ὅτι δ' οὐχ ὁμοία ἡ ἀπάθεια τοῦ
αἰσθητικοῦ καὶ τοῦ νοητικοῦ, φανερὸν ἐπὶ τῶν
αἰσθητηρίων καὶ τῆς αἰσθήσεως. ἡ μὲν γὰρ
429 b αἴσθησις οὐ δύναται αἰσθάνεσθαι ἐκ τοῦ σφόδρα
αἰσθητοῦ, οἷον ψόφου ἐκ τῶν μεγάλων ψόφων, οὐδ'

with which the soul knows and thinks, we have to consider what is its distinguishing characteristic, and how thinking comes about. If it is analogous to perceiving, it must be either a process in which the soul is acted upon by what is thinkable, or something else of a similar kind. This part, then, must (although impassive) be receptive of the form of an object, *i.e.*, must be potentially the same as its object, although not identical with it : as the sensitive is to the sensible, so must mind be to the thinkable. It is necessary then that mind, since it thinks all things, should be uncontaminated, as Anaxagoras says, in order that it may be in control, that is, that it may know ; for the intrusion of anything foreign hinders and obstructs it. Hence the mind, too, can have no characteristic except its capacity to receive. That part of the soul, then, which we call mind (by mind I mean that part by which the soul thinks and forms judgements) has no actual existence until it thinks. So it is unreasonable to suppose that it is mixed with the body ; for in that case it would become somehow qualitative, *e.g.*, hot or cold, or would even have some organ, as the sensitive faculty has ; but in fact it has none. It has been well said that the soul is the place of forms, except that this does not apply to the soul as a whole, but only in its thinking capcity, and the forms occupy it not actually but only potentially. But that the perceptive and thinking faculties are not alike in their impassivity is obvious if we consider the sense organs and sensation. For the sense loses sensation under the stimulus of a too violent sensible object ; *e.g.*, of sound immediately after loud sounds, and

ἐκ τῶν ἰσχυρῶν χρωμάτων καὶ ὀσμῶν οὔτε ὁρᾶν
οὔτε ὀσμᾶσθαι· ἀλλ' ὁ νοῦς ὅταν τι νοήσῃ σφόδρα
νοητόν, οὐχ ἧττον νοεῖ τὰ ὑποδεέστερα, ἀλλὰ καὶ
5 μᾶλλον· τὸ μὲν γὰρ αἰσθητικὸν οὐκ ἄνευ σώματος,
ὁ δὲ χωριστός. ὅταν δ' οὕτως ἕκαστα γένηται
ὡς ἐπιστήμων λέγεται ὁ κατ' ἐνέργειαν (τοῦτο δὲ
συμβαίνει, ὅταν δύνηται ἐνεργεῖν δι' αὐτοῦ), ἔστι
μὲν[1] καὶ τότε δυνάμει πως, οὐ μὴν ὁμοίως καὶ
πρὶν μαθεῖν ἢ εὑρεῖν· καὶ αὐτὸς δὲ αὑτὸν τότε
10 δύναται νοεῖν.

’Επεὶ δ' ἄλλο ἐστὶ τὸ μέγεθος καὶ τὸ μεγέθει
εἶναι καὶ ὕδωρ καὶ ὕδατι εἶναι (οὕτω δὲ καὶ ἐφ'
ἑτέρων πολλῶν, ἀλλ' οὐκ ἐπὶ πάντων· ἐπ' ἐνίων
γὰρ ταὐτόν ἐστι) τὸ σαρκὶ εἶναι καὶ σάρκα ἢ ἄλλῳ
ἢ ἄλλως ἔχοντι κρίνει· ἡ γὰρ σὰρξ οὐκ ἄνευ τῆς
ὕλης, ἀλλ' ὥσπερ τὸ σιμὸν τόδε ἐν τῷδε. τῷ
15 μὲν οὖν αἰσθητικῷ τὸ θερμὸν καὶ τὸ ψυχρὸν κρίνει,
καὶ ὧν λόγος τις ἡ σάρξ· ἄλλῳ δὲ ἤτοι χωριστῷ
ἢ ὡς ἡ κεκλασμένη ἔχει πρὸς αὑτὴν ὅταν ἐκταθῇ,
τὸ σαρκὶ εἶναι κρίνει. πάλιν δ' ἐπὶ τῶν ἐν ἀφαιρέσει ὄντων τὸ εὐθὺ ὡς τὸ σιμόν· μετὰ συνεχοῦς
20 γάρ· τὸ δὲ τί ἦν εἶναι, εἰ ἔστιν ἕτερον τὸ εὐθεῖ
εἶναι καὶ τὸ εὐθύ, ἄλλο[2]· ἔστω γὰρ δυάς. ἑτέρῳ
ἄρα ἢ ἑτέρως ἔχοντι κρίνει. καὶ ὅλως ἄρα ὡς

---

[1] ἔστι μὲν ὁμοίως Bekker.
[2] ἄλλο TVX, Bonitz : ἄλλῳ.

neither seeing nor smelling is possible just after strong colours and scents ; but when mind thinks the highly intelligible, it is not less able to think of slighter things, but even more able ; for the faculty of sense is not apart from the body, whereas the mind is separable. But when the mind has become the several groups of its objects, as the learned man when active is said to do (and this happens, when he can exercise his function by himself), even then the mind is in a sense potential, though not quite in the same way as before it learned and discovered ; moreover the mind is then capable of thinking itself.

Since magnitude is not the same as the essence of magnitude, nor water the same as the essence of water (and so too in many other cases, but not in all, because in some cases there is no difference), we judge flesh and the essence of flesh either by different faculties, or by the same faculty in different relations ; for flesh cannot exist without its matter, but like " snub-nosed " implies a definite form in a definite matter. Now it is by the sensitive faculty that we judge hot and cold, and all qualities whose due proportion constitutes flesh ; but it is by a different sense, either quite distinct, or related to it in the same way as a bent line to itself when pulled out straight, that we judge the essence of flesh. Again, among abstract objects " straight " is like " snub-nosed," for it is always combined with extension ; but its essence, if " straight " and " straightness " are not the same, is something different ; let us call it duality. Therefore we judge it by another faculty, or by the same faculty in a different relation. And speaking gener-

ally, as objects are separable from their matter so also are the corresponding faculties of the mind.

One might raise the question : if the mind is a simple thing, and not liable to be acted upon, and has nothing in common with anything else, as Anaxagoras says, how will it think, if thinking is a form of being acted upon ? For it is when two things have something in common that we regard one as acting and the other as acted upon. And our second problem is whether the mind itself can be an object of thought. [a] For either mind will be present in all other objects (if, that is, mind is an object of thought in itself and not in virtue of something else, and what is thought is always identical in form), or else it will contain some common element, which makes it an object of thought like other things. Or there is the explanation which we have given before of the phrase " being acted upon in virtue of some common element," that mind is potentially identical with the objects of thought but is actually nothing, until it thinks. What the mind thinks must be in it in the same sense as letters are on a tablet which bears no actual writing; this is just what happens in the case of the mind. It is also itself thinkable, just like other objects of thought. [b] For in the case of things without matter that which thinks and that which is thought are the same ; for speculative knowledge is the same as its object. (We must consider why mind does not always think.) In things which have matter, each of the objects of thought is only potentially present. Hence while material objects will not have mind in them (for it is apart from their matter that mind is potentially identical with them) mind will still have the capacity of being thought.

Marginal notes:
Two problems :
(1) How does the mind think?
(2) Is the mind an object of thought ?

430 a

10 V. Ἐπεὶ δ' ὥσπερ ἐν ἁπάσῃ τῇ φύσει ἐστί τι
τὸ μὲν ὕλη ἑκάστῳ γένει (τοῦτο δὲ ὃ πάντα
δυνάμει ἐκεῖνα), ἕτερον δὲ τὸ αἴτιον καὶ ποιητικόν,
τῷ ποιεῖν πάντα, οἷον ἡ τέχνη πρὸς τὴν ὕλην
πέπονθεν, ἀνάγκη καὶ ἐν τῇ ψυχῇ ὑπάρχειν ταύτας
τὰς διαφοράς. καὶ ἔστιν ὁ μὲν τοιοῦτος νοῦς τῷ
15 πάντα γίνεσθαι, ὁ δὲ τῷ πάντα ποιεῖν, ὡς ἕξις
τις, οἷον τὸ φῶς· τρόπον γάρ τινα καὶ τὸ φῶς
ποιεῖ τὰ δυνάμει ὄντα χρώματα ἐνεργείᾳ χρώματα.
καὶ οὗτος ὁ νοῦς χωριστὸς καὶ ἀπαθὴς καὶ ἀμιγὴς
τῇ οὐσίᾳ ὢν ἐνεργείᾳ. ἀεὶ γὰρ τιμιώτερον τὸ
ποιοῦν τοῦ πάσχοντος καὶ ἡ ἀρχὴ τῆς ὕλης. τὸ δ'
20 αὐτό ἐστιν ἡ κατ' ἐνέργειαν ἐπιστήμη τῷ πράγ-
ματι· ἡ δὲ κατὰ δύναμιν χρόνῳ προτέρα ἐν τῷ ἑνί,
ὅλως δὲ οὐ χρόνῳ· ἀλλ' οὐχ ὁτὲ μὲν νοεῖ ὁτὲ δ' οὐ
νοεῖ. χωρισθεὶς δ' ἐστὶ μόνον τοῦθ' ὅπερ ἐστί, καὶ
τοῦτο μόνον ἀθάνατον καὶ ἀΐδιον (οὐ μνημονεύομεν
δέ, ὅτι τοῦτο μὲν ἀπαθές, ὁ δὲ παθητικὸς νοῦς
25 φθαρτός), καὶ ἄνευ τούτου οὐθὲν νοεῖ.

VI. Ἡ μὲν οὖν τῶν ἀδιαιρέτων νόησις ἐν τούτοις,
περὶ ἃ οὐκ ἔστι τὸ ψεῦδος· ἐν οἷς δὲ καὶ τὸ ψεῦδος
καὶ τὸ ἀληθές, σύνθεσίς τις ἤδη νοημάτων ὥσπερ
ἓν ὄντων, καθάπερ Ἐμπεδοκλῆς ἔφη '' ᾗ πολλῶν
30 μὲν κόρσαι ἀναύχενες ἐβλάστησαν,'' ἔπειτα συν-
τίθεσθαι τῇ φιλίᾳ—, οὕτω καὶ ταῦτα κεχωρισμένα
συντίθεται, οἷον τὸ ἀσύμμετρον καὶ ἡ διάμετρος,
430 b ἂν δὲ γενομένων[1] ἢ ἐσομένων, τὸν χρόνον προσ-

---

[1] γινομένων VWX, Bekker, Trendelenburg.

---

[a] *Sc.*, its previous activity.

V. Since in every class of objects, just as in the whole of nature, there is something which is their matter, *i.e.*, which is potentially all the individuals, and something else which is their cause or agent in that it makes them all—the two being related as an art to its material—these distinct elements must be present in the soul also. Mind in the passive sense is such because it becomes all things, but mind has another aspect in that it makes all things ; this is a kind of positive state like light ; for in a sense light makes potential into actual colours. Mind in this sense is separable, impassive and unmixed, since it is essentially an activity ; for the agent is always superior to the patient, and the originating cause to the matter. Actual knowledge is identical with its object. Potential is prior in time to actual knowledge in the individual, but in general it is not prior in time. Mind does not think intermittently. When isolated it is its true self and nothing more, and this alone is immortal and everlasting (we do not remember [a] because, while mind in this sense cannot be acted upon, mind in the passive sense is perishable), and without this nothing thinks.

VI. The thinking of indivisible objects of thought occurs among things concerning which there can be no falsehood ; where truth and falsehood are possible there is implied a compounding of thoughts into a fresh unity, as Empedocles said,[b] " where without necks the heads of many grew," and then were joined together by Love—, so also these separate entities are combined, as for instance " incommensurable " and " diagonal." But if the thinking is concerned with things past or future, then we take into account

Mind active and passive.

Individual and combined concepts.

[b] Fr. 57.

171

εννοῶν καὶ συντιθείς. τὸ γὰρ ψεῦδος ἐν συνθέσει ἀεί· καὶ γὰρ ἂν τὸ λευκὸν μὴ λευκόν, τὸ
μὴ λευκὸν συνέθηκεν. ἐνδέχεται δὲ καὶ διαίρεσιν
φάναι πάντα. ἀλλ᾽ οὖν ἔστι γε οὐ μόνον τὸ ψεῦδος
5 ἢ ἀληθές, ὅτι λευκὸς Κλέων ἐστίν, ἀλλὰ καὶ ὅτι
ἦν ἢ ἔσται. τὸ δὲ ἓν ποιοῦν, τοῦτο ὁ νοῦς ἕκαστον.

Τὸ δ᾽ ἀδιαίρετον ἐπεὶ διχῶς, ἢ δυνάμει ἢ
ἐνεργείᾳ, οὐθὲν κωλύει νοεῖν τὸ ἀδιαίρετον, ὅταν
νοῇ τὸ μῆκος (ἀδιαίρετον γὰρ ἐνεργείᾳ), καὶ ἐν
χρόνῳ ἀδιαιρέτῳ· ὁμοίως γὰρ ὁ χρόνος διαιρετὸς
10 καὶ ἀδιαίρετος τῷ μήκει. οὔκουν ἔστιν εἰπεῖν εἰ
τῷ ἡμίσει τί ἐνόει[1] ἑκατέρῳ· οὐ γάρ ἐστιν, ἂν
μὴ διαιρεθῇ, ἀλλ᾽ ἢ δυνάμει. χωρὶς δ᾽ ἑκάτεροι
νοῶν τῶν ἡμίσεων διαιρεῖ καὶ τὸν χρόνον ἅμα·
τότε δ᾽ οἱονεὶ μήκη. εἰ δ᾽ ὡς ἐξ ἀμφοῖν, καὶ ἐν
τῷ χρόνῳ τῷ ἐπ᾽ ἀμφοῖν.

15 Τὸ δὲ μὴ κατὰ ποσὸν ἀδιαίρετον ἀλλὰ τῷ εἴδει
νοεῖ ἐν ἀδιαιρέτῳ χρόνῳ καὶ ἀδιαιρέτῳ τῆς ψυχῆς·
κατὰ συμβεβηκὸς δέ, καὶ οὐχ ᾗ ἐκεῖνα διαιρετά,
ᾧ νοεῖ καὶ ἐν ᾧ χρόνῳ, ἀλλ᾽ ᾗ ἀδιαίρετα· ἔνεστι
γὰρ κἂν τούτοις τι ἀδιαίρετον, ἀλλ᾽ ἴσως οὐ
χωριστόν, ὃ ποιεῖ ἕνα τὸν χρόνον καὶ τὸ μῆκος.
20 καὶ τοῦθ᾽ ὁμοίως ἐν ἅπαντί ἐστι τῷ συνεχεῖ καὶ
χρόνῳ καὶ μήκει. ἡ δὲ στιγμὴ καὶ πᾶσα διαίρεσις,
καὶ τὸ οὕτως ἀδιαίρετον, δηλοῦται ὥσπερ ἡ

---

[1] ἐνόει L, Torstrik : ἐννοεῖ.

and include the notion of time. For falsehood always lies in the process of combination, for if a man calls white not-white, he has combined the notion not-white. It is equally possible to say that all these cases involve division. At any rate it is not merely true or false to say that Cleon is white, but also that he was or will be. The principle which unifies is in every case the mind.

Since the term indivisible has two senses—potential or actual—there is nothing to prevent the mind from thinking of the indivisible when it thinks of length (which is in actuality undivided), and that in indivisible time. Time is also both divisible and indivisible in the same sense as length. So it is impossible to say what it was thinking in each half of the time ; for the half has no existence, except potentially, unless the whole is divided. But by thinking each half separately, mind divides the time as well ; in which case the halves are treated as separate units of length. But if the line is thought of as the sum of two halves, it is also thought of in a time which covers both half periods.

But when the object of thought is not quantitatively but qualitatively indivisible, the mind thinks of it in indivisible time, and by an indivisible activity of the soul ; but incidentally this whole is divisible, not in the sense in which the activity and the time are divisible, but in the sense in which they are indivisible ; for there is an indivisible element even in these, though perhaps incapable of separate existence, which makes the time and the length one. And this is equally true of every continuous thing whether time or length. Points and all divisions and everything indivisible in this sense are apprehended in the same

173

430 b

στέρησις. καὶ ὅμοιος ὁ λόγος ἐπὶ τῶν ἄλλων,
οἷον πῶς τὸ κακὸν γνωρίζει ἢ τὸ μέλαν· τῷ
ἐναντίῳ γάρ πως γνωρίζει. δεῖ δὲ δυνάμει εἶναι
τὸ γνωρίζον καὶ ἐνεῖναι ἐν αὐτῷ. εἰ δέ τινι μή
25 ἐστιν ἐναντίον,[1] αὐτὸ ἑαυτὸ γινώσκει καὶ ἐνεργείᾳ
ἐστὶ καὶ χωριστόν. ἔστι δ' ἡ μὲν φάσις τι κατά
τινος, ὥσπερ ἡ κατάφασις, καὶ ἀληθὴς ἢ ψευδὴς
πᾶσα· ὁ δὲ νοῦς οὐ πᾶς, ἀλλ' ὁ τοῦ τί ἐστι κατὰ
τὸ τί ἦν εἶναι ἀληθής, καὶ οὐ τὶ κατά τινος· ἀλλ'
30 ὥσπερ τὸ ὁρᾶν τοῦ ἰδίου ἀληθές, εἰ δ' ἄνθρωπος
τὸ λευκὸν ἢ μή, οὐκ ἀληθὲς ἀεί, οὕτως ἔχει ὅσα
ἄνευ ὕλης.

431 a   VII. Τὸ δ' αὐτό ἐστιν ἡ κατ' ἐνέργειαν ἐπιστήμη
τῷ πράγματι. ἡ δὲ κατὰ δύναμιν χρόνῳ προτέρα
ἐν τῷ ἑνί, ὅλως δὲ οὐδὲ χρόνῳ· ἔστι γὰρ ἐξ ἐντελεχείᾳ ὄντος πάντα τὰ γιγνόμενα. φαίνεται δὲ
5 τὸ μὲν αἰσθητὸν ἐκ δυνάμει ὄντος τοῦ αἰσθητικοῦ
ἐνεργείᾳ ποιοῦν· οὐ γὰρ πάσχει οὐδ' ἀλλοιοῦται.
διὸ ἄλλο εἶδος τοῦτο κινήσεως· ἡ γὰρ κίνησις τοῦ
ἀτελοῦς ἐνέργεια ἦν, ἡ δ' ἁπλῶς ἐνέργεια ἑτέρα
ἡ τοῦ τετελεσμένου. τὸ μὲν οὖν αἰσθάνεσθαι
ὅμοιον τῷ φάναι μόνον καὶ νοεῖν· ὅταν δὲ ἡδὺ
10 ἢ λυπηρόν, οἷον καταφᾶσα ἢ ἀποφᾶσα, διώκει
ἢ φεύγει· καὶ ἔστι τὸ ἥδεσθαι καὶ λυπεῖσθαι τὸ ἐνεργεῖν τῇ αἰσθητικῇ μεσότητι πρὸς τὸ ἀγαθὸν
ἢ κακόν, ᾗ τοιαῦτα. καὶ ἡ φυγὴ δὲ καὶ ἡ
ὄρεξις τοῦτο ἡ κατ' ἐνέργειαν, καὶ οὐχ ἕτερον τὸ

---

[1] ἐναντίον τῶν αἰτίων Bekker.

[a] 417 b 2 sqq.

way as privations. And the same explanation applies in all other cases ; e.g., how the mind cognizes evil or black ; for it recognizes them, in a sense, by their contraries. The cognizing agent must be potentially one contrary, and contain the other. But if there is anything which has no contrary, it is self-cognizant, actual and separately existent. Assertion, like affirmation, states an attribute of a subject, and is always either true or false ; but this is not always so with the mind : the thinking of the definition in the sense of the essence is always true and is not an instance of predication ; but just as while the seeing of a proper object is always true, the judgement whether the white object is a man or not is not always true, so it is with every object abstracted from its matter.

VII. Knowledge when actively operative is identical with its object. In the individual potential knowledge has priority in time, but generally it is not prior even in time ; for everything comes out of that which actually is. And clearly the sensible object makes the sense-faculty actually operative from being only potential ; it is not acted upon, nor does it undergo change of state ; and so, if it is motion, it is motion of a distinct kind ; for motion, as we saw,[a] is an activity of the imperfect, but activity in the absolute sense, that is activity of the perfected, is different. Sensation, then, is like mere assertion and thinking ; when an object is pleasant or unpleasant, the soul pursues or avoids it, thereby making a sort of assertion or negation. To feel pleasure or pain is to adopt an attitude with the sensitive mean towards good or bad as such. This is what avoidance or appetite, when actual, really means, and the faculties

The practical intellect in operation.

431 a

ὀρεκτικὸν καὶ φευκτικόν, οὔτ' ἀλλήλων οὔτε τοῦ
αἰσθητικοῦ· ἀλλὰ τὸ εἶναι ἄλλο. τῇ δὲ διανοητικῇ
15 ψυχῇ τὰ φαντάσματα οἷον αἰσθήματα ὑπάρχει.
ὅταν δὲ ἀγαθὸν ἢ κακὸν φήσῃ ἢ ἀποφήσῃ, φεύγει
ἢ διώκει. διὸ οὐδέποτε νοεῖ ἄνευ φαντάσματος ἡ
ψυχή, ὥσπερ δὲ ὁ ἀὴρ τὴν κόρην τοιανδὶ ἐποίησεν,
αὐτὴ δ' ἕτερον, καὶ ἡ ἀκοὴ ὡσαύτως· τὸ δὲ
ἔσχατον ἕν, καὶ μία μεσότης· τὸ δ' εἶναι αὐτῇ
20 πλείω.

Τίνι δ' ἐπικρίνει τί διαφέρει γλυκὺ καὶ θερμόν,
εἴρηται μὲν καὶ πρότερον, λεκτέον δὲ καὶ ὧδε.
ἔστι γὰρ ἕν τι, οὕτω δὲ καὶ ὡς ὅρος. καὶ ταῦτα
ἓν τῷ ἀνάλογον καὶ τῷ ἀριθμῷ ὂν[1] ἔχει πρὸς
ἑκάτερον ὡς ἐκεῖνα πρὸς ἄλληλα· τί γὰρ διαφέρει
25 τὸ ἀπορεῖν πῶς τὰ μὴ[2] ὁμογενῆ κρίνει ἢ τἀναντία,
οἷον λευκὸν καὶ μέλαν ; ἔστω δὴ ὡς τὸ Α τὸ
λευκὸν πρὸς τὸ Β τὸ μέλαν, τὸ Γ πρὸς τὸ Δ [ὡς
ἐκεῖνα πρὸς ἄλληλα][3]· ὥστε καὶ ἐναλλάξ. εἰ δὴ
τὰ ΓΔ ἑνὶ εἴη ὑπάρχοντα, οὕτως ἕξει ὥσπερ καὶ
τὰ ΑΒ τὸ αὐτὸ μὲν καὶ ἓν τὸ δ' εἶναι οὐ τὸ
431 b αὐτό, κἀκεῖνα[4] ὁμοίως. ὁ δ' αὐτὸς λόγος καὶ εἰ
τὸ μὲν Α τὸ γλυκὺ εἴη, τὸ δὲ Β τὸ λευκόν.

Τὰ μὲν οὖν εἴδη τὸ νοητικὸν ἐν τοῖς φαντάσμασι
νοεῖ, καὶ ὡς ἐν ἐκείνοις ὥρισται αὐτῷ τὸ διωκτὸν
καὶ φευκτόν, καὶ ἐκτὸς τῆς αἰσθήσεως, ὅταν ἐπὶ
5 τῶν φαντασμάτων ᾖ, κινεῖται, οἷον αἰσθανόμενος

---

[1] ὂν Freudenthal : ὄν.   [2] μὴ om. TVW, Bekker.
[3] secl. Biehl.   [4] κἀκεῖνα Pacius : κἀκεῖνο.

---

[a] 426 b 12 sqq.   [b] Sc., to pursue or avoidance.

of appetite or avoidance are not really different from each other, or from the sensitive faculty, though their actual essence is different. Now for the thinking soul images take the place of direct perceptions; and when it asserts or denies that they are good or bad, it avoids or pursues them. Hence the soul never thinks without a mental image. The process is just like that in which air affects the eye in a particular way, and the eye again affects something else; and similarly with hearing. The last thing to be affected is a single entity and a single mean, although it has more than one aspect.

We have explained before [a] what part of the soul distinguishes between sweet and hot, but some further details must now be added. It is a unity, but in the sense just described, i.e., as a point of connexion. The faculties which it connects, being analogically and numerically one, are related to one another just as their sensible objects are. It makes no difference whether we ask how the soul distinguishes things which are not of the same class, or contraries like white and black. Suppose that as A (white) is to B (black), so is C to D. Then alternando C is to A as D is to B. If then C and D belong to one subject, they will stand in the same relation as A and B; A and B are one and the same, though their being has different aspects, and so it is with C and D. The same also holds good if we take A as sweet and B as white.

So the thinking faculty thinks the forms in mental images, and just as in the sphere of sense what is to be pursued and avoided is defined for it, so also outside sensation, when it is occupied with mental images, is moved.[b] For instance in perceiving a beacon

177

431 b

τὸν φρυκτὸν ὅτι πῦρ, τῇ κοινῇ γνωρίζει, ὁρῶν κινούμενον, ὅτι πολέμιος. ὁτὲ δὲ τοῖς ἐν τῇ ψυχῇ φαν-
τάσμασιν ἢ νοήμασιν ὥσπερ ὁρῶν λογίζεται καὶ
βουλεύεται τὰ μέλλοντα πρὸς τὰ παρόντα· καὶ ὅταν
εἴπῃ ὡς ἐκεῖ τὸ ἡδὺ ἢ λυπηρόν, ἐνταῦθα φεύγει ἢ
10 διώκει, καὶ ὅλως ἐν πράξει. καὶ τὸ ἄνευ δὲ πράξεως, τὸ ἀληθὲς καὶ τὸ ψεῦδος ἐν τῷ αὐτῷ γένει
ἐστὶ τῷ ἀγαθῷ καὶ κακῷ· ἀλλὰ τῷ γε ἁπλῶς διαφέρει καὶ τινί. τὰ δὲ ἐν ἀφαιρέσει λεγόμενα νοεῖ
ὥσπερ ἂν εἰ τὸ σιμόν, ᾗ μὲν σιμόν, οὐ κεχωρισ-
15 μένως, ᾗ δὲ κοῖλον, εἴ τις ἐνόει ἐνεργείᾳ, ἄνευ τῆς
σαρκὸς ἂν ἐνόει ἐν ᾗ τὸ κοῖλον. οὕτω τὰ μαθη-
ματικὰ οὐ κεχωρισμένα ὡς κεχωρισμένα νοεῖ, ὅταν
νοῇ ἐκεῖνα. ὅλως δὲ ὁ νοῦς ἐστιν ὁ κατ᾽ ἐνέργειαν
τὰ πράγματα νοῶν. ἆρα δ᾽ ἐνδέχεται τῶν κεχω-
ρισμένων τι νοεῖν ὄντα αὐτὸν μὴ κεχωρισμένον
μεγέθους, ἢ οὔ, σκεπτέον ὕστερον.

20 VIII. Νῦν δὲ περὶ ψυχῆς τὰ λεχθέντα συγκεφα-
λαιώσαντες, εἴπωμεν πάλιν ὅτι ἡ ψυχὴ τὰ ὄντα πώς
ἐστι πάντα. ἢ γὰρ αἰσθητὰ τὰ ὄντα ἢ νοητά, ἔστι
δ᾽ ἡ ἐπιστήμη μὲν τὰ ἐπιστητά πως, ἡ δ᾽ αἴσθησις
τὰ αἰσθητά· πῶς δὲ τοῦτο, δεῖ ζητεῖν. τέμνεται
25 οὖν ἡ ἐπιστήμη καὶ ἡ αἴσθησις εἰς τὰ πράγματα,
ἡ μὲν δυνάμει εἰς τὰ δυνάμει, ἡ δ᾽ ἐντελεχείᾳ εἰς
τὰ ἐντελεχείᾳ. τῆς δὲ ψυχῆς τὸ αἰσθητικὸν καὶ
τὸ ἐπιστημονικὸν δυνάμει ταῦτά[1] ἐστι, τὸ μὲν
ἐπιστητὸν τὸ δὲ αἰσθητόν. ἀνάγκη δ᾽ ἢ αὐτὰ ἢ

---

[1] ταῦτά E², Sophonias et vetus translatio : ταὐτόν.

a man recognizes that it is fire ; then seeing it moving he knows that it signifies an enemy. But sometimes by means of the images or thoughts in the soul, just as if it were seeing, it calculates and plans for the future in view of the present ; and when it makes a statement, as in sensation it asserts that an object is pleasant or unpleasant, in this case it avoids or pursues ; and so generally in action. What does not involve action, *i.e.*, the true or false, belongs to the same sphere as what is good or evil ; but they differ in having respectively a universal and a particular reference. Abstract objects, as they are called, the mind thinks as if it were thinking the snub-nosed ; *qua* snub-nosed, it would not be thought of apart from flesh, but *qua* hollow, if it were actually so conceived, it would be thought of apart from the flesh in which the hollowness resides. So when mind thinks the objects of mathematics, it thinks them as separable though actually they are not. In general, the mind when actively thinking is identical with its objects. Whether it is possible for the mind to think of un-extended objects when it is not itself unextended, must be considered later.

VIII. Now summing up what we have said about Summary the soul, let us assert once more that in a sense the soul is all existing things. What exists is either sensible or intelligible ; and in a sense knowledge is the knowable and sensation the sensible. We must consider in what sense this is so. Both knowledge and sensation are divided to correspond to their objects, the potential to the potential, and the actual to the actual. The sensitive and cognitive faculties of the soul are potentially these objects, *viz.*, the sensible and the knowable. These faculties, then, must be identical either with the objects themselves

431 b

τὰ εἴδη εἶναι. αὐτὰ μὲν γὰρ δὴ οὔ· οὐ γὰρ ὁ
432 a λίθος ἐν τῇ ψυχῇ, ἀλλὰ τὸ εἶδος· ὥστε ἡ ψυχὴ
ὥσπερ ἡ χείρ ἐστιν· καὶ γὰρ ἡ χεὶρ ὄργανόν ἐστιν
ὀργάνων, καὶ ὁ νοῦς εἶδος εἰδῶν καὶ ἡ αἴσθησις
εἶδος αἰσθητῶν. ἐπεὶ δὲ οὐδὲ πρᾶγμα οὐθέν ἐστι
παρὰ τὰ μεγέθη, ὡς δοκεῖ, τὰ αἰσθητὰ κεχωρισ-
5 μένον, ἐν τοῖς εἴδεσι τοῖς αἰσθητοῖς τὰ νοητά ἐστι,
τά τε ἐν ἀφαιρέσει λεγόμενα, καὶ ὅσα τῶν αἰσθητῶν
ἕξεις καὶ πάθη. καὶ διὰ τοῦτο οὔτε μὴ αἰσθανόμενος μηθὲν οὐθὲν ἂν μάθοι οὐδὲ ξυνείη· ὅταν τε
θεωρῇ, ἀνάγκη ἅμα φάντασμά τι θεωρεῖν· τὰ γὰρ
10 φαντάσματα ὥσπερ αἰσθήματά ἐστι, πλὴν ἄνευ
ὕλης. ἔστι δ᾽ ἡ φαντασία ἕτερον φάσεως καὶ
ἀποφάσεως· συμπλοκὴ γὰρ νοημάτων ἐστὶ τὸ
ἀληθὲς ἢ ψεῦδος. τὰ δὲ πρῶτα νοήματα τίνι
διοίσει τοῦ μὴ φαντάσματα εἶναι; ἢ οὐδὲ τἆλλα
φαντάσματα, ἀλλ᾽ οὐκ ἄνευ φαντασμάτων.

15 IX. Ἐπεὶ δὲ ἡ ψυχὴ κατὰ δύο ὥρισται δυνάμεις
ἡ τῶν ζῴων, τῷ τε κριτικῷ, ὃ διανοίας ἔργον ἐστὶ
καὶ αἰσθήσεως, καὶ ἔτι τῷ κινεῖν τὴν κατὰ τόπον
κίνησιν, περὶ μὲν αἰσθήσεως καὶ νοῦ διωρίσθω
τοσαῦτα, περὶ δὲ τοῦ κινοῦντος, τί ποτέ ἐστι τῆς
20 ψυχῆς, σκεπτέον, πότερον ἕν τι μόριον αὐτῆς
χωριστὸν ὂν ἢ μεγέθει ἢ λόγῳ, ἢ πᾶσα ἡ ψυχή,

180

or with their forms. Now they are not identical with the objects; for the stone does not exist in the soul, but only the form of the stone. The soul, then, acts like a hand; for the hand is an instrument which employs instruments, and in the same way the mind is a form which employs forms, and sense is a form which employs the forms of sensible objects. But since apparently nothing has a separate existence, except sensible magnitudes, the objects of thought —both the so-called abstractions of mathematics and all states and affections of sensible things—reside in the sensible forms. And for this reason as no one could ever learn or understand anything without the exercise of perception, so even when we think speculatively, we must have some mental picture of which to think; for mental images are similar to objects perceived except that they are without matter. But imagination is not the same thing as assertion and denial; for truth and falsehood involve a combination of notions. How then will the simplest notions differ from mental pictures? Surely neither these simple notions nor any others are mental pictures, but they cannot occur without such mental pictures. *Thought and feeling are not identical but interdependent.*

IX. The soul in living creatures is distinguished by two functions, the judging capacity which is a function of the intellect and of sensation combined, and the capacity for exciting movement in space. We have completed our account of sense and mind, and must now consider what it is in the soul that excites movement; whether it is a part separable from the soul itself, either in extension or only in defini- *The relation of the soul to movement.*

181

tion, or whether it is the whole soul ; and if it is a part, whether it is a special part beyond those usually described, and of which we have given an account, or whether it is one of them. A problem at once arises : in what sense should we speak of parts of the soul, and how many are there ? For in one sense they seem to be infinite, and not confined to those which some thinkers describe, when they attempt analysis, as calculative, emotional, and desiderative, or, as others have it, rational and irrational. When we consider the distinctions according to which they classify, we shall find other parts exhibiting greater differences than those of which we have already spoken ; for instance the nutritive part, which belongs both to plants and to all living creatures, and the sensitive part, which one could not easily assign either to the rational or irrational part ; and also the imaginative part, which appears to be different in essence from them all, but which is extremely difficult to identify with, or to distinguish from any one of them, if we are to suppose that the parts of the soul are separate. Beyond these again is the appetitive part, which in both definition and capacity would seem to be different from them all. And it is surely unreasonable to split this up ; for there is will in the calculative, and desire and passion in the irrational part ; and if the soul is divided into three, appetite will be found in each.

Moreover, to come to the point with which our inquiry is now concerned, what is it that makes the living creature move in space ? The generative and nutritive faculties, which all share, would seem responsible for movement in the sense of growth and

Parts of the soul.

Movement cannot be associated with any one part of the soul.

183

432 b

γεννητικὸν καὶ θρεπτικόν· περὶ δὲ ἀναπνοῆς καὶ
ἐκπνοῆς καὶ ὕπνου καὶ ἐγρηγόρσεως ὕστερον ἐπι-
σκεπτέον· ἔχει γὰρ καὶ ταῦτα πολλὴν ἀπορίαν.
ἀλλὰ περὶ τῆς κατὰ τόπον κινήσεως, τί τὸ κινοῦν
τὸ ζῷον τὴν πορευτικὴν κίνησιν, σκεπτέον. ὅτι
15 μὲν οὖν οὐχ ἡ θρεπτικὴ δύναμις, δῆλον· ἀεί τε
γὰρ ἕνεκά του ἡ κίνησις αὕτη, καὶ ἢ μετὰ φαντασίας ἢ ὀρέξεώς ἐστιν· οὐθὲν γὰρ μὴ ὀρεγόμενον
ἢ φεῦγον κινεῖται ἀλλ' ἢ βίᾳ. ἔτι κἂν τὰ φυτὰ
κινητικὰ ἦν, κἂν εἶχέ τι μόριον ὀργανικὸν πρὸς
τὴν κίνησιν ταύτην. ὁμοίως δὲ οὐδὲ τὸ αἰσθη-
20 τικόν· πολλὰ γάρ ἐστι τῶν ζῴων ἃ αἴσθησιν μὲν
ἔχει, μόνιμα δ' ἐστὶ καὶ ἀκίνητα διὰ τέλους. εἰ
οὖν ἡ φύσις μήτε ποιεῖ μάτην μηθὲν μήτε ἀπο-
λείπει τι τῶν ἀναγκαίων, πλὴν ἐν τοῖς πηρώμασι
καὶ ἐν τοῖς ἀτελέσιν (τὰ δὲ τοιαῦτα τῶν ζῴων
τέλεια καὶ οὐ πηρώματά ἐστιν· σημεῖον δ' ὅτι
25 ἔστι γεννητικὰ καὶ ἀκμὴν ἔχει καὶ φθίσιν)—ὥστ'
εἶχεν ἂν καὶ τὰ ὀργανικὰ μέρη τῆς πορείας. ἀλλὰ
μὴν οὐδὲ τὸ λογιστικὸν καὶ ὁ καλούμενος νοῦς
ἐστὶν ὁ κινῶν· ὁ μὲν γὰρ θεωρητικὸς οὐθὲν νοεῖ
πρακτόν, οὐδὲ λέγει περὶ φευκτοῦ καὶ διωκτοῦ
οὐθέν, ἡ δὲ κίνησις ἢ φεύγοντός τι ἢ διώκοντός
30 τί ἐστιν. ἀλλ' οὐδ' ὅταν θεωρῇ τι τοιοῦτον, ἤδη
κελεύει φεύγειν ἢ διώκειν, οἷον πολλάκις διανοεῖται
φοβερόν τι ἢ ἡδύ, οὐ κελεύει δὲ φοβεῖσθαι, ἡ δὲ
433 a καρδία κινεῖται, ἂν δ' ἡδύ, ἕτερόν τι μόριον. ἔτι

---

[a] *i.e.* if movement in space were due to the sensitive faculty.

184

decay, as this movement belongs to them all ; later on we shall have to consider inspiration and expiration, and sleep and waking ; for these also present considerable difficulty. But now, about movement in space, we must consider what it is that causes the living animal to exhibit a travelling movement. Obviously it is not the nutritive faculty ; for this movement always has an object in view, and is combined with imagination or appetite ; for nothing moves except under compulsion, unless it is seeking or avoiding something. Besides, plants would be capable of locomotion, and would have some part instrumental towards this movement. Nor is it the sensitive faculty ; for there are many living creatures which have feeling, but are stationary, and do not move throughout their existence. Then seeing that nature does nothing in vain, and omits nothing essential, except in maimed or imperfect animals (and the sort of animal under consideration is perfect and not maimed ; this is proved by the fact that they propagate their species and have a prime and decline), they would also have parts instrumental to progression.[a] Nor is the calculative faculty, which is called mind, the motive principle, for the speculative mind thinks of nothing practical, and tells us nothing about what is to be avoided or pursued ; but movement is characteristic of one who is either avoiding or pursuing something. Even when the mind contemplates such an object, it does not directly suggest avoidance or pursuit ; e.g., it often thinks of something fearful or pleasant without suggesting fear. It is the heart which is moved,—or if the object is pleasant, some other part. Further, even when the mind orders and

185

thought urges avoidance or pursuit, there is no movement, but action is prompted by desire, *e.g.*, in the absence of self-control. Speaking generally, we see that the man possessing knowledge of the healing art is not always healing, so that there is some other factor which causes action in accordance with knowledge, and not knowledge itself. Finally, it is not appetite which is responsible for movement ; for the self-controlled, though they may crave and desire, do not do these things for which they have an appetite, but follow their reason.

X. These two then, appetite and mind, are clearly capable of causing movement if, that is, one regards imagination as some sort of thinking process ; for men often follow their imaginations contrary to knowledge, and in living creatures other than man there is neither thinking nor calculation, but only imagination. Both of these, then, mind and appetite, are productive of movement in space. But the mind in question is that which makes its calculations with an end in view, that is, the practical mind : it differs from the speculative mind in the end that it pursues. And every appetite is directed towards an end ; for the thing at which appetite aims is the starting-point of the practical mind, and the last step of the practical mind is the beginning of the action. So these two, appetite and practical thought, seem reasonably considered as the producers of movement ; for the object of appetite produces movement, and therefore thought produces movement, because the object of appetite is its beginning. Imagination, too, when it starts movement, never does so without appetite. That which moves, then, is a single faculty, that of appetite. If there were two movers, mind as well as appetite, they would produce movement in virtue of a common characteristic. But, as things are, mind is never seen to produce movement without appetite (for will is a form of appetite, and when movement accords with calculation, it accords also with choice), but appetite produces movement contrary to calculation; for desire is a form of appetite. Now mind is always right; but appetite and imagination may be right or wrong. Thus the object of appetite always produces movement, but this may be either the real or the apparent good; and not every good can excite movement, but only practical good. Practical good is that which is capable of being otherwise.

It is clear, then, that movement is caused by such a faculty of the soul as we have described, *viz.*, that which is called appetite. But those who divide up the parts of the soul, if they divide and distinguish them by their functions, get a great many parts : nutritive, sensitive, intelligent, deliberate and appetitive as well ; for these differ from one another more than the desiderative does from the emotional. Now appetites may conflict, and this happens wherever reason and desire are opposed, and this occurs in creatures which have a sense of time (for the mind advises us to resist with a view to the future, while desire only looks to the present ; for what is momentarily pleasant seems to be absolutely pleasant and absolutely good, because desire cannot look to the future). Thus while that which causes movement is specifically one, *viz.*, the faculty of appetite *qua* appetitive, or ultimately the object of appetite (for this, though unmoved, causes

Appetite produces movement.

189

433 b

κινούμενον τῷ νοηθῆναι ἢ φαντασθῆναι), ἀριθμῷ
δὲ πλείω τὰ κινοῦντα.

Ἐπειδὴ δ' ἐστὶ τρία, ἓν μὲν τὸ κινοῦν, δεύτερον
δ' ᾧ κινεῖ, τρίτον τὸ κινούμενον· τὸ δὲ κινοῦν
15 διττόν, τὸ μὲν ἀκίνητον, τὸ δὲ κινοῦν καὶ κινού-
μενον· ἔστι δὲ τὸ μὲν ἀκίνητον τὸ πρακτὸν ἀγαθόν,
τὸ δὲ κινοῦν καὶ κινούμενον τὸ ὀρεκτικόν (κινεῖται
γὰρ τὸ κινούμενον ᾗ ὀρέγεται, καὶ ἡ ὄρεξις κίνησίς[1]
τίς ἐστιν ᾗ ἐνέργεια), τὸ δὲ κινούμενον τὸ ζῷον·
ᾧ δὲ κινεῖ ὀργάνῳ ἡ ὄρεξις, ἤδη τοῦτο σωματικόν
20 ἐστιν· διὸ ἐν τοῖς κοινοῖς σώματος καὶ ψυχῆς ἔργοις
θεωρητέον περὶ αὐτοῦ. νῦν δὲ ὡς ἐν κεφαλαίῳ
εἰπεῖν τὸ κινοῦν ὀργανικῶς ὅπου ἀρχὴ καὶ τελευτὴ
τὸ αὐτό, οἷον ὁ γιγγλυμός· ἐνταῦθα γὰρ τὸ κυρτὸν
καὶ κοῖλον τὸ μὲν τελευτὴ τὸ δ' ἀρχή· διὸ τὸ μὲν
25 ἠρεμεῖ τὸ δὲ κινεῖται, λόγῳ μὲν ἕτερα ὄντα, μεγέθει
δ' ἀχώριστα· πάντα γὰρ ὤσει καὶ ἕλξει κινεῖται.
διὸ δεῖ ὥσπερ ἐν κύκλῳ μένειν τι, καὶ ἐντεῦθεν
ἄρχεσθαι τὴν κίνησιν. ὅλως μὲν οὖν, ὥσπερ
εἴρηται, ᾗ ὀρεκτικὸν τὸ ζῷον, ταύτῃ αὑτοῦ κινη-
τικόν· ὀρεκτικὸν δὲ οὐκ ἄνευ φαντασίας· φαντασία
30 δὲ πᾶσα ἢ λογιστικὴ ἢ αἰσθητική. ταύτης μὲν
οὖν καὶ τὰ ἄλλα ζῷα μετέχει.

XI. Σκεπτέον δὲ καὶ περὶ τῶν ἀτελῶν, τί τὸ
434 a κινοῦν ἐστίν, οἷς ἁφὴ μόνον ὑπάρχει αἴσθησις,

[1] ὄρεξις κίνησίς] κίνησις ὄρεξίς EL, Bekker.

movement by being thought of or imagined), the things which cause movement are numerically many.

But movement involves three factors : first the moving cause, secondly the means by which it produces movement, and thirdly the thing moved. The moving cause is of two kinds ; one is unmoved and the other both moves and is moved. The former is the practical good, while that which both moves and is moved is the appetite (for that which is moved is moved *qua* influenced by appetite, and appetite *qua* actual is a kind of movement), and the thing moved is the animal. The instrument by which appetite causes movement belongs already to the physical sphere ; so it must be considered among the functions common to body and soul. But for the present we may say briefly that the motive instrument is found where a beginning and end coincide, as in a ball-and-socket joint. For there the convex surface (the ball) and the concave surface (the socket) are respectively the end and the beginning of the movement ; consequently the latter is at rest while the former moves. They are distinct in definition, but spatially inseparable ; for all movement consists of pushing and pulling ; so that, as in a wheel, one point must remain fixed, and from it the movement must be initiated. Speaking generally then, as has been said, in so far as the living creature is capable of appetite, it is also capable of self-movement ; but it is not capable of appetite without imagination, and all imagination involves either calculation or sensation. This latter all other living creatures share besides man.

XI. We must now consider what the moving principle is in the case of those imperfect animals, whose only sensation is that of touch, and whether it is or

How movement takes place.

How is movement possible in the lower

191

is not possible for them to have imagination and desire. For it is evident that they are liable to pain and pleasure. If they have these they must also have desire. But in what sense could they have imagination ? Perhaps, just as their movements are indeterminate, so they also have imagination and desire, but only indeterminately. Imagination in the form of sense is found, as we have said, in all animals, but deliberative imagination only in the calculative ; for to decide whether one shall do this or that calls at once for calculation, and one must measure by a single standard ; for one pursues the *greater* good. This implies the ability to combine several images into one. This is why imagination is thought not to involve opinion, because it does not involve opinion which is based on inference, whereas opinion involves imagination. Hence appetite does not imply capacity for deliberation. Sometimes it overcomes the will and sways it, as one sphere moves another ; or appetite influences appetite, when the subject lacks self-control (but in nature the upper sphere always controls and moves the lower) ; thus we now have three modes of movement. The cognitive faculty is not moved but remains still. Since one premiss or statement is universal and the other particular (for the one asserts that a man in such a position should do such a thing, but the other asserts that this present act is such a thing and that I am a

man in such a position), it is surely this latter opinion
which causes movement, not the universal. Or perhaps it is both, but the universal tends to remain at
rest, and the other does not.

XII. Every living thing, then, must have the nutri- The soul
tive soul, and in fact has a soul from its birth until its and life.
death ; for what has been born must have growth, a
highest point of development, and decay, and these
things are impossible without food. The nutritive
faculty must then exist in all things which grow and
decay. But sensation is not necessarily present in all
living things. Those whose bodies are uncompounded
cannot have a sense of touch, nor can those which are
incapable of receiving forms without their matter. But
an animal must have sensation, if it is a fact that nature
does nothing in vain. For all provisions of nature are
means to an end, or must be regarded as coincidental to
such means. Any body capable of moving from place
to place, if it had no sensation, would be destroyed, and
would not reach the end which is its natural function ;
for how could it be nourished ? Stationary living
things can draw their food from the source from which
they were born, but it is not possible for a body to
possess a soul and a mind capable of judgement without also having sensation, if that body is not stationary
but produced by generation ; nor even if it is un-
generated.¹ For why should it not² have sensation ?
Either for the good of the soul or for that of the body,
but in fact neither alternative is true ; for the soul
will not think any better, and the body will be no

---

¹ οὔτε . . . ζῷον secl. Torstrik.
² οὐχ om. LSX, Bekker.

434 b

μᾶλλον δι' ἐκεῖνο. οὐθὲν ἄρα ἔχει ψυχὴν σῶμα
μὴ μόνιμον ἄνευ αἰσθήσεως.

Ἀλλὰ μὴν εἴγε αἴσθησιν ἔχει, ἀνάγκη τὸ σῶμα
10 εἶναι ἢ ἁπλοῦν ἢ μικτόν. οὐχ οἷόν τε δὲ ἁπλοῦν·
ἁφὴν γὰρ οὐχ ἕξει, ἔστι δὲ ἀνάγκη ταύτην ἔχειν.
τοῦτο δὲ ἐκ τῶνδε δῆλον. ἐπεὶ γὰρ τὸ ζῷον
σῶμα ἔμψυχόν ἐστι, σῶμα δὲ ἅπαν ἁπτόν, ἁπτὸν
δὲ τὸ αἰσθητὸν ἁφῇ, ἀνάγκη καὶ τὸ τοῦ ζῴου
σῶμα ἁπτικὸν εἶναι, εἰ μέλλει σώζεσθαι τὸ ζῷον.
15 αἱ γὰρ ἄλλαι αἰσθήσεις δι' ἑτέρων αἰσθάνονται,
οἷον ὄσφρησις ὄψις ἀκοή· ἁπτόμενον δέ, εἰ μὴ
ἕξει αἴσθησιν, οὐ δυνήσεται τὰ μὲν φεύγειν τὰ δὲ
λαβεῖν. εἰ δὲ τοῦτο, ἀδύνατον ἔσται σώζεσθαι τὸ
ζῷον. διὸ καὶ ἡ γεῦσίς ἐστιν ὥσπερ ἁφή τις·
τροφῆς γάρ ἐστιν, ἡ δὲ τροφὴ τὸ σῶμα τὸ ἁπτόν.
20 ψόφος δὲ καὶ χρῶμα καὶ ὀσμὴ οὐ τρέφει, οὐδὲ
ποιεῖ οὔτ' αὔξησιν οὔτε φθίσιν. ὥστε καὶ τὴν
γεῦσιν ἀνάγκη ἁφὴν εἶναί τινα, διὰ τὸ τοῦ ἁπτοῦ
καὶ θρεπτικοῦ αἴσθησιν εἶναι. αὗται μὲν οὖν
ἀναγκαῖαι τῷ ζῴῳ, καὶ φανερὸν ὅτι οὐχ οἷόν τε
ἄνευ ἁφῆς εἶναι ζῷον.

25 Αἱ δὲ ἄλλαι τοῦ τε εὖ ἕνεκα καὶ γένει ζῴων
ἤδη οὐ τῷ τυχόντι, ἀλλὰ τισίν, οἷον τῷ πορευτικῷ
ἀνάγκη ὑπάρχειν· εἰ γὰρ μέλλει σώζεσθαι, οὐ
μόνον δεῖ ἁπτόμενον αἰσθάνεσθαι ἀλλὰ καὶ ἄποθεν.
τοῦτο δ' ἂν εἴη, εἰ διὰ τοῦ μεταξὺ αἰσθητικὸν εἴη

better, for not having sensation. No, body, then, which is not stationary possesses a soul without sensation.

Further, if it does possess sensation, the body must be either simple or compound. But it cannot be simple ; for in that case it will have no sense of touch, and this is indispensable to it. This is obvious from the following considerations. For since the living animal is a body possessing soul, and every body is tangible, and tangible means perceptible by touch, it follows that the body of the animal must have the faculty of touch if the animal is to survive. For the other senses, such as smell, vision and hearing, perceive through the medium of something else ; but the animal when it touches, if it has no sensation, will not be able to avoid some things and seize others. In that case it will be impossible for the animal to survive. This is why taste is a kind of touch ; for it relates to food, and food is a tangible body. Sound, colour and smell supply no food, nor do they produce growth and decay. Hence taste must be some kind of touch, because it is the perception of what is tangible and nutritive. These two senses then, are essential to the animal, and it is obvious that an animal cannot exist without a sense of touch.

The other senses are means to well-being ; they do not belong to any class of living creatures taken at random, but only to certain ones, e.g., they are essential to the animal which is capable of locomotion; for if it is to survive, not only must it perceive when in contact, but also from a distance. And this will occur only if it can perceive through a medium, the

197

medium being affected and set in motion by the sensible object, and the animal itself by the medium. For just as that which produces movement in space causes change up to a certain point, and that which has given an impulse causes something else to give one also, and the movement takes place through a medium ; and as the first mover impels without being impelled, while the last in the series is impelled without impelling, but the medium both impels and is impelled, and there may be many media : so it is in the case of alteration, except that the subject suffers alteration without changing place. If one were to dip something into wax, the movement would occur in the wax just so far as one dipped it ; stone would not be moved at all, but water would be to a great distance. But it is air that is moved, acting and being acted upon to the greatest extent, so long as it remains a constant unity. This is why in the case of reflection it is better to suppose, not that sight proceeds from the eye and is reflected, but rather that the air, so long as it remains a unity, is affected by the shape and colour. Now on a smooth surface it is a unity ; and so it in its turn sets the sight in motion, just as if the impression on the wax extended right through to the other side.

XIII. It is obvious that the body of an animal cannot consist of a single element such as fire or air. For without a sense of touch it is impossible to have any other sensation ; for every body possessing soul has the faculty of touch, as has been said.[a] Now except for earth, all the other elements would become sense organs, but they all produce sensation by means of something else, that is through media.

Touch is the most elementary and indispensable faculty.

---

[a] 434 b 10 *sqq.*

435 a

μεταξύ· ἡ δ' ἁφὴ τῷ αὐτῶν ἅπτεσθαί ἐστιν, διὸ
καὶ τοὔνομα τοῦτο ἔχει. καίτοι καὶ τὰ ἄλλα
αἰσθητήρια ἁφῇ αἰσθάνεται, ἀλλὰ δι' ἑτέρου· αὕτη
20 δὲ δοκεῖ μόνη δι' αὑτῆς, ὥστε τῶν μὲν τοιούτων
στοιχείων οὐθὲν ἂν εἴη σῶμα τοῦ ζῴου. οὐδὲ δὴ
γήινον. πάντων γὰρ ἡ ἁφὴ τῶν ἁπτῶν ἐστὶν
ὥσπερ μεσότης, καὶ δεκτικὸν τὸ αἰσθητήριον οὐ
μόνον ὅσαι διαφοραὶ γῆς εἰσίν, ἀλλὰ καὶ θερμοῦ
καὶ ψυχροῦ καὶ τῶν ἄλλων ἁπτῶν ἁπάντων. καὶ
25 διὰ τοῦτο τοῖς ὀστοῖς καὶ ταῖς θριξὶ καὶ τοῖς
435 b τοιούτοις μορίοις οὐκ αἰσθανόμεθα, ὅτι γῆς ἐστίν.
καὶ τὰ φυτὰ διὰ τοῦτο οὐδεμίαν ἔχει αἴσθησιν, ὅτι
γῆς ἐστίν· ἄνευ δὲ ἁφῆς οὐδεμίαν οἷόν τε ἄλλην
ὑπάρχειν, τοῦτο δὲ τὸ αἰσθητήριον οὐκ ἔστιν οὔτε
γῆς οὔτε ἄλλου τῶν στοιχείων οὐδενός. φανερὸν
5 τοίνυν ὅτι ἀνάγκη μόνης ταύτης στερισκόμενα τῆς
αἰσθήσεως τὰ ζῷα ἀποθνήσκειν· οὔτε γὰρ ταύτην
ἔχειν οἷόν τε μὴ ζῷον, οὔτε ζῷον ὂν ἄλλην ἔχειν
ἀνάγκη πλὴν ταύτης. καὶ διὰ τοῦτο τὰ μὲν ἄλλα
αἰσθητὰ ταῖς ὑπερβολαῖς οὐ διαφθείρει τὸ ζῷον,
οἷον χρῶμα καὶ ψόφος καὶ ὀσμή, ἀλλὰ μόνον τὰ
10 αἰσθητήρια, ἂν μὴ κατὰ συμβεβηκός, οἷον ἂν ἅμα
τῷ ψόφῳ ὦσις γένηται καὶ πληγή, καὶ ὑπὸ ὁρα-
μάτων καὶ ὀσμῆς ἕτερα κινεῖται, ἃ τῇ ἁφῇ φθείρει.
καὶ ὁ χυμὸς δὲ ᾗ ἅμα συμβαίνει ἁπτικὸν εἶναι,
ταύτῃ φθείρει. ἡ δὲ τῶν ἁπτῶν ὑπερβολή, οἷον
θερμῶν καὶ ψυχρῶν καὶ σκληρῶν, ἀναιρεῖ τὸ ζῷον·
15 παντὸς μὲν γὰρ αἰσθητοῦ ὑπερβολὴ ἀναιρεῖ τὸ
αἰσθητήριον, ὥστε καὶ τὸ ἁπτὸν τὴν ἁφήν, ταύτῃ
δὲ ὥρισται τὸ ζῆν· ἄνευ γὰρ ἁφῆς δέδεικται ὅτι

But touch occurs by direct contact with its objects, and that is why it has its name. The other sense organs perceive by contact too, but through a medium; touch alone seems to perceive immediately. Thus no one of these elements could compose the animal body. Nor could earth. For touch is a kind of mean between all tangible qualities, and its organ is receptive not only of all the different qualities of earth, but also of hot, cold, and all other tangible qualities. This is why we do not perceive by our bones and hair, and such parts of the body, because they are composed of earth. And for this reason plants have no sensation, because they are composed of earth. Without touch there can be no other sense, and the organ of touch is composed neither of earth nor of any other single element. It is obvious, then, that deprived of this one sense alone, animals must die; for it is impossible for anything but an animal to possess this, nor need an animal possess any sense but this. And this explains another fact. Other sensibles, such as colour, sound and smell, do not destroy the animal by excess, but only the sense organs; except incidentally, as for instance when a thrust or blow is delivered at the same time as the sound, or when by the objects of sight or smell other things are set in motion, which destroy by contact. Flavour, again, destroys only in so far as it is at the same time tactile. But the excess of tangible qualities, such as heat, cold, and hardness, destroys the animal. For excess in any sensible quality destroys the organ; and so the tangible also destroys touch. But this is the distinguishing characteristic of life, for it has been shown that without touch an animal cannot exist. Hence excess in tangible qualities destroys not only the sense organ, but also the animal, because touch is the one sense which the animal must possess. The animal possesses the other senses, as has been said,[a] not for mere existence but for well-being; for instance the animal has sight in order that it may see, because it lives in air or water, or generally in a transparent medium; and it has taste because of what is sweet and bitter, in order that it may perceive these qualities in food, and may feel desire and be set in motion; and hearing that it may have significant sounds made to it, and a tongue that it may make significant sounds to another animal.

[a] 434 b 24.

ON SENSE AND SENSIBLE
OBJECTS

connected with divisibility, including the possibility of simultaneous perception (chs. vi.-vii.).

Senses and Elements. The introductory section states the subjects to be treated ; points out the contributions of the several senses to existence, well-being, knowledge and intelligence ; and then turns to examine the generally accepted theory that each of the sense-organs corresponds to and is mainly if not entirely composed of one of the four elements. At first sight this is scarcely distinguishable from the doctrine that " like is known by like," which has already been rejected in the *De Anima.* It is true that there the main objection is to the constitution of the soul itself out of the corporeal elements, but the whole idea seems inconsistent with Aristotle's own view that prior to perception the sense-organ is only potentially like its object. It may be urged that in the present context the sense-organs are made to correspond in their material composition not to their respective objects but to the media of their respective senses—the point being that in this way they will be better able to pass on sensations to the central sense-organ ; thus water is chosen for the eye because it is transparent ; so is air, but water fits the observed characteristics of the eye better, and air, being the best medium for sound, is needed for the ear. So far so good, but difficulties arise in the case of the other organs. Fire, to which the nose is referred, is no doubt a concomitant of smoky vapour, but it is in so sense the medium of smell ; and the explicit qualification that the nose is only potentially (whereas smelling is actually) fire does not seem to help in achieving consistency, for the eye is water not potentially but actually. Finally,

f earth is at least the main constituent of flesh, and
flesh is the medium of touch, the organ of touch
might well be supposed to consist of earth ; but this
view seems to be flatly contradicted by the opening
paragraphs of *De Anima* III. xiii. The inference is
that Aristotle has been unsuccessful in an attempt
to harmonize a current opinion with his own theory.

Nature of Sensible Objects. The chief point which
emerges from Aristotle's discussion of the objects of
sense is that they have objective reality. In his
account of *Colour* he first amplifies the theory of
transparency outlined in *De Anima* II. vii. There the
question was how vision takes place, and " the transparent " was primarily treated as the medium of
vision. (Strictly speaking, the transparent is only
the potential receptacle of light. In the absence of
an illuminant it is darkness—a purely negative state.
The corresponding positive state, *i.e.*, the actualization
of the transparent by the presence of an illuminant,
is light. Hence Aristotle speaks of the medium of
vision sometimes as air or water, as having the quality
of transparency, sometimes as transparency, as being
potentially illuminable, and sometimes as light itself.)
The relation of transparency to colour was only
briefly stated in 418 b 27-32. There we were told
that the transparent is also the vehicle of colour.
Now the inference is drawn that all visible objects
must be transparent in some degree, since only colour
is visible. The colour is displayed at the external
limit of the transparent, but must be supposed to
extend throughout it. White and black correspond
to the positive and negative determinations of the
transparent as light and darkness ; the problem is :
How are the other " intermediate " colours pro-

duced? Aristotle rejects two theories and adopts a third—that they are caused by a chemical mixture of white and black in varying proportions. He can hardly be blamed for not having anticipated the comparatively modern theory of different-coloured rays (though it was a major error to reject the doctrine of " emanations " held by Alcmaeon, Empedocles and Plato, which at least made light travel and might well have led to important discoveries), but his own doctrine could hardly have survived a few experiments with pigments. It must be admitted, however, that Greek views of the blending of colours are generally perplexing ; cf. Plato, Timaeus 67 E 6 ff.

Aristotle's restriction of the species of colour (as also of flavour and smell) to seven was probably due to analogy with the seven notes of the heptachord ; but if it was a legacy of Pythagoreanism it does not appear to have reached Aristotle through Plato (cf. Timaeus, loc. cit.).

Flavours. It has been seen in the De Anima (II. x.) that the process of tasting is impossible without the presence of liquid. Aristotle now considers more closely what part is played in the process by water. Water does not contain flavours in itself ; it is merely the solvent vehicle of taste-particles contained in dry constituents of food. Actual food is never absolutely dry ; nutriment requires the presence of moisture. Hence it is inferred that flavour is a property of food. Sweetness is the positive flavour, and sweet stuff is the purest form of nutriment. Other flavours represent intermediate stages between sweet and its contrary, bitter, which is negative. These diversities of flavour are due to natural modifications (corresponding to the artificial process of seasoning) whose purpose is to give variety and to regulate nutrition.

As of colours, so of flavours there are seven distinct species (the figure is not arrived at without some manipulation) ; these represent different numerical ratios of sweet to bitter. The flavours which give most pleasure are those whose ratios are " definite," which presumably means those which, like the musical concords, are expressible in simple integers.

Smells. Chapter v. repeats much which has been already laid down in *De Anima* II. x., including the close analogy between smell and taste, and the curious theory of " nose-lids " ; the re-statement of the latter, however, clears up some confusions and omissions. The objects of the two senses are ultimately the same, *viz.*, stuff endowed with flavour, and both sensations require the presence of moisture ; but whereas in taste the flavoured substance must be itself moistened and in contact with the tongue, no extended medium being necessary, in smell the flavoured substance is (relatively) dry, and its quality is transmitted without direct contact through the moist extended medium of air or water.

If the objects of taste and smell are ultimately the same, it is natural that the species of taste and smell should correspond (443 b 9). This has scarcely been stated, however, when Aristotle, as if by an afterthought, introduces an entirely different class of smells, perceptible only by man, and having nothing to do with food : these are scents or perfumes, pleasurable in themselves (whereas the former class, which exist merely to attract animals to their food, are pleasurable only incidentally, and cease to please when hunger is satisfied), and salutary in their

211

operation, since the heat which they excite tends to counteract the excessive coolness of the human brain.

Here Aristotle's teleology seems to run away with him, since he does not mention (and apparently forgets) the class of essentially disagreeable smells until 444 b 29. The function of these is glossed over : we are told that animals are insensitive to such smells although (like human beings) they are affected by any injurious quality in the substances which emit them. It is difficult to explain away the inconsistency which makes " scents " beneficial *qua* scents, but " stinks " injurious only incidentally.

Problems of Divisibility. The discussions which form the two last chapters bring out some points of importance. The first problem—Are sensible qualities infinitely divisible ?—is satisfactorily answered in the affirmative, with the qualification that they are so potentially (and, it should have been added, incidentally, since it is in fact the coloured object, not the colour itself, that is divisible). The claim that the same argument explains why the species of sensible qualities are limited appears to be false. This view rests on the unproved assumption that the intermediates between contrary extremes are limited in number. The evidence cited by G. R. T. Ross from *An. Post.* 82 a 21 ff. is, as he concedes, inapplicable here. Aristotle is probably influenced (*a*) by his belief in the fixity of biological species, (*b*) by an instinctive dislike of the indeterminate, (*c*) by the analogy of the musical scale (see above, p. 210).

In answer to the question whether sensation is divisible as a process, Aristotle admits the possibility in the case of hearing and smelling, but reaffirms his conviction that sight is instantaneous, *i.e.*, is not a process at all.

The last serious problem—whether more than one object can be perceived simultaneously—is treated at some length. The argument, partly because of textual uncertainties, is not clear in all its details, but it is finally decided that homogeneous objects can be perceived simultaneously as a blend or compound by their proper sense, and heterogeneous objects can be perceived simultaneously as separate by the common sense-faculty, which is essentially one but conceptually analysable in relation to its objects. Thus the doctrine of *De Anima* III is confirmed.

ΠΕΡΙ ΑΙΣΘΗΣΕΩΣ
ΚΑΙ ΑΙΣΘΗΤΩΝ

I. Ἐπεὶ δὲ περὶ ψυχῆς καθ᾽ αὑτὴν διώρισται
καὶ περὶ τῶν δυνάμεων ἑκάστης κατὰ μόριον
αὑτῆς, ἐχόμενόν ἐστι ποιήσασθαι τὴν ἐπίσκεψιν
περὶ τῶν ζῴων καὶ τῶν ζωὴν ἐχόντων ἁπάντων,
5 τίνες εἰσὶν ἴδιαι καὶ τίνες κοιναὶ πράξεις αὐτῶν.
τὰ μὲν οὖν εἰρημένα περὶ ψυχῆς ὑποκείσθω, περὶ
δὲ τῶν λοιπῶν λέγωμεν, καὶ πρῶτον περὶ τῶν
πρώτων. φαίνεται δὲ τὰ μέγιστα, καὶ τὰ κοινὰ
καὶ τὰ ἴδια τῶν ζῴων, κοινὰ τῆς ψυχῆς ὄντα καὶ
τοῦ σώματος, οἷον αἴσθησις καὶ μνήμη καὶ θυμὸς
10 καὶ ἐπιθυμία καὶ ὅλως ὄρεξις, καὶ πρὸς τούτοις
ἡδονή τε καὶ λύπη· καὶ γὰρ ταῦτα σχεδὸν ὑπάρχει πᾶσι τοῖς ζῴοις. πρὸς δὲ τούτοις τὰ μὲν
πάντων ἐστὶ τῶν μετεχόντων ζωῆς κοινά, τὰ δὲ
τῶν ζῴων ἐνίοις. τυγχάνουσι δὲ τούτων τὰ
μέγιστα τέτταρες οὖσαι συζυγίαι τὸν ἀριθμόν, οἷον
15 ἐγρήγορσις καὶ ὕπνος, καὶ νεότης καὶ γῆρας, καὶ
ἀναπνοὴ καὶ ἐκπνοή, καὶ ζωὴ καὶ θάνατος· περὶ
ὧν θεωρητέον, τί τε ἕκαστον αὐτῶν, καὶ διὰ τίνας
αἰτίας συμβαίνει.

Φυσικοῦ δὲ καὶ περὶ ὑγιείας καὶ νόσου τὰς
πρώτας ἰδεῖν ἀρχάς· οὔτε γὰρ ὑγίειαν οὔτε νόσον

ON SENSE AND SENSIBLE
OBJECTS

I. SINCE we have now dealt in detail with the soul by itself, and with each of its several faculties, our next task is to consider animals and all things possessed of life, and to discover what are their peculiar and what are their common activities. All that has already been said about the soul is to be assumed, but let us now discuss the remaining questions, dealing first of all with those which naturally come first. The most important characteristics of animals, whether common or peculiar, are clearly those which belong to both soul and body, such as sensation, memory, passion, desire, and appetite generally, and in addition to these pleasure and pain ; for these belong to almost all living creatures. Besides these there are some which are common to all things that have a share in life, and others which are peculiar to certain animals. The most important of these are the four pairs, namely waking and sleep, youth and age, inhalation and exhalation, life and death ; we have now to investigate what each of these is, and what are the reasons for their occurrence.

It is further the duty of the natural philosopher to study the first principles of disease and health ; for neither health nor disease can be properties of

Our subject is the activities of the living animal.

Disease and health.

436 a

οἷόν τε γίνεσθαι τοῖς ἐστερημένοις ζωῆς. διὸ
20 σχεδὸν τῶν τε περὶ φύσεως οἱ πλεῖστοι καὶ τῶν
ἰατρῶν οἱ φιλοσοφωτέρως τὴν τέχνην μετιόντες,
436 b οἱ μὲν τελευτῶσιν εἰς τὰ περὶ ἰατρικῆς, οἱ δ' ἐκ
τῶν περὶ φύσεως ἄρχονται περὶ τῆς ἰατρικῆς.

"Ότι δὲ τὰ λεχθέντα κοινὰ τῆς τε ψυχῆς ἐστὶ
καὶ τοῦ σώματος, οὐκ ἄδηλον. πάντα γὰρ τὰ
μὲν μετ' αἰσθήσεως συμβαίνει, τὰ δὲ δι' αἰσθήσεως·
5 ἔνια δὲ τὰ μὲν πάθη ταύτης ὄντα τυγχάνει, τὰ δ'
ἕξεις, τὰ δὲ φυλακαὶ καὶ σωτηρίαι, τὰ δὲ φθοραὶ
καὶ στερήσεις. ἡ δ' αἴσθησις ὅτι διὰ σώματος
γίνεται τῇ ψυχῇ, δῆλον καὶ διὰ τοῦ λόγου καὶ
τοῦ λόγου χωρίς.

Ἀλλὰ περὶ μὲν αἰσθήσεως καὶ τοῦ αἰσθάνεσθαι,
τί ἐστι καὶ διὰ τί συμβαίνει τοῖς ζῴοις τοῦτο τὸ
10 πάθος, εἴρηται πρότερον ἐν τοῖς περὶ ψυχῆς. τοῖς
δὲ ζῴοις, ᾗ μὲν ζῷον ἕκαστον, ἀνάγκη ὑπάρχειν
αἴσθησιν· τούτῳ γὰρ τὸ ζῷον εἶναι καὶ μὴ ζῷον
διορίζομεν. ἰδίᾳ δ' ἤδη καθ' ἕκαστον ἡ μὲν ἁφὴ
καὶ γεῦσις ἀκολουθεῖ πᾶσιν ἐξ ἀνάγκης, ἡ μὲν
15 ἁφὴ διὰ τὴν εἰρημένην αἰτίαν ἐν τοῖς περὶ ψυχῆς,
ἡ δὲ γεῦσις διὰ τὴν τροφήν· τὸ γὰρ ἡδὺ διακρίνει
καὶ τὸ λυπηρὸν αὕτη περὶ τὴν τροφήν, ὥστε τὸ
μὲν φεύγειν τὸ δὲ διώκειν, καὶ ὅλως ὁ χυμός ἐστι
τοῦ θρεπτικοῦ μορίου πάθος. αἱ δὲ διὰ τῶν ἔξωθεν
αἰσθήσεις τοῖς πορευτικοῖς αὐτῶν, οἷον ὄσφρησις
20 καὶ ἀκοὴ καὶ ὄψις, πᾶσι μὲν τοῖς ἔχουσι σωτηρίας
ἕνεκεν ὑπάρχουσιν, ὅπως διώκωσί τε προαισθανό-
μενα τὴν τροφὴν καὶ τὰ φαῦλα καὶ τὰ φθαρτικὰ

---

things deprived of life. Hence one may say that most natural philosophers, and those physicians who take a scientific interest in their art, have this in common : the former end by studying medicine, and the latter base their medical theories on the principles of natural science.

It is obvious that the characteristics already mentioned belong to both soul and body. For all of them either appear in conjunction with sensation or arise through sensation : some again are affections of sensations and some are positive states ; some again tend to guard and preserve life, and others to destroy and extinguish it. That sensation is produced in the soul through the medium of the body is obvious on theoretical grounds and also apart from theory.

Connexion of body and soul.

Now we have already explained, in our work *On the Soul*,[a] what sensation and sentience are, and why this affection appears among animals. Every animal *qua* animal must have sensation. For it is by this that we differentiate between what is and what is not an animal. As for the various individual senses, touch and taste are necessarily present in all animals, touch for the reason given in our work *On the Soul*,[b] and taste on account of nutrition ; for it is taste which discriminates between pleasant and unpleasant in food, so that the one is avoided and the other pursued; and speaking generally flavour is an affection of the nutritive element. But those senses which act through external media, such as smell, hearing and vision, belong to such animals as are capable of locomotion. To all those which possess them they are a means of preservation, in order that they may be aware of their food before they pursue it, and may avoid what is inferior or destructive, while in those

Sensation is essential to the living creature.

437 a φεύγωσι, τοῖς δὲ καὶ φρονήσεως τυγχάνουσι τοῦ
εὖ ἕνεκα· πολλὰς γὰρ εἰσαγγέλλουσι διαφοράς, ἐξ
ὧν ἥ τε τῶν νοητῶν ἐγγίνεται φρόνησις καὶ ἡ
τῶν πρακτῶν.

Αὐτῶν δὲ τούτων πρὸς μὲν τὰ ἀναγκαῖα κρείτ-
5 των ἡ ὄψις καὶ καθ᾽ αὑτήν, πρὸς δὲ νοῦν καὶ κατὰ
συμβεβηκὸς ἡ ἀκοή. διαφορὰς μὲν γὰρ πολλὰς
εἰσαγγέλλει καὶ παντοδαπὰς ἡ τῆς ὄψεως δύναμις
διὰ τὸ πάντα τὰ σώματα μετέχειν χρώματος, ὥστε
καὶ τὰ κοινὰ διὰ ταύτης αἰσθάνεσθαι μάλιστα
(λέγω δὲ κοινὰ σχῆμα, μέγεθος, κίνησιν, ἀριθμόν)·
10 ἡ δ᾽ ἀκοὴ τὰς τοῦ ψόφου διαφορὰς μόνον, ὀλίγοις
δὲ καὶ τὰς τῆς φωνῆς. κατὰ συμβεβηκὸς δὲ πρὸς
φρόνησιν ἡ ἀκοὴ πλεῖστον συμβάλλεται μέρος. ὁ
γὰρ λόγος αἴτιός ἐστι τῆς μαθήσεως ἀκουστὸς
ὤν, οὐ καθ᾽ αὑτὸν ἀλλὰ κατὰ συμβεβηκός· ἐξ
ὀνομάτων γὰρ σύγκειται, τῶν δ᾽ ὀνομάτων ἕκαστον
15 σύμβολόν ἐστιν. διόπερ φρονιμώτεροι τῶν ἐκ
γενετῆς ἐστερημένων εἰσὶν ἑκατέρας τῆς αἰσθήσεως
οἱ τυφλοὶ τῶν ἐνεῶν καὶ κωφῶν.

II. Περὶ μὲν οὖν τῆς δυνάμεως ἣν ἔχει τῶν αἰσθήσεων ἑκάστη, πρότερον εἴρηται. τοῦ δὲ σώμα-
20 τος ἐν οἷς ἐγγίγνεσθαι πέφυκεν αἰσθητηρίοις, νῦν[1]
μὲν ζητοῦσι κατὰ τὰ στοιχεῖα τῶν σωμάτων· οὐκ
εὐποροῦντες δὲ πρὸς τέτταρα πέντ᾽ οὔσας συνάγειν,
γλίχονται περὶ τῆς πέμπτης. ποιοῦσι δὲ πάντες
τὴν ὄψιν πυρὸς διὰ τὸ πάθους τινὸς ἀγνοεῖν τὴν

[1] νῦν EMY: ἔνιοι.

that have intelligence also these senses exist for the sake of well-being ; for they inform us of many differences, from which arises understanding both of the objects of thought and of the affairs of practical life.

Of these faculties, for the mere necessities of life and in itself, sight is the more important, but for the mind and indirectly hearing is the more important. For the faculty of sight informs us of many differences of all kinds, because all bodies have a share of colour, so that it is chiefly by this medium that we perceive the common sensibles. (By these I mean shape, magnitude, movement and number.) But hearing only conveys differences of sound, and to a few animals differences of voice. Indirectly, hearing makes the largest contribution to wisdom. For discourse, which is the cause of learning, is so because it is audible ; but it is audible not in itself but indirectly, because speech is composed of words, and each word is a rational symbol. Consequently, of those who have been deprived of one sense or the other from birth, the blind are more intelligent than the deaf and the dumb. *Sight and hearing.*

II. Concerning the capacity which each of these senses has, we have already spoken. As for the parts of the body in which, as their organs, the several senses are naturally engendered, modern thinkers seek to refer them to the elements of which bodies are composed. But finding it difficult to adjust the five senses to the four elements, they are seriously concerned about the fifth. They all make vision consist of fire, because they do not understand the reason of one of the peculiarities of vision. When the eye *Are the senses composed of the elements ?*

437 a

αἰτίαν· θλιβομένου γὰρ καὶ κινουμένου τοῦ ὀφ

25 θαλμοῦ φαίνεται πῦρ ἐκλάμπειν· τοῦτο δ' ἐν τ

σκότει πέφυκε συμβαίνειν, ἢ τῶν βλεφάρων ἐπι

κεκαλυμμένων· γίνεται γὰρ καὶ τότε σκότος. ἔχε

δ' ἀπορίαν τοῦτο καὶ ἑτέραν. εἰ γὰρ μὴ ἔστ

λανθάνειν αἰσθανόμενον καὶ ὁρῶντα ὁρώμενόν τι

ἀνάγκη ἄρ' αὐτὸν ἑαυτὸν ὁρᾶν τὸν ὀφθαλμόν.

30 τί οὖν ἠρεμοῦντι τοῦτ' οὐ συμβαίνει; τὰ δ' αἴτι

τούτου, καὶ τῆς ἀπορίας καὶ τοῦ δοκεῖν πῦρ εἶνα

τὴν ὄψιν, ἐντεῦθεν ληπτέον. τὰ γὰρ λεῖα πέφυκε

ἐν τῷ σκότει λάμπειν, οὐ μέντοι φῶς γε ποιεῖ, το

437 b δ' ὀφθαλμοῦ τὸ καλούμενον μέλαν καὶ μέσον λεῖο

φαίνεται. φαίνεται δὲ τοῦτο κινουμένου τοῦ ὄμ

ματος διὰ τὸ συμβαίνειν ὥσπερ δύο γίνεσθαι

ἕν. τοῦτο δ' ἡ ταχυτὴς ποιεῖ τῆς κινήσεως, ὥστ

δοκεῖν ἕτερον εἶναι τὸ ὁρῶν καὶ τὸ ὁρώμενον. δι

5 καὶ οὐ γίνεται, ἂν μὴ ταχέως καὶ ἐν σκότει τοῦτ

συμβῇ· τὸ γὰρ λεῖον ἐν τῷ σκότει πέφυκε λάμπειν

οἷον κεφαλαὶ ἰχθύων τινῶν καὶ ὁ τῆς σηπίας θολὸς

καὶ βραδέως μεταβάλλοντος τοῦ ὄμματος οὐ συμ

βαίνει ὥστε δοκεῖν ἅμα ἓν καὶ δύο εἶναι τό θ' ὁρῶ

10 καὶ τὸ ὁρώμενον· ἐκείνως δ' αὐτὸς αὐτὸν ὁρᾶ

ὀφθαλμός, ὥσπερ καὶ ἐν τῇ ἀνακλάσει, ἐπεὶ εἴ γ

πῦρ ἦν, καθάπερ Ἐμπεδοκλῆς φησὶ καὶ ἐν τ

Τιμαίῳ γέγραπται, καὶ συνέβαινε τὸ ὁρᾶν ἐξιόντο

ὥσπερ ἐκ λαμπτῆρος τοῦ φωτός, διὰ τί οὐ καὶ ἐ

τῷ σκότει ἑώρα ἂν ἡ ὄψις; τὸ δ' ἀποσβέννυσθα

---

[a] i.e. seeing sparks.

[b] A. appears to be thinking of what we call persistence o
vision. If the movement is sufficiently rapid, for instance
in a vibrating string, we shall appear to see not one string
in successive positions but two strings each stationary in th
two extreme positions.

220

is pressed and moved, fire seems to flash out.[a] This naturally happens in the dark, or when eyes are closed ; for then, too, there is darkness. But this only raises another difficulty. For unless we suppose that it is possible for a sentient subject to see a visible object without knowing it, the eye must on this theory see itself. Why then does this not happen when the eye is at rest ? The reason for this, and the solution of our difficulty, and the cause of the theory that vision is fire, must be found in the following considerations. It is always smooth surfaces that shine in the dark, though they do not create light ; and the centre of the eye which men call the " black " of the eye is clearly smooth. The phenomenon occurs when the eye is moved because then the effect is as though one thing became two. This is due to the rapidity of the movement,[b] which causes the seeing subject to appear different from the object seen. Hence the phenomenon does not occur, unless the movement is rapid and in the dark ; for it is in the dark that a smooth surface appears to shine, for instance the heads of certain fishes, and the dark fluid of the cuttlefish ; when the movement of the eye is slow, it is impossible that the seeing organ and the object seen should appear to be both one and two at the same moment. But in the other case (when the movement is rapid) the eye sees itself just as in reflection ; if the eye were actually fire, as Empedocles says, and as is stated in the *Timaeus*,[c] and if vision occurred when light issues from the eye as from a lantern, why should not vision be equally possible in the dark ? It is quite

Difficulties of this theory.

[c] *Tim.* 45 c.

221

437 b

15 φάναι ἐν τῷ σκότει ἐξιοῦσαν, ὥσπερ ὁ Τίμαιος
λέγει, κενόν ἐστι παντελῶς· τίς γὰρ ἀπόσβεσις
φωτός ἐστιν; σβέννυται γὰρ ἢ ὑγρῷ ἢ ψυχρῷ τὸ
θερμὸν καὶ ξηρόν, οἷον δοκεῖ τό τ᾿ ἐν τοῖς ἀνθρα-
κώδεσιν εἶναι πῦρ καὶ ἡ φλόξ, ὧν τῷ φωτὶ οὐδέτερον φαίνεται ὑπάρχον. εἰ δ᾿ ἄρα ὑπάρχει μὲν
20 ἀλλὰ διὰ τὸ ἠρέμα λανθάνει ἡμᾶς, ἔδει μεθ᾿ ἡμέραν
τε καὶ ἐν τῷ ὕδατι ἀποσβέννυσθαι τὸ φῶς, καὶ ἐν
τοῖς πάγοις μᾶλλον γίνεσθαι σκότον· ἡ γοῦν φλὸξ
καὶ τὰ πεπυρωμένα σώματα πάσχει τοῦτο· νῦν δ᾿
οὐδὲν συμβαίνει τοιοῦτον. Ἐμπεδοκλῆς δ᾿ ἔοικε
25 νομίζοντι ὁτὲ μὲν ἐξιόντος τοῦ φωτός, ὥσπερ
εἴρηται πρότερον, βλέπειν· λέγει γοῦν οὕτως.

ὡς δ᾿ ὅτε τις πρόοδον νοέων ὡπλίσσατο λύχνον,
χειμερίην διὰ νύκτα πυρὸς σέλας αἰθομένοιο,
ἅψας παντοίων ἀνέμων λαμπτῆρας ἀμοργούς,
οἵ τ᾿ ἀνέμων μὲν πνεῦμα διασκιδνᾶσιν ἀέντων,
30 φῶς δ᾿ ἔξω διαθρῶσκον, ὅσον ταναώτερον ἦεν,
λάμπεσκεν κατὰ βηλὸν ἀτειρέσιν ἀκτίνεσσιν·
ὡς δὲ τότ᾿ ἐν μήνιγξιν ἐεργμένον ὠγύγιον πῦρ

438 a
λεπτῇσιν ὀθόνῃσι λοχάζετο κύκλοπα κούρην·
αἱ δ᾿ ὕδατος μὲν βένθος ἀπέστεγον ἀμφινάοντος,
πῦρ δ᾿ ἔξω δίεσκον,[1] ὅσον ταναώτερον ἦεν.

ὁτὲ μὲν οὖν οὕτως ὁρᾶν φησιν, ὁτὲ δὲ ταῖς ἀπορ-
5 ροίαις ταῖς ἀπὸ τῶν ὁρωμένων. Δημόκριτος δ᾿
ὅτι μὲν ὕδωρ εἶναί φησι, λέγει καλῶς, ὅτι δ᾿ οἴεται
τὸ ὁρᾶν εἶναι τὴν ἔμφασιν, οὐ καλῶς· τοῦτο μὲν
γὰρ συμβαίνει ὅτι τὸ ὄμμα λεῖον, καὶ ἔστιν οὐκ
ἐν ἐκείνῳ ἀλλ᾿ ἐν τῷ ὁρῶντι· ἀνάκλασις γὰρ τὸ
πάθος. ἀλλὰ καθόλου περὶ τῶν ἐμφαινομένων καὶ

---

[1] δίεσκον P : διαθρῶσκον.

---

[a] *Tim.* 45 D.   [b] Empedocles' unifying and creative force.

futile to say, as the *Timaeus* [a] does, that on its emergence from the eye it is extinguished in the dark; for what meaning can we attach to this extinguishing of light? That which is hot and dry, such as a coal fire or a flame is held to be, is extinguished by wet or cold; but heat and dryness are clearly not attributes of light. If they are, but to so slight a degree that we do not notice it, the light should be extinguished in the daytime when it rains, and darkness should occur more commonly in frosty weather. Flame and bodies on fire do show this phenomenon; but no such thing occurs in the other case. Empedocles seems sometimes to imagine that one sees because light issues from the eye, as we have said before; at any rate he says: *Empedocles' theory of vision.*

" As when a man, thinking to sally forth, furnishes him with a lamp, a glow of fire blazing through the stormy night, to protect it against all winds fits thereto screens, which scatter the breath of the winds as they blow; and leaping forth, because it is more tenuous, the light shines over his threshold with tireless rays, so did Love [b] surround the web-enclosed primeval fire, even the round pupil, with fine membranes; and these shut out the depth of surrounding water, but let the fire pass through, because it was more tenuous." [c]

At times, then, he explains vision in this way, but at other times he accounts for it by emanations from objects seen. Democritus is right when he says that the eye is water, but wrong when he supposes vision to be mere mirroring. The image is visible to the eye because the eye is smooth; it exists not in the eye, but in the observer; for the phenomenon is only reflection. It seems, however, that there was *Democritus on vision.*

[c] Frag. 84 (Diels).

438 a

10 ἀνακλάσεως οὐδέν πω δῆλον ἦν, ὡς ἔοικε. ἄτοπον
δὲ καὶ τὸ μὴ ἐπελθεῖν αὐτῷ ἀπορῆσαι διὰ τί ὁ
ὀφθαλμὸς ὁρᾷ μόνον, τῶν δ' ἄλλων οὐδὲν ἐν οἷς
ἐμφαίνεται τὰ εἴδωλα. τὸ μὲν οὖν τὴν ὄψιν εἶναι
ὕδατος ἀληθὲς μέν, οὐ μέντοι συμβαίνει τὸ ὁρᾶν
15 ᾗ ὕδωρ ἀλλ' ᾗ διαφανές· ὃ καὶ ἐπὶ τοῦ ἀέρος κοινόν
ἐστιν. ἀλλ' εὐφυλακτότερον καὶ ἐπιληπτότερον¹ τὸ
ὕδωρ τοῦ ἀέρος· διόπερ ἡ κόρη καὶ τὸ ὄμμα ὕδατός
ἐστιν. τοῦτο δὲ καὶ ἐπ' αὐτῶν τῶν ἔργων δῆλον·
φαίνεται γὰρ ὕδωρ τὸ ἐκρέον διαφθειρομένων, καὶ
20 ἔν γε τοῖς πάμπαν ἐμβρύοις τῇ ψυχρότητι ὑπερβάλλον καὶ τῇ λαμπρότητι. καὶ τὸ λευκὸν τοῦ
ὄμματος ἐν τοῖς ἔχουσιν αἷμα πῖον καὶ λιπαρόν·
ὅπερ διὰ τοῦτ' ἐστί, πρὸς τὸ διαμένειν τὸ ὑγρὸν
ἄπηκτον. καὶ διὰ τοῦτο τοῦ σώματος ἀρριγότατον
ὁ ὀφθαλμός ἐστιν· οὐδεὶς γάρ πω τὸ ἐντὸς τῶν
25 βλεφάρων ἐρρίγωσεν. τῶν δ' ἀναίμων σκληρό-
δερμοι οἱ ὀφθαλμοί εἰσι, καὶ τοῦτο ποιεῖ τὴν
σκέπην.

Ἄλογον δὲ ὅλως τὸ ἐξιόντι τινὶ τὴν ὄψιν ὁρᾶν,
καὶ ἀποτείνεσθαι μέχρι τῶν ἄστρων, ἢ μέχρι τινὸς
ἐξιοῦσαν συμφύεσθαι, καθάπερ λέγουσί τινες.
τούτου μὲν γὰρ βέλτιον τὸ ἐν ἀρχῇ συμφύεσθαι τοῦ
ὄμματος. ἀλλὰ καὶ τοῦτο εὔηθες· τό τε γὰρ συμ-
30 φύεσθαι τί ἐστι φωτὶ πρὸς φῶς; ἢ πῶς οἷόν θ'

438 b ὑπάρχειν; οὐ γὰρ τῷ τυχόντι συμφύεται τὸ τυχόν.
τό τ' ἐντὸς τῷ ἐκτὸς πῶς; ἡ γὰρ μήνιγξ μεταξύ
ἐστιν. περὶ μὲν οὖν τοῦ ἄνευ φωτὸς μὴ ὁρᾶν
εἴρηται ἐν ἄλλοις·ᵃ ἀλλ' εἴτε φῶς εἴτ' ἀὴρ ἐστι τὸ
μεταξὺ τοῦ ὁρωμένου καὶ τοῦ ὄμματος, ἡ διὰ

¹ ἐπιληπτότερον scripsi : εὐπιλητότερον.

ᵃ 418 b 3.

as yet no clear general theory about mirrored objects and reflection. But it is strange that it never occurred to him to wonder why only the eye sees, and none of the other things in which images appear do so. It is true that the eye consists of water, but it has the power of vision not because it is water, but because it is transparent ; an attribute which it shares with air. But water is more easily controlled and confined than air ; hence the pupil or eye proper is composed of water. This is obvious from the observed facts. When the eyes decay, what exudes is clearly water, and this, in mere embryos, is exceedingly cold and shining. And the white of the eye in animals which have blood is fat and oily ; this is so in order that the moisture may remain unfrozen. For this reason the eye is the part of the body least sensitive to cold ; for no one has ever felt cold inside his eyelids. The eyes of bloodless animals have a hard skin, which gives them similar protection.

In general it is unreasonable to suppose that seeing occurs by something issuing from the eye ; that the ray of vision reaches as far as the stars, or goes to a certain point and there coalesces with the object, as some think. It would be better to suppose that coalescence occurs in the very heart of the eye. But even this is foolish ; what is the meaning of light coalescing with light ? How can it occur ? For coalescence is not between any chance objects. And how could the light inside coalesce with that outside ? For the membrane is between them. Elsewhere [a] we have shown the impossibility of vision without light ; but whether light or air is the medium between the visible object and the eye, the motion through

Vision as an emanation from the eye.

225

438 b

τούτου κίνησίς ἐστιν ἡ ποιοῦσα τὸ ὁρᾶν. καὶ
εὐλόγως τὸ ἐντός ἐστιν ὕδατος· διαφανὲς γὰρ τὸ
ὕδωρ. ὁρᾶται δὲ ὥσπερ καὶ ἔξω οὐκ ἄνευ φωτός,
οὕτω καὶ ἐντός· διαφανὲς ἄρα δεῖ εἶναι. καὶ ἀνάγκη
ὕδωρ εἶναι, ἐπειδὴ οὐκ ἀήρ. οὐ γὰρ ἐπὶ τοῦ
ἐσχάτου ὄμματος ἡ ψυχὴ ἢ τῆς ψυχῆς τὸ αἰσθητήριόν ἐστιν, ἀλλὰ δῆλον ὅτι ἐντός· διόπερ ἀνάγκη
διαφανὲς εἶναι καὶ δεκτικὸν φωτὸς τὸ ἐντὸς τοῦ
ὄμματος. καὶ τοῦτο καὶ ἐπὶ τῶν συμβαινόντων
δῆλον· ἤδη γάρ τισι πληγεῖσιν ἐν πολέμῳ παρὰ τὸν
κρόταφον οὕτως ὥστ' ἐκτμηθῆναι τοὺς πόρους τοῦ
ὄμματος, ἔδοξε γενέσθαι σκότος ὥσπερ λύχνου
ἀποσβεσθέντος, διὰ τὸ οἷον λαμπτῆρά τινα ἀπο-
τμηθῆναι τὸ διαφανές, τὴν καλουμένην κόρην.

Ὥστ' εἴπερ τούτων τι συμβαίνει, καθάπερ λέγομεν, φανερὸν ὡς δεῖ τοῦτον τὸν τρόπον ἀποδιδόναι καὶ προσάπτειν ἕκαστον τῶν αἰσθητηρίων ἑνὶ
τῶν στοιχείων. τοῦ μὲν ὄμματος τὸ ὁρατικὸν
ὕδατος ὑποληπτέον, ἀέρος δὲ τὸ τῶν ψόφων αἰσθητικόν, πυρὸς δὲ τὴν ὄσφρησιν (ὃ γὰρ ἐνεργείᾳ ἡ
ὄσφρησις, τοῦτο δυνάμει τὸ ὀσφραντικόν· τὸ γὰρ
αἰσθητὸν ἐνεργεῖν ποιεῖ τὴν αἴσθησιν, ὥσθ' ὑπάρχειν
ἀνάγκη αὐτὴν δυνάμει πρότερον· ἡ δ' ὀσμὴ καπνώδης τίς ἐστιν ἀναθυμίασις, ἡ δ' ἀναθυμίασις ἡ
καπνώδης ἐκ πυρός· διὸ καὶ τῷ περὶ τὸν ἐγκέφαλον τόπῳ τὸ τῆς ὀσφρήσεως αἰσθητήριόν ἐστιν
ἴδιον· δυνάμει γὰρ θερμὴ ἡ τοῦ ψυχροῦ ὕλη ἐστίν·
καὶ ἡ τοῦ ὄμματος γένεσις τὸν αὐτὸν ἔχει τρόπον·
ἀπὸ τοῦ ἐγκεφάλου γὰρ συνέστηκεν· οὗτος γὰρ
ὑγρότατος καὶ ψυχρότατος τῶν ἐν τῷ σώματι

439 a

μορίων ἐστίν), τὸ δ' ἁπτικὸν γῆς· τὸ δὲ γευστικὸν
εἶδός τι ἁφῆς ἐστιν. καὶ διὰ τοῦτο πρὸς τῇ καρδίᾳ

this medium is what produces vision. And it is natural that what is within should consist of water; for water is transparent. And just as there is no vision outside without light, so also within; there must be a transparency. And this, since it is not air, must be water. For the soul or the sense organ of the soul does not reside in the surface of the eye, but must evidently be within; consequently the part within the eye must be transparent and receptive of light. This is clear from what actually occurs; for it is a fact that when in war men have been struck on the temple so as to sever the channels from the eye, darkness has seemed to fall on them as if a lamp has failed, because the transparent substance, called the pupil, has been cut off, like a lamp screen.

If the facts are at all as we have described, evidently the following is the only method by which we can allot and adapt each of the sense organs to one of the elements. We must suppose the seeing part of the eye to consist of water, that which is sensitive to sound of air, smell of fire (not the organ of smell; for the organ of smell is potentially what the sense of smell is on actualization; for since the sense is actualized by its object, it must pre-exist potentially). Now odour is a kind of smoky vapour, and a smoky vapour arises from fire. Hence the sense organ of smell is proper to the region about the brain; for the matter of what is cold is potentially hot. The same applies to the genesis of the eye (for it is developed from the brain, which is the wateriest and coldest of all parts of the body), and the tactual organ of earth. The faculty of taste is a form of touch. For this reason

Senses and the elements.

227

the sense organ of both taste and touch is near the heart. For the heart is the antithesis of the brain, and is the hottest of all parts of the body. So much by way of description of the parts of the body which have perceptive faculties.

III. The sensible organs corresponding to each of the sense organs, *viz.*, colour, sound, smell, flavour and touch, have been treated generally in the treatise *On the Soul*, where their function is explained, and the effect of their actualization in respect of the several sense organs ; but we have now to consider how we are to describe any of them, *i.e.*, to answer the question what is sound, or colour, or smell, or flavour, and similarly with regard to touch. Let us Colour. deal with colour first.

Each of these terms is used in two senses : as actual or potential. We have explained in the treatise *On the Soul* the sense in which actual colour and sound are identical with or different from the actual sensations, that is, seeing or hearing.[a] Now let us explain what each of them must be to produce the sensation in full actuality. In that treatise we have already said [b] of light, that it is, indirectly, the colour of the transparent ; for whenever there is a fiery element in the transparent, its presence is light, while its absence is darkness. What we call " transparent " is not peculiar to air or water or any other body so described, but a common nature or potency, which is not separable but resides in these bodies and in all others, to a greater or less extent ; hence just as every body must have some bound, so must this. The nature of light resides in the transparent when undefined ; but clearly the transparent which inheres

[a] *De An.* 425 b 26.   [b] 418 a 26.

439 a

σώμασι διαφανοῦς τὸ ἔσχατον, ὅτι μὲν εἴη ἄν τι,
δῆλον, ὅτι δὲ τοῦτ' ἐστὶ τὸ χρῶμα, ἐκ τῶν συμ-
30 βαινόντων φανερόν. τὸ γὰρ χρῶμα ἢ ἐν τῷ πέρατί
ἐστιν ἢ πέρας· διὸ καὶ οἱ Πυθαγόρειοι τὴν ἐπι-
φάνειαν χροιὰν ἐκάλουν. ἔστι μὲν γὰρ ἐν τῷ τοῦ
σώματος πέρατι, ἀλλ' οὔ τι τὸ τοῦ σώματος πέρας,
ἀλλὰ τὴν αὐτὴν φύσιν δεῖ νομίζειν, ἥπερ καὶ ἔξω
439 b χρωματίζεται, ταύτην καὶ ἐντός. φαίνεται δὲ καὶ
ἀὴρ καὶ ὕδωρ χρωματιζόμενα· καὶ γὰρ ἡ αὐγὴ
τοιοῦτόν ἐστιν. ἀλλ' ἐκεῖ μὲν διὰ τὸ ἐν ἀορίστῳ
οὐ τὴν αὐτὴν ἐγγύθεν καὶ προσιοῦσι καὶ πόρρωθεν
5 ἔχει χροιὰν οὔθ' ὁ ἀὴρ οὔθ' ἡ θάλαττα· ἐν δὲ τοῖς
σώμασιν ἐὰν μὴ τὸ περιέχον ποιῇ τὸ μεταβάλλειν,
ὥρισται καὶ ἡ φαντασία τῆς χρόας. δῆλον ἄρα
ὅτι τὸ αὐτὸ κἀκεῖ κἀνθάδε δεκτικὸν τῆς χρόας
ἐστίν. τὸ ἄρα διαφανὲς καθ' ὅσον ὑπάρχει ἐν τοῖς
σώμασιν (ὑπάρχει δὲ μᾶλλον καὶ ἧττον ἐν πᾶσι)
10 χρώματος ποιεῖ μετέχειν. ἐπεὶ δ' ἐν πέρατι ἡ
χρόα, τούτου ἂν ἐν πέρατι εἴη. ὥστε χρῶμα ἂν
εἴη τὸ τοῦ διαφανοῦς ἐν σώματι ὡρισμένῳ πέρας.
καὶ αὐτῶν δὲ τῶν διαφανῶν, οἷον ὕδατος καὶ εἴ
τι ἄλλο τοιοῦτον, καὶ ὅσοις φαίνεται χρῶμα ἴδιον
ὑπάρχειν κατὰ τὸ ἔσχατον, ὁμοίως πᾶσιν ὑπάρχει.
15 ἔστι μὲν οὖν ἐνεῖναι ἐν τῷ διαφανεῖ τοῦθ' ὅπερ
καὶ ἐν τῷ ἀέρι ποιεῖ φῶς, ἔστι δὲ μή, ἀλλ'
ἐστερῆσθαι. ὥσπερ οὖν ἐκεῖ τὸ μὲν φῶς τὸ δὲ
σκότος, οὕτως ἐν τοῖς σώμασιν ἐγγίνεται τὸ λευκὸν
καὶ τὸ μέλαν.

in bodies must have a bound, and it is plain from the facts that this bound is colour; for colour either is in the limit or else is the limit itself. This is why the Pythagoreans called the surface of a body its colour. Colour lies at the limit of the body, but this limit is not a real thing; we must suppose that the same nature which exhibits colour outside, also exists within. Air and water obviously have colour; for their brightness is of the nature of colour. But in their case because the colour resides in something undefined, air and sea do not show the same colour near at hand and to those who approach them as they have at a distance. But in bodies, unless the surrounding envelope causes a change, even the appearance of the colour is defined. Hence clearly it is the same thing that is receptive of the colour, both in the one case and in the other. It is then the transparent, in proportion as it exists in bodies (and it exists in them all to a greater or less extent), which causes them to share in colour. But since colour resides in the limit, it must lie in the limit of the transparent. Hence colour will be the limit of the transparent in a defined body. Both in objects actually transparent, such as water, etc., and in all those things which seem to have at the limit a characteristic colour, the colour always inheres in the bounding surface. That which in air causes light may be present in the transparent,[a] or may not be present, the body being deprived of it. Thus the same conditions which in air produce light and darkness in bodies produce white and black.

[a] *Sc.*, which is present in a defined body.

439 b

Περὶ δὲ τῶν ἄλλων χρωμάτων ἤδη διελομένους
20 ποσαχῶς ἐνδέχεται γίγνεσθαι λεκτέον. ἐνδέχεται
μὲν γὰρ παρ' ἄλληλα τιθέμενα τὸ λευκὸν καὶ τὸ
μέλαν, ὥσθ' ἑκάτερον μὲν εἶναι ἀόρατον διὰ
σμικρότητα, τὸ δ' ἐξ ἀμφοῖν ὁρατὸν οὕτω γίνεσθαι.
τοῦτο γὰρ οὔτε λευκὸν οἷόν τε φαίνεσθαι οὔτε
μέλαν· ἐπεὶ δ' ἀνάγκη μέν τι ἔχειν χρῶμα, τούτων
25 δ' οὐδέτερον δυνατόν, ἀνάγκη μικτόν τι εἶναι καὶ
εἶδός τι χρόας ἕτερον. ἔστι μὲν οὖν οὕτως ὑπολαβεῖν πλείους εἶναι χρόας παρὰ τὸ λευκὸν καὶ τὸ
μέλαν, πολλὰς δὲ τῷ λόγῳ· τρία γὰρ πρὸς δύο,
καὶ τρία πρὸς τέτταρα, καὶ κατ' ἄλλους ἀριθμοὺς
ἔστι παρ' ἄλληλα κεῖσθαι (τὰ δ' ὅλως κατὰ μὲν
30 λόγον μηδένα, καθ' ὑπεροχὴν δέ τινα καὶ ἔλλειψιν
ἀσύμμετρον), καὶ τὸν αὐτὸν δὴ τρόπον ἔχειν ταῦτα
ταῖς συμφωνίαις· τὰ μὲν γὰρ ἐν ἀριθμοῖς εὐλογί-
στοις χρώματα, καθάπερ ἐκεῖ τὰς συμφωνίας, τὰ
440 a ἥδιστα τῶν χρωμάτων εἶναι δοκοῦντα, οἷον τὸ
ἁλουργὸν καὶ φοινικοῦν καὶ ὀλίγ' ἄττα τοιαῦτα,
δι' ἥνπερ αἰτίαν καὶ αἱ συμφωνίαι ὀλίγαι, τὰ δὲ
μὴ ἐν ἀριθμοῖς τἆλλα χρώματα· ἢ καὶ πάσας τὰς
5 χρόας ἐν ἀριθμοῖς εἶναι, τὰς μὲν τεταγμένας τὰς
δὲ ἀτάκτους, καὶ αὐτὰς ταύτας, ὅταν μὴ καθαραὶ
ὦσι, διὰ τὸ μὴ ἐν ἀριθμοῖς εἶναι τοιαύτας γίνεσθαι.

Εἷς μὲν οὖν τρόπος τῆς γενέσεως τῶν χρωμάτων
οὗτος, εἷς δὲ τὸ φαίνεσθαι δι' ἀλλήλων, οἷον ἐνίοτε
οἱ γραφῆς ποιοῦσιν, ἑτέραν χρόαν ἐφ' ἑτέραν
ἐναργεστέραν ἐπαλείφουσιν, ὥσπερ ὅταν ἐν ὕδατί
10 τι ἢ ἐν ἀέρι βούλωνται ποιῆσαι φαινόμενον, καὶ

We must now speak of the other colours and explain the various ways in which they may arise. One possibility is that white and black particles alternate in such a way that while each by itself is invisible because of its smallness, the compound of the two is visible. This cannot appear either as white or as black ; but since it must have some colour, and cannot have either of these, it must evidently be some kind of mixture, *i.e.*, some other kind of colour. It is thus possible to believe that there are more colours than just white and black, and that their number is due to the proportion of their components ; for these may be grouped in the ratio of three to two, or three to four, or in other numerical ratios (or they may be in no expressible ratio, but in an incommensurable relation of excess or defect), so that these colours are determined like musical intervals. For on this view the colours that depend on simple ratios, like the concords in music, are regarded as the most attractive, *e.g.*, purple and red and a few others like them—few for the same reason that the concords are few—, while the other colours are those which have no numerical ratios ; or it may be that all are expressible in numbers, but while some are regular in ratio, others are not ; and the latter, when they are not pure, have this character because they are not in a pure numerical ratio.

This is one way of accounting for colours. Another theory is that they appear through one another, as sometimes painters produce them, when they lay a colour over another more vivid one, *e.g.*, when they want to make a thing show through water or mist ;

The formation of colours other than white and black.

Colours due to strips of white and black in different proportions.

Colour due to superposition.

just as the sun appears white when seen directly, but red when seen through fog and smoke. But on this view too the multiplicity of colours will be explained in the same way as before ; for there will be some definite ratio between the superimposed colours and those below, and others again will not be in any expressible ratio.

But to say, as the old philosophers did, that colours are emanations from objects and are visible on this account, is unreasonable ; for in any case they would have to explain sensation by contact, so that it would be better to say at once that sensation is caused because the sensible object sets in motion the medium of the sensation, that is by contact and not by emanations. On the theory of alternate particles we must assume not only invisible magnitude but also imperceptible time, if we are not to notice that the stimuli arrive successively, and if the particles are to give a single impression by appearing to us simultaneously. In the other case we need not do so ; the upper colour will affect the medium differently according as it is itself unaffected or affected by the underlying colour. Hence it appears as a different colour, neither white nor black. Thus if no magnitude can be invisible, but every magnitude is visible from some distance, this second theory would account for the blending of colour. Even on the former view the particles might appear as a compound colour—at a distance, but only so ; for we are to show later that no magnitude can be invisible.

But a mixture of bodies occurs, not merely, as some people think, by the alternation of their smallest particles, which are imperceptible to sense, but by

Colours as emanations from objects.

Mixture.

αἴσθησιν, ἀλλ' ὅλως πάντῃ πάντως, ὥσπερ ἐν τοῖς
περὶ μίξεως εἴρηται καθόλου περὶ πάντων. ἐκείνως
5 μὲν γὰρ μίγνυται ταῦτα μόνον ὅσα ἐνδέχεται
διελεῖν εἰς τὰ ἐλάχιστα, καθάπερ ἀνθρώπους ἵππους
ἢ τὰ σπέρματα· τῶν μὲν γὰρ ἀνθρώπων ἄνθρωπος
ἐλάχιστος, τῶν δ' ἵππων ἵππος· ὥστε τῇ τούτων
παρ' ἄλληλα θέσει τὸ πλῆθος μέμικται τῶν συν-
αμφοτέρων· ἄνθρωπον δὲ ἕνα ἑνὶ ἵππῳ οὐ λέγομεν
10 μεμῖχθαι. ὅσα δὲ μὴ διαιρεῖται εἰς τὸ ἐλάχιστον,
τούτων οὐκ ἐνδέχεται μίξιν γενέσθαι τὸν τρόπον
τοῦτον ἀλλὰ τῷ πάντῃ μεμῖχθαι, ἅπερ καὶ μάλιστα
μίγνυσθαι πέφυκεν. πῶς δὲ τοῦτο γίγνεσθαι
δυνατόν, ἐν τοῖς περὶ μίξεως εἴρηται πρότερον.
ἀλλ' ὅτι ἀνάγκη μιγνυμένων καὶ τὰς χρόας μίγνυ-
15 σθαι, δῆλον, καὶ ταύτην τὴν αἰτίαν εἶναι κυρίαν τοῦ
πολλὰς εἶναι χροίας, ἀλλὰ μὴ τὴν ἐπιπόλασιν μηδὲ
τὴν παρ' ἄλληλα θέσιν· οὐ γὰρ πόρρωθεν μὲν
ἐγγύθεν δ' οὒ φαίνεται μία χρόα τῶν μεμιγμένων,
ἀλλὰ πάντοθεν. πολλαὶ δ' ἔσονται χρόαι διὰ τὸ
κατὰ πολλοὺς λόγους ἐνδέχεσθαι μίγνυσθαι ἀλ-
20 λήλοις τὰ μιγνύμενα, καὶ τὰ μὲν ἐν ἀριθμοῖς τὰ
δὲ καθ' ὑπεροχὴν μόνον. καὶ τἆλλα δὴ τὸν αὐτὸν
τρόπον ὅνπερ ἐπὶ τῶν παρ' ἄλληλα τιθεμένων
χρωμάτων ἢ ἐπιπολῆς, ἐνδέχεται λέγειν καὶ περὶ
τῶν μιγνυμένων· διὰ τίνα δ' αἰτίαν εἴδη τῶν
χρωμάτων ἐστὶν ὡρισμένα καὶ οὐκ ἄπειρα, καὶ
25 χυμῶν καὶ ψόφων, ὕστερον ἐροῦμεν.

IV. Τί μὲν οὖν ἐστὶ χρῶμα καὶ διὰ τίν' αἰτίαν
πολλαὶ χροιαί εἰσιν, εἴρηται· περὶ δὲ ψόφου καὶ
φωνῆς εἴρηται πρότερον ἐν τοῖς περὶ ψυχῆς· περὶ

complete interfusion of all their parts, as we have said in our discussion of mixtures in general.[a] On their view mixture is only possible in the case of those things which can be divided into minimal parts—*e.g.*, men, horses or kinds of seeds ; for a man is the smallest unit of men, and a horse of horses ; so that when these are placed alternately, the whole number becomes a mixture of both ; but we do not say that one man is mixed with one horse. But of things which are not divisible into their smallest units there can be no mixture in this sense, but only a complete fusion, which is the most natural form of mixture. How this can occur has been discussed previously in our discussion of mixture.[a] But it is clear that colours must be mixed when the bodies in which they occur are mixed, and that this is the real reason why there are many colours ; it is not due either to overlaying or to alternation ; for it is not from afar only (but not from near at hand) that the colour of mixed bodies seems uniform, but from all distances. The multiplicity of colours will be due to the fact that the components may be combined in various ratios, some being numerical and some merely expressing preponderance. All that we said of colours which are due to alternation or overlaying applies equally to those which are due to mixture. Why the possible forms of colour are limited and not unlimited, which is also true of flavours and sounds, we will discuss later on.[b]

IV. We have now explained what colour is, and why there are many colours. We have previously discussed sound and voice in our treatise *On the*

Smell and flavour.

---

[a] *Cf. De Gen. et Cor.* i. 10.   [b] Ch. vi.

440 b

δὲ ὀσμῆς καὶ χυμοῦ νῦν λεκτέον. σχεδὸν γάρ
30 ἐστι τὸ αὐτὸ πάθος, οὐκ ἐν τοῖς αὐτοῖς δ' ἐστὶν
ἑκάτερον αὐτῶν. ἐναργέστερον δ' ἐστὶν ἡμῖν τὸ
τῶν χυμῶν γένος ἢ τὸ τῆς ὀσμῆς. τούτου δ'
441 a αἴτιον ὅτι χειρίστην ἔχομεν τῶν ἄλλων ζῴων τὴν
ὄσφρησιν καὶ τῶν ἐν ἡμῖν αὐτοῖς αἰσθήσεων, τὴν
δ' ἁφὴν ἀκριβεστάτην τῶν ἄλλων ζῴων· ἡ δὲ γεῦσις ἁφή τις ἐστίν.

Ἡ μὲν οὖν τοῦ ὕδατος φύσις βούλεται ἄχυμος
5 εἶναι· ἀνάγκη δ' ἢ ἐν αὐτῷ τὸ ὕδωρ ἔχειν τὰ
γένη τῶν χυμῶν ἀναίσθητα διὰ μικρότητα, καθάπερ
Ἐμπεδοκλῆς φησίν, ἢ ὕλην τοιαύτην ἐνεῖναι οἷον
πανσπερμίαν χυμῶν, καὶ ἅπαντα μὲν ἐξ ὕδατος
γίγνεσθαι, ἄλλα δ' ἐξ ἄλλου μέρους, ἢ μηδεμίαν
ἔχοντος διαφορὰν τοῦ ὕδατος τὸ ποιοῦν αἴτιον
10 εἶναι, οἷον εἰ τὸ θερμὸν καὶ τὸν ἥλιον φαίη τις
τούτων δ', ὡς μὲν Ἐμπεδοκλῆς λέγει, λίαν εὐ-
σύνοπτον τὸ ψεῦδος· ὁρῶμεν γὰρ μεταβάλλοντας
ὑπὸ τοῦ θερμοῦ τοὺς χυμοὺς ἀφαιρουμένων τῶν
περικαρπίων εἰς τὸν ἥλιον καὶ πυρουμένων, ὡς
οὐ τῷ ἐκ τοῦ ὕδατος ἕλκειν τοιούτους γινομένους
ἀλλ' ἐν αὐτῷ τῷ περικαρπίῳ μεταβάλλοντας, κα
15 ἐξικμαζομένους δὲ καὶ κειμένους, διὰ τὸν χρόνον
αὐστηροὺς ἐκ γλυκέων καὶ πικροὺς καὶ παντοδαποὺς γινομένους, καὶ ἑψομένους, εἰς πάντα τὰ
γένη τῶν χυμῶν ὡς εἰπεῖν μεταβάλλοντας. ὁμοίως
δὲ καὶ τὸ πανσπερμίας εἶναι τὸ ὕδωρ ὕλην ἀδύνα-
20 τον· ἐκ τοῦ αὐτοῦ γὰρ ὁρῶμεν ὡς[1] τροφῆς γινο-

---

[1] ὡς ἐκ τῆς αὐτῆς LSU : ὡς ἐκ τῆς P.

238

Soul.[a] We have now to consider smell and flavour. These two are almost the same affection, though they do not occur in the same circumstances. The class of flavours is more easily detected than that of smells. The reason for this is that our sense of smell is inferior to that of all other living creatures, and also inferior to all the other senses we possess, while our sense of touch is more accurate than that of any other living creature ; and taste is a form of touch.

Now the nature of water tends to be tasteless ; we must explain the facts in one of three ways : (1) Water may possess within itself all kinds of flavours, which are imperceptible because of their small quantity, as Empedocles suggests. (2) Water may contain matter such that it comprises the seeds of all flavours, that is to say that all flavours arise from water, some from one part and some from another. (3) Differences of flavour may not reside in water, but be caused by some external agent; *e.g.*, one might suggest heat or the sun as the cause. Now of these three theories, the first—that of Empedocles—is palpably false ; for we find flavours changing under heat, when pericarpal fruits are picked and dried in the sun or at a fire ; which shows that the juices do not owe their nature to drawing flavour from water,[b] but change in the fruit itself, and when they are extracted and left to lie, in time they become, instead of sweet, harsh or bitter, or assume various other tastes ; and when they ferment, change into almost every kind of flavour. Similarly that water should be a material comprising the seeds of all flavours is impossible ; for we see different kinds of flavours generated from the same water, this being their food.

Theories of taste.

(1) Empedocles.

(2) Water is not the origin of all flavours.

---

[a] *De An.* 419 b, 420 a. [b] *Sc.*, in the earth.

441 a
μένους ἑτέρους χυμούς. λείπεται δὴ τῷ πάσχειν
τι τὸ ὕδωρ μεταβάλλειν. ὅτι μὲν τοίνυν οὐχ ὑπὸ
τῆς τοῦ θερμοῦ δυνάμεως λαμβάνει ταύτην τὴν
δύναμιν ἣν καλοῦμεν χυμόν, φανερόν· λεπτότατον
γὰρ τῶν πάντων ὑγρῶν τὸ ὕδωρ ἐστί, καὶ αὐτοῦ
τοῦ ἐλαίου. ἀλλ' ἐπεκτείνεται ἐπὶ πλεῖον τὸ
25 ὕδατος τὸ ἔλαιον διὰ τὴν γλισχρότητα. τὸ δ'
ὕδωρ ψαθυρόν ἐστι· διὸ καὶ χαλεπώτερον φυλάξαι
ἐν τῇ χειρὶ ὕδωρ ἢ ἔλαιον. ἐπεὶ δὲ θερμαινόμενον
οὐδὲν φαίνεται παχυνόμενον τὸ ὕδωρ αὐτὸ μόνον,
δῆλον ὅτι ἑτέρα τις ἂν εἴη αἰτία· οἱ γὰρ χυμοὶ
πάντες πάχος ἔχουσι μᾶλλον· τὸ δὲ θερμὸν συν-
30 αίτιον. φαίνονται δ' οἱ χυμοὶ ὅσοιπερ καὶ ἐν τοῖς
441 b περικαρπίοις, οὗτοι ὑπάρχοντες καὶ ἐν τῇ γῇ. διὸ
καὶ πολλοί φασι τῶν ἀρχαίων φυσιολόγων τοιοῦτον
εἶναι τὸ ὕδωρ δι' οἵας ἂν γῆς πορεύηται. καὶ τοῦτο
δῆλόν ἐστιν ἐπὶ τῶν ἁλμυρῶν ὑδάτων μάλιστα·
οἱ γὰρ ἅλες γῆς τι εἶδός εἰσιν. καὶ τὰ διὰ τῆς
5 τέφρας διηθούμενα πικρᾶς οὔσης πικρὸν ποιεῖ τὸν
χυμόν. εἰσί τε κρῆναι πολλαὶ αἱ μὲν πικραί, αἱ
δ' ὀξεῖαι, αἱ δὲ παντοδαποὺς ἔχουσαι χυμοὺς
ἄλλους. εὐλόγως δ' ἐν τοῖς φυομένοις τὸ τῶν χυμῶν γίνεται γένος μάλιστα. πάσχειν γὰρ πέφυκε
10 τὸ ὑγρόν, ὥσπερ καὶ τἆλλα, ὑπὸ τοῦ ἐναντίου·
ἐναντίον δὲ τὸ ξηρόν. διὸ καὶ ὑπὸ τοῦ πυρὸς
πάσχει τι· ξηρὰ γὰρ ἡ τοῦ πυρὸς φύσις. ἀλλ'
ἴδιον τοῦ πυρὸς τὸ θερμόν ἐστι, γῆς δὲ τὸ ξηρόν,
ὥσπερ εἴρηται ἐν τοῖς περὶ στοιχείων. ᾗ μὲν οὖν
πῦρ καὶ ᾗ γῆ, οὐδὲν πέφυκε ποιεῖν ἢ πάσχειν,
οὐδ' ἄλλο οὐδέν· ᾗ δ' ὑπάρχει ἐναντιότης ἐν
15 ἑκάστῳ, ταύτῃ πάντα καὶ ποιοῦσι καὶ πάσχουσιν.
ὥσπερ οὖν οἱ ἐναποπλύνοντες ἐν τῷ ὑγρῷ τὰ

The remaining solution is that the water changes by (3) Water is affected in being affected in some way. Now it is clear that it some way. does not acquire this quality which we call taste from the action of heat; for water is the thinnest of all liquids, thinner even than oil. Oil, because of its viscosity, will spread over a larger surface than water, whereas water is volatile; consequently it is more difficult to hold water in the hand than oil. But since water by itself when heated shows no sign of thickening, it is clear that there must be some other reason; for all flavours tend to have density. Still, heat is a contributory cause. Now the flavours found in pericarpal fruits are clearly also present in the earth. Hence many of the old natural philosophers hold that water is assimilated to the earth through which it passes. This is quite obvious in the case of salt springs; for salt is a form of earth. Water which is filtered through ashes, which are bitter, has a bitter flavour. There are the many springs, too, which are bitter or acid or have various other flavours. The range of flavours is best represented in plants, as is natural. For the wet, like everything else, is by nature affected only by its contrary, *viz.*, the dry. This is why it is affected to some extent by fire; for the nature of fire is dry. But the special property of fire is heat, dryness being that of earth, as has been said in our discussion of the elements.[a] Thus *qua* How the fire and earth they cannot naturally produce or suffer water is affected. an effect; nothing produces or suffers an effect except in so far as it contains some element of contrariety. So just as those who steep colours or flavours in liquid

[a] *De Gen. et Corr.* 329 a.

441 b

χρώματα καὶ τοὺς χυμοὺς τοιοῦτον ἔχειν ποιοῦσι
τὸ ὕδωρ, οὕτω καὶ ἡ φύσις τὸ ξηρὸν καὶ τὸ
γεῶδες, καὶ διὰ τοῦ ξηροῦ καὶ γεώδους διηθοῦσα
καὶ κινοῦσα τῷ θερμῷ ποιόν τι τὸ ὑγρὸν παρασκευάζει. καὶ ἔστι τοῦτο χυμὸς τὸ γιγνόμενον
20 ὑπὸ τοῦ εἰρημένου ξηροῦ πάθος ἐν τῷ ὑγρῷ τῆς
γεύσεως τῆς κατὰ δύναμιν ἀλλοιωτικὸν εἰς ἐνέρ-
γειαν· ἄγει γὰρ τὸ αἰσθητικὸν εἰς τοῦτο δυνάμει
προϋπάρχον· οὐ γὰρ κατὰ τὸ μανθάνειν ἀλλὰ κατὰ
τὸ θεωρεῖν ἐστι τὸ αἰσθάνεσθαι.

Ὅτι δ' οὐ παντὸς ξηροῦ ἀλλὰ τοῦ τροφίμου οἱ
25 χυμοὶ ἢ πάθος εἰσὶν ἢ στέρησις, δεῖ λαβεῖν ἐντεῦθεν,
ὅτι οὔτε τὸ ξηρὸν ἄνευ τοῦ ὑγροῦ οὔτε τὸ ὑγρὸν
ἄνευ τοῦ ξηροῦ· τροφὴ γὰρ οὐδὲν αὐτῶν τοῖς
ζῴοις, ἀλλὰ τὸ μεμιγμένον. καὶ ἔστι τῆς προσ-
φερομένης τροφῆς τοῖς ζῴοις τὰ μὲν ἁπτὰ τῶν
αἰσθητῶν αὔξησιν ποιοῦντα καὶ φθίσιν· τούτων
30 μὲν γὰρ αἴτιον ἢ θερμὸν καὶ ψυχρὸν τὸ προσφερό-
442 a μενον· ταῦτα γὰρ ποιεῖ καὶ αὔξησιν καὶ φθίσιν·
τρέφει δὲ ᾗ γευστὸν τὸ προσφερόμενον· πάντα
γὰρ τρέφεται τῷ γλυκεῖ, ἢ ἁπλῶς ἢ μεμιγμένως.
δεῖ μὲν οὖν διορίζειν περὶ τούτων ἐν τοῖς περὶ
γενέσεως, νῦν δ' ὅσον ἀναγκαῖον ἅψασθαι αὐτῶν.
5 τὸ γὰρ θερμὸν αὐξάνει καὶ δημιουργεῖ τὴν τροφήν,
καὶ τὸ μὲν κοῦφον ἕλκει, τὸ δ' ἁλμυρὸν καὶ πικρὸν
καταλείπει διὰ βάρος. ὃ δὴ ἐν τοῖς ἔξω σώμασι
ποιεῖ τὸ ἔξω θερμόν, τοῦτο τὸ¹ ἐν τῇ φύσει τῶν
ζῴων καὶ φυτῶν· διὸ τρέφεται τῷ γλυκεῖ. συμμίγ-
νυνται δ' οἱ ἄλλοι χυμοὶ εἰς τὴν τροφὴν τὸν αὐτὸν
10 τρόπον τῷ ἁλμυρῷ καὶ ὀξεῖ, ἀντὶ ἡδύσματος. ταῦτα

¹ τὸ om. ELMY.

ᵃ Cf. De An. 417 b 19.

cause the water to assume these colours or flavours, so also nature treats what is dry and earthy : causing water, by the agency of heat, to percolate through what is dry and earthy, she invests the liquid with a certain quality. And this, *viz.*, the affection produced in the liquid by the aforesaid dry, capable of altering potential into actual taste, is flavour. For it brings the sensitive faculty, which already exists potentially, to actuality ; for active sensation is analogous not to the acquisition of knowledge, but to the exercise of it.[a]

That flavours are not an affection nor a privative state of everything dry, but only of dry food, can be deduced from the fact that no dry without wet, or wet without dry, is nutrient ; for no element by itself, but only a composite product, serves as food for animals. Now of the sensible elements in the food assimilated by animals, the tangible cause growth and decay ; which are caused by the assimilated food *qua* hot and cold ; for these effect growth and decay. But the food assimilated nourishes *qua* tastable ; for everything is nourished by the sweet, either isolated or in combination. The details of this must be discussed in the treatise *On Generation*,[b] but for the moment need only be referred to as far as is essential. Heat expands and modifies the food, and extracts from it what is light, leaving behind what is harsh and bitter owing to its weight. The function performed by external heat in external bodies is performed by their natural heat in animals and plants ; this is how they are nourished by the sweet. The other flavours are blended with food in the same way as we use salt and vinegar, for seasoning. These are required to

[b] *De Gen. et Corr.* 1. 5.

442 a

δὲ διὰ τὸ ἀντισπᾶν τῷ λίαν τρόφιμον εἶναι τὸ
γλυκὺ καὶ ἐπιπολαστικόν.

Ὥσπερ δὲ τὰ χρώματα ἐκ λευκοῦ καὶ μέλανος
μίξεώς ἐστιν, οὕτως οἱ χυμοὶ ἐκ γλυκέος καὶ
πικροῦ. καὶ κατὰ λόγον δὴ τῷ μᾶλλον καὶ ἧττον
15 ἕκαστοί εἰσιν, εἴτε κατ' ἀριθμούς τινας τῆς μίξεως
καὶ κινήσεως,[1] εἴτε καὶ ἀορίστως. οἱ δὲ τὴν ἡδονὴν
ποιοῦντες μιγνύμενοι, οὗτοι ἐν ἀριθμοῖς.

Μόνος μὲν οὖν λιπαρὸς ὁ τοῦ[2] γλυκέος ἐστὶ
χυμός, τὸ δ' ἁλμυρὸν καὶ πικρὸν σχεδὸν τὸ αὐτό, ὁ
δὲ αὐστηρὸς καὶ δριμὺς καὶ στρυφνὸς καὶ ὀξὺς ἀνὰ
20 μέσον. σχεδὸν γὰρ ἴσα καὶ τὰ τῶν χυμῶν εἴδη καὶ
τὰ τῶν χρωμάτων ἐστίν. ἑπτὰ γὰρ ἀμφοτέρων
εἴδη, ἄν τις τιθῇ, ὥσπερ εὔλογον, τὸ φαιὸν μέλαν τι
εἶναι· λείπεται γὰρ τὸ ξανθὸν μὲν τοῦ λευκοῦ εἶναι
ὥσπερ τὸ λιπαρὸν τοῦ γλυκέος, τὸ φοινικοῦν δὲ
καὶ ἁλουργὸν καὶ πράσινον καὶ κυανοῦν μεταξὺ
25 τοῦ λευκοῦ καὶ μέλανος, τὰ δ' ἄλλα μικτὰ ἐκ
τούτων. καὶ ὥσπερ τὸ μέλαν στέρησις ἐν τῷ
διαφανεῖ τοῦ λευκοῦ, οὕτω τὸ ἁλμυρὸν καὶ πικρὸν
τοῦ γλυκέος ἐν τῷ τροφίμῳ ὑγρῷ. διὸ καὶ ἡ
τέφρα τῶν κατακαομένων πικρὰ πάντων· ἐξ-
ίκμασται γὰρ τὸ πότιμον ἐξ αὐτῶν.

30 Δημόκριτος δὲ καὶ οἱ πλεῖστοι τῶν φυσιολόγων,
ὅσοι λέγουσι περὶ αἰσθήσεως, ἀτοπώτατόν τι
442 b ποιοῦσιν· πάντα γὰρ τὰ αἰσθητὰ ἁπτὰ ποιοῦσιν.
καίτοι εἰ οὕτω τοῦτ' ἔχει, δῆλον ὡς καὶ τῶν ἄλλων
αἰσθήσεων ἑκάστη ἁφή τις ἐστίν· τοῦτο δ' ὅτι
ἀδύνατον, οὐ χαλεπὸν συνιδεῖν. ἔτι δὲ τοῖς κοινοῖς
5 τῶν αἰσθήσεων πασῶν χρῶνται ὡς ἰδίοις· μέγεθος
γὰρ καὶ σχῆμα καὶ τὸ τραχὺ καὶ τὸ λεῖον, ἔτι

[1] κινήσεως E (Biehl): κινήσεις.

counteract the tendency of the sweet to be over-nutritious and to lie undigested in the stomach.

As colours come from a mixing of white and black, so do flavours from a mixing of sweet and bitter. The several colours exhibit varying proportions, whether the ratio of their mixture and stimulative effect is exactly numerical or indefinite. Those which when mingled give pleasure are all in numerical ratios. Flavours and colours are analogous.

Only the flavour of the sweet is rich, and the salt is virtually the same as the bitter ; between these extremes lie the harsh, the pungent, the astringent and the acid. The kinds of flavour are roughly equal in number to those of colours. There are seven of each, if, as is natural, one regards grey as a variety of black (the alternative is to class yellow with white, as rich with sweet) ; red, purple, green and blue are colours intermediate between white and black, and the rest are combinations of these. And just as black is a privation of white in the transparent, so the salt or bitter is a privation of the sweet in nutrient moisture. This is why the ash of everything burned is bitter ; for the drinkable moisture has been evaporated from it. The analogy continued.

Democritus and most of the natural philosophers who treat of sensation produce a most unreasonable hypothesis ; for they make all sensible objects objects of touch. And yet it is obvious that, if this be so, each of the other senses is a kind of touch. Now it is not difficult to see that this is impossible. Again they treat perceptible objects which are common to all the senses as if they were peculiar to one ; for size, shape, roughness and smoothness, and also Some thinkers refer all senses to touch.

---

² ἀριθμοῖς μόνον. ὁ μὲν οὖν λιπαρὸς τοῦ κτλ. LSUP.

442 b

δὲ τὸ ὀξὺ καὶ τὸ ἀμβλὺ τὸ ἐν τοῖς ὄγκοις κοιν
τῶν αἰσθήσεών ἐστιν, εἰ δὲ μὴ πασῶν, ἀλλ᾽ ὄψεώ
γε καὶ ἁφῆς. διὸ καὶ περὶ μὲν τούτων ἀπατῶντα
περὶ δὲ τῶν ἰδίων οὐκ ἀπατῶνται, οἷον ὄψις περ
10 χρώματος καὶ ἀκοὴ περὶ ψόφων. οἱ δὲ τὰ ἴδι
εἰς ταῦτα ἀνάγουσιν, ὥσπερ Δημόκριτος· τὸ γὰ
λευκὸν καὶ τὸ μέλαν τὸ μὲν τραχύ φησιν εἶναι τ
δὲ λεῖον, εἰς δὲ τὰ σχήματα ἀνάγει τοὺς χυμούς
καίτοι ἢ οὐδεμιᾶς ἢ μᾶλλον τῆς ὄψεως τὰ κοιν
γνωρίζειν. εἰ δ᾽ ἄρα τῆς γεύσεως μᾶλλον, τ
15 γοῦν ἐλάχιστα τῆς ἀκριβεστάτης ἐστὶν αἰσθήσεω
διακρίνειν περὶ ἕκαστον γένος, ὥστε ἐχρῆν τὴ
γεῦσιν καὶ τῶν ἄλλων κοινῶν αἰσθάνεσθαι μάλιστ
καὶ τῶν σχημάτων εἶναι κριτικωτάτην. ἔτι τ
μὲν αἰσθητὰ πάντα ἔχει ἐναντίωσιν, οἷον ἐν χρώ
ματι τῷ μέλανι τὸ λευκὸν καὶ ἐν χυμοῖς τῷ γλυκε
20 τὸ πικρόν· σχῆμα δὲ σχήματι οὐ δοκεῖ εἶνα
ἐναντίον· τίνι γὰρ τῶν πολυγώνων τὸ¹ περιφερὲ
ἐναντίον; ἔτι ἀπείρων ὄντων τῶν σχημάτω
ἀναγκαῖον καὶ τοὺς χυμοὺς εἶναι ἀπείρους· διὰ τ
γὰρ ὁ μὲν ἂν ποιήσειεν αἴσθησιν, ὁ δ᾽ οὐκ ἂ
ποιήσειεν; καὶ περὶ μὲν τοῦ γευστοῦ καὶ χυμο
25 εἴρηται· τὰ γὰρ ἄλλα πάθη τῶν χυμῶν οἰκεία
ἔχει τὴν σκέψιν ἐν τῇ φυσιολογίᾳ τῇ περὶ τῶ
φυτῶν.

V. Τὸν αὐτὸν δὲ τρόπον δεῖ νοῆσαι καὶ περὶ τὰ
ὀσμάς· ὅπερ γὰρ ποιεῖ ἐν τῷ ὑγρῷ τὸ ξηρόν
τοῦτο ποιεῖ ἐν ἄλλῳ γένει τὸ ἔγχυμον ὑγρόν, ε
30 ἀέρι καὶ ὕδατι ὁμοίως. κοινὸν δὲ κατὰ τούτων νῦ
443 a μὲν λέγομεν τὸ διαφανές, ἔστι δ᾽ ὀσφραντὸν οὐχ
ἣ διαφανές, ἀλλ᾽ ἣ πλυντικὸν ἢ ῥυπτικὸν ἐγχύμου

---

¹ τὸ E, om. ceteri codd.

sharpness and bluntness, as found in solid bodies, are common, if not to all the senses, at least to sight and touch. Hence the senses are liable to error about common sensibles, but not about their proper sensibles ; *e.g.*, vision is not in error about colour, nor hearing about sound. But these thinkers reduce proper to common sensibles, as Democritus does. For he says that white and black are rough and smooth respectively, and he refers flavours to shapes. And yet surely if any single sense can recognize common sensibles, it should rather be sight. If we attribute such a function to taste, then since it is the mark of the keenest sense to recognize the smallest differences in each class, taste ought, besides being most discriminative of shapes, to be most perceptive of the other common sensibles too. Again all sensible objects exhibit contrariety, *e.g.*, that of white to black in colours, or bitter to sweet in flavours. But no figure appears contrary to any other ; to what polygon is a sphere contrary ? Again, as figures are infinite in number, flavours must also be infinite ; for why should one flavour produce sensation and not another? This finishes our discussion of taste and flavour ; the other affections of flavour have their proper place of inquiry in the *Natural History of Plants.*[a]

V. We must regard smells in the same way ; for Smell. the effect which the dry produces in the wet is also produced by flavoured liquid in another sphere, in air and water alike. We have just said that in these transparence is a common property, but the object is smellable not *qua* transparent, but because it is capable of washing or cleansing the flavoured dryness ;

---

[a] No such treatise by Aristotle has come down to us.

443 a

ξηρότητος· οὐ γὰρ μόνον ἐν ἀέρι ἀλλὰ καὶ ἐν ὕδατι
τὸ τῆς ὀσφρήσεώς ἐστιν. δῆλον δ' ἐπὶ τῶν ἰχθύων
καὶ τῶν ὀστρακοδέρμων· φαίνονται γὰρ ὀσφραινό-
5 μενα οὔτε ἀέρος ὄντος ἐν τῷ ὕδατι (ἐπιπολάζει γὰρ
ὁ ἀήρ, ὅταν ἐγγένηται) οὔτ' αὐτὰ ἀναπνέοντα. εἰ
οὖν τις θείη καὶ τὸν ἀέρα καὶ τὸ ὕδωρ ἄμφω ὑγρά,
εἴη ἂν ἡ ἐν ὑγρῷ τοῦ ἐγχύμου ξηροῦ φύσις ὀσμή,
καὶ ὀσφραντὸν τὸ τοιοῦτον. ὅτι δ' ἀπ' ἐγχύμου[1]
10 ἐστὶ τὸ πάθος, δῆλον ἐκ τῶν ἐχόντων καὶ μὴ
ἐχόντων ὀσμήν· τά τε γὰρ στοιχεῖα ἄοσμα, οἷον πῦρ
ἀὴρ ὕδωρ γῆ, διὰ τὸ τά τε ξηρὰ αὐτῶν καὶ τὰ ὑγρὰ
ἄχυμα εἶναι, ἐὰν μή τι μιγνύμενον ποιῇ. διὸ καὶ
ἡ θάλαττα ἔχει ὀσμήν· ἔχει γὰρ χυμὸν καὶ ξηρό-
τητα. καὶ ἅλες μᾶλλον λίτρου ὀσμώδεις· δηλοῖ δὲ
τὸ ἐξικμαζόμενον ἐξ αὐτῶν ἔλαιον· τὸ δὲ λίτρον γῆς
15 ἐστὶ μᾶλλον. ἔτι λίθος μὲν ἄοσμον, ἄχυμον γάρ,
τὰ δὲ ξύλα ὀσμώδη, ἔγχυμα γάρ· καὶ τούτων τὰ
ὑδατώδη ἧττον. ἔτι τῶν μεταλλευομένων χρυσὸς
ἄοσμον, ἄχυμον γάρ, ὁ δὲ χαλκὸς καὶ ὁ σίδηρος
ὀσμώδη. ὅταν δ' ἐκκαυθῇ τὸ ὑγρόν, ἀοσμότεραι αἱ
20 σκωρίαι γίγνονται πάντων. ἄργυρος δὲ καὶ κατ-
τίτερος τῶν μὲν μᾶλλον ὀσμώδη τῶν δ' ἧττον·
ὑδατώδη γάρ.

Δοκεῖ δ' ἐνίοις ἡ καπνώδης ἀναθυμίασις εἶναι
ὀσμή, οὖσα κοινὴ γῆς τε καὶ ἀέρος. [καὶ πάντες
ἐπιφέρονται ἐπὶ τοῦτο περὶ ὀσμῆς·][2] διὸ καὶ Ἡράκλειτος οὕτως εἴρηκεν, ὡς εἰ πάντα τὰ ὄντα καπ-

---

[1] ἅπαν χυμοῦ EMY.
[2] καὶ ... ὀσμῆς damn. Thurot, secl. Biehl.

for this phenomenon of smelling occurs not only in air, but also in water. This is obvious in the case of fishes and hard-shelled creatures ; for they evidently have the power of smelling, though there is no air in water (for whenever it is generated in water it rises to the surface), nor do such creatures inhale. If, then, we assume that water and air are both moist, smell will be the nature which the flavoured dry shows in a wet medium, and what is so conditioned will be an object of smell. That the effect is due to the possession of flavour is obvious from a consideration of those things which have, and those which have not smell. The elements, *viz.*, fire, air, water and earth, are odourless, because both those which are dry and those which are wet have no flavour, unless they form a combination. That is why the sea has a smell ; for it has flavour and a dry ingredient. Salt has more smell than sodium carbonate : the oil extracted from the former proves this ; but sodium carbonate belongs more to earth. Stone again has no smell because it has no taste, but woods have a smell because they have also taste ; wet woods have less smell than dry. In the case of the metals gold has no smell because it has no taste, but bronze and iron have smell. But when the moisture is burned out of them, the slag of all of them has less smell. Silver and tin have more smell than gold, and less than bronze and iron ; for they contain water.

Some people think that smell is a smoky vapour, which is partly earth and partly air. [Indeed all incline to this view about smell.] It is with this idea that Heracleitus has said that, if everything that exists became smoke, the nose would be the organ

The nature of smell.

249

443 a

25 νὸς γένοιτο, ῥῖνες ἂν διαγνοῖεν, ἐπὶ δὲ τὴν ὀσμὴν
πάντες¹ ἐπιφέρονται οἱ μὲν ὡς ἀτμίδα, οἱ δ' ὡς
ἀναθυμίασιν, οἱ δ' ὡς ἄμφω ταῦτα. ἔστι δ' ἡ μὲν
ἀτμὶς ὑγρότης τις, ἡ δὲ καπνώδης ἀναθυμίασις,
ὥσπερ εἴρηται, κοινὸν ἀέρος καὶ γῆς· καὶ συνίσταται
ἐκ μὲν ἐκείνης ὕδωρ, ἐκ δὲ ταύτης γῆς τι εἶδος.
30 ἀλλ' οὐδέτερον τούτων ἔοικεν· ἡ μὲν γὰρ ἀτμὶς
ἐστιν ὕδατος, ἡ δὲ καπνώδης ἀναθυμίασις ἀδύνατος
ἐν ὕδατι γενέσθαι. ὀσμᾶται δὲ καὶ τὰ ἐν τῷ ὕδατι,
443 b ὥσπερ εἴρηται πρότερον. ἔτι ἡ ἀναθυμίασις ὁμοίως
λέγεται ταῖς ἀπορροίαις. εἰ οὖν μηδ' ἐκείνη καλῶς,
οὐδ' αὕτη καλῶς.

Ὅτι μὲν οὖν ἐνδέχεται ἀπολαύειν τὸ ὑγρὸν καὶ τὸ
ἐν τῷ πνεύματι καὶ τὸ ἐν τῷ ὕδατι καὶ πάσχειν τι
5 ὑπὸ τῆς ἐγχύμου ξηρότητος, οὐκ ἄδηλον· καὶ γὰρ
ὁ ἀὴρ ὑγρὸν τὴν φύσιν ἐστίν. ἔτι δ' εἴπερ² ὁμοίως
ἐν τοῖς ὑγροῖς ποιεῖ καὶ ἐν τῷ ἀέρι οἷον ἀποπλυνό-
μενον τὸ ξηρόν, φανερὸν ὅτι² δεῖ ἀνάλογον εἶναι τὰς
ὀσμὰς τοῖς χυμοῖς. ἀλλὰ μὴν τοῦτό γε ἐπ' ἐνίων
10 συμβέβηκεν· καὶ γὰρ δριμεῖαι καὶ γλυκεῖαί εἰσιν
ὀσμαὶ καὶ αὐστηραὶ καὶ στρυφναὶ καὶ λιπαραί, καὶ
τοῖς πικροῖς τὰς σαπρὰς ἄν τις ἀνάλογον εἴποι.
διὸ ὥσπερ ἐκεῖνα δυσκατάποτα, τὰ σαπρὰ δυσανά-
πνευστά ἐστιν. δῆλον ἄρα ὅτι ὅπερ ἐν τῷ ὕδατι ὁ
15 χυμός, τοῦτ' ἐν τῷ ἀέρι καὶ ὕδατι ἡ ὀσμή. καὶ διὰ
τοῦτο τὸ ψυχρὸν καὶ ἡ πῆξις καὶ τοὺς χυμοὺς ἀμβλύνει
καὶ τὰς ὀσμὰς ἀφανίζει· τὸ γὰρ θερμὸν τὸ κινοῦν
καὶ δημιουργοῦν ἀφανίζουσιν ἡ ψύξις καὶ ἡ πῆξις.

Εἴδη δὲ τοῦ ὀσφραντοῦ δύο ἐστίν· οὐ γὰρ ὥσπερ
τινές φασιν, οὐκ ἔστιν εἴδη τοῦ ὀσφραντοῦ, ἀλλ'
ἔστιν. διοριστέον δὲ πῶς ἔστι καὶ πῶς οὐκ ἔστιν·

---

¹ ἐπὶ τοῦτο post πάντες add. Christ.

to perceive it. All tend to regard smell in this way ; some as vapour, some as smoke, and some as a mixture of the two. Vapour is a kind of moisture, but smoky exhalation, as we have said, is a compound of air and earth ; the former when condensed becomes water, but the latter becomes a kind of earth. But probably smell is neither of these ; for vapour consists of water, and smoky exhalation cannot occur in water. Yet water-creatures have a sense of smell, as has been said before. Further the smoky exhalation theory is like the theory of emanations. If the latter is unsound, so also is the former.

It is obvious that it is possible that the moisture both in air and in water absorbs the nature of and is affected by flavoured dryness ; for air too has a moist nature. Again, if the dry produces in liquids and air alike an effect as of something dissolved away, clearly smells must be analogous to flavours. Moreover this is certainly so in some cases ; for smells like flavours are pungent, sweet, harsh, astringent and rich, and one could call the fetid analogous to the bitter. Hence as these flavours are unpleasant to drink, so fetid smells are unpleasant to inhale. It is clear then that smell in air and in water is the same thing as flavour in water alone. This is why cold and freezing dulls flavours, and causes smells to disappear ; for cold and freezing counteract the heat which excites and develops flavour. *Analogy between smell and taste.*

There are two species of smellable objects ; for it is untrue to say, as some do, that there are no species of the smellable, for there are. But we must distinguish in what sense there are such species, and *Objects smelt are of different kinds.*

---

² ἔτι δ' εἴπερ] ὅτι δ' M, Bekker.
³ φανερὸν ὅτι] φανερόν. ἔτι EMY, Bekker.

443 b

τὸ μὲν γάρ ἐστι κατὰ τοὺς χυμοὺς τεταγμένον
20 αὐτῶν, ὥσπερ εἴπομεν, καὶ τὸ ἡδὺ καὶ τὸ λυπηρὸν
κατὰ συμβεβηκὸς ἔχουσιν· διὰ γὰρ τὸ τοῦ θρεπτικοῦ
πάθη εἶναι, ἐπιθυμούντων μὲν ἡδεῖαι αἱ ὀσμαὶ
τούτων εἰσί, πεπληρωμένοις δὲ καὶ μηδὲν δεομένοις
οὐχ ἡδεῖαι, οὐδ' ὅσοις μὴ καὶ ἡ τροφὴ ἡ ἔχουσα
25 τὰς ὀσμὰς ἡδεῖα, οὐδὲ τούτοις. ὥστε αὗται μέν,
καθάπερ εἴπομεν, κατὰ συμβεβηκὸς ἔχουσι τὸ ἡδὺ
καὶ λυπηρόν, διὸ καὶ πάντων εἰσὶ κοιναὶ τῶν ζῴων·
αἱ δὲ καθ' αὑτὰς ἡδεῖαι τῶν ὀσμῶν εἰσίν, οἷον αἱ
τῶν ἀνθῶν· οὐδὲν γὰρ μᾶλλον οὐδ' ἧττον πρὸς τὴν
τροφὴν παρακαλοῦσιν, οὐδὲ συμβάλλεται πρὸς
30 ἐπιθυμίαν οὐδέν, ἀλλὰ τοὐναντίον μᾶλλον· ἀληθὲς
γὰρ ὅπερ Εὐριπίδην σκώπτων εἶπε Στράττις,
" ὅταν φακῆν ἕψητε, μὴ 'πιχεῖν[1] μύρον." οἱ δὲ νῦν
444 a μιγνύντες εἰς τὰ πόματα τὰς τοιαύτας δυνάμεις
βιάζονται τῇ συνηθείᾳ τὴν ἡδονήν, ἕως ἂν ἐκ δύ'
αἰσθήσεων γένηται τὸ ἡδὺ ὡς ἓν καὶ ἀπὸ μιᾶς.
τοῦτο μὲν οὖν τὸ ὀσφραντὸν ἴδιον ἀνθρώπου ἐστίν,
5 ἡ δὲ κατὰ τοὺς χυμοὺς τεταγμένη καὶ τῶν ἄλλων
ζῴων, ὥσπερ εἴρηται πρότερον· κἀκείνων μέν, διὰ
τὸ κατὰ συμβεβηκὸς ἔχειν τὸ ἡδύ, διήρηται τὰ
εἴδη κατὰ τοὺς χυμούς, ταύτης δ' οὐκέτι, διὰ τὸ
τὴν φύσιν αὑτῆς εἶναι καθ' αὑτὴν ἡδεῖαν ἢ λυπηράν.

Αἴτιον δὲ τοῦ ἴδιον εἶναι ἀνθρώπου τὴν τοιαύτην
ὀσμὴν διὰ τὴν ἕξιν[2] τὴν περὶ τὸν ἐγκέφαλον.
10 ψυχροῦ γὰρ ὄντος τὴν φύσιν τοῦ ἐγκεφάλου, καὶ
τοῦ αἵματος τοῦ περὶ αὐτὸν ἐν τοῖς φλεβίοις ὄντος
λεπτοῦ μὲν καὶ καθαροῦ, εὐψύκτου δέ (διὸ καὶ ἡ
τῆς τροφῆς ἀναθυμιάσις ψυχομένη διὰ τὸν τόπον τὰ
νοσηματικὰ ῥεύματα ποιεῖ), τοῖς ἀνθρώποις πρὸς

---

¹ ἐπιχεῖν Bekker.　　　　² ἕξιν EMY : ψύξιν.

in what sense there are not. There is one kind of smells which is ranged parallel with the corresponding flavours, as we have said, and to these pleasantness and unpleasantness are incidental ; for because they are affections of nutritive substance, these smells are pleasant when we are hungry, but to those who are sated and require nothing they are not pleasant ; nor is the smell pleasant to those to whom the food having the smell is unpleasant. So, as we have said, these smells are pleasant or unpleasant only inciden-tally, and thus too they are common to all animals. The other kind of smells are pleasant in themselves, *e.g.*, those of flowers ; for they have no effect, great or small, as incitements to eat, nor do they contribute anything to desire, but rather the opposite ; for what Strattis said in deriding Euripides is true, " When you make pea-soup don't put perfume in it." Those who " doctor " drinks in this way develop an acquired taste for them, until pleasure arises from two senses as if it were a single pleasure from one. This kind of smell-percept is peculiar to man, but those which correspond to flavours are perceptible to all other animals, as has been said before : the latter, because their sweetness is incidental, can be divided into classes, according to flavours, but the former cannot, because the sweetness, or the reverse, is part of its essential nature.

The reason why the former kind of smell is peculiar to man is to be found in the conditions prevailing around the brain. For the brain is naturally cold, and the blood around it in the veins is thin and pure and easily cooled. (This is why the fumes of food, when they are chilled by the coldness of that region, cause catarrh.) This kind of smell, then, has been

The function of the sense of smell.

444 a

15 βοήθειαν ὑγιείας γέγονε τὸ τοιοῦτον εἶδος τῆς
ὀσμῆς· οὐδὲν γὰρ ἄλλο ἔργον ἐστὶν αὐτῆς ἢ τοῦτο.
τοῦτο δὲ ποιεῖ φανερῶς· ἡ μὲν γὰρ τροφὴ ἡδεῖα
οὖσα, καὶ ἡ ξηρὰ καὶ ἡ ὑγρά, πολλάκις νοσώδης
ἐστίν, ἡ δ' ἀπὸ τῆς ὀσμῆς τῆς καθ' αὑτὴν εὐώδους
ὁπωσοῦν ἔχουσιν ὠφέλιμος ὡς εἰπεῖν ἀεί. καὶ διὰ
20 τοῦτο γίνεται διὰ τῆς ἀναπνοῆς, οὐ πᾶσιν ἀλλὰ τοῖς
ἀνθρώποις καὶ τῶν ἐναίμων οἷον τοῖς τετράποσι
καὶ ὅσα μετέχει μᾶλλον τῆς τοῦ ἀέρος φύσεως·
ἀναφερομένων γὰρ τῶν ὀσμῶν πρὸς τὸν ἐγκέφαλον
διὰ τὴν ἐν αὐταῖς τῆς θερμότητος κουφότητα,
ὑγιεινοτέρως ἔχει τὰ περὶ τὸν τόπον τοῦτον· ἡ γὰρ
25 τῆς ὀσμῆς δύναμις θερμὴ τὴν φύσιν ἐστίν.

Κατακέχρηται δ' ἡ φύσις τῇ ἀναπνοῇ ἐπὶ δύο,
ὡς ἔργῳ μὲν ἐπὶ τὴν εἰς τὸν θώρακα βοήθειαν, ὡς
παρέργῳ δ' ἐπὶ τὴν ὀσμήν· ἀναπνέοντος γὰρ ὥσπερ
ἐκ παρόδου ποιεῖται διὰ τῶν μυκτήρων τὴν κίνησιν.
ἴδιον δὲ τῆς τοῦ ἀνθρώπου φύσεώς ἐστι τὸ τῆς
30 ὀσμῆς τῆς τοιαύτης γένος διὰ τὸ πλεῖστον ἐγκέφαλον καὶ ὑγρότατον ἔχειν τῶν ζῴων ὡς κατὰ
μέγεθος· διὰ γὰρ τοῦτο καὶ μόνον ὡς εἰπεῖν αἰσθάνεται τῶν ζῴων ἄνθρωπος καὶ χαίρει ταῖς τῶν
ἀνθῶν καὶ ταῖς τῶν τοιούτων ὀσμαῖς· σύμμετρος

444 b

γὰρ αὐτῶν ἡ θερμότης καὶ ἡ κίνησις πρὸς τὴν
ὑπερβολὴν τῆς ἐν τῷ τόπῳ ὑγρότητος καὶ ψυχρότη-
τός ἐστιν. τοῖς δ' ἄλλοις ὅσα πλεύμονα ἔχει διὰ
τὸ ἀναπνεῖν τοῦ ἑτέρου γένους τῆς ὀσμῆς τὴν
αἴσθησιν ἀποδέδωκεν ἡ φύσις, ὅπως μὴ δύο αἰσθη-
5 τήρια ποιῇ. ἀπόχρη γάρ, ἐπείπερ καὶ ὡς ἀναπνέουσιν, ὥσπερ τοῖς ἀνθρώποις ἀμφοτέρων τῶν
ὀσφραντῶν, τούτοις τῶν ἑτέρων μόνων ὑπάρχουσα

evolved for man to preserve his health; this is its only function. And it obviously performs it; for sweet food, whether dry or moist, is often unhealthy, whereas that which has a smell which is in itself pleasant is nearly always beneficial to persons in any state of health. And for this reason smell is conveyed by inhalation, not in all animals, but in man and some sanguineous animals, *e.g.*, quadrupeds and those which have a larger share in the nature of air; for as the smells rise to the brain because of the lightness of the heat contained in them, the parts of the body in this region are the healthier for these smells; for the potency of smell is naturally hot.

Nature has employed respiration for two purposes, first, and as its main function, for the aid of the chest, and in a secondary sense for the purpose of smell; for when a creature inhales, smell effects its stimulus of the nostrils as though from a side entrance. Smell of this type is peculiar to man because he has the largest and moistest brain, in proportion to his size, of all animals; hence too alone, one may say, among animals man is conscious of and enjoys the smell of flowers and the like; for the heat and stimulation produced by these balance the excess of moisture and coldness in that region of the body. But nature has allotted perception of the second class of smell to all the other animals which have lungs through respiration, to avoid making two separate sense organs; for in respiration the animals have sufficient means for the perception of one kind of smells, just as men have for the perception of both kinds. But

Smell and breathing.

255

444 b

ἡ αἴσθησις. τὰ δὲ μὴ ἀναπνέοντα, ὅτι μὲν ἔχει
αἴσθησιν τοῦ ὀσφραντοῦ, φανερόν· καὶ γὰρ ἰχθύες
καὶ τὸ τῶν ἐντόμων γένος πᾶν ἀκριβῶς καὶ πόρρω-
10 θεν αἰσθάνεται, διὰ τὸ θρεπτικὸν εἶδος τῆς ὀσμῆς,
ἀπέχοντα πολὺ τῆς οἰκείας τροφῆς, οἷον αἵ τε
μέλιτται [ποιοῦσι πρὸς τὸ μέλι]¹ καὶ τὸ τῶν μικρῶν
μυρμήκων γένος, οὓς καλοῦσί τινες κνῖπας, καὶ
τῶν θαλαττίων αἱ πορφύραι, καὶ πολλὰ τῶν ἄλλων
τῶν τοιούτων ζῴων ὀξέως αἰσθάνεται τῆς τροφῆς
15 διὰ τὴν ὀσμήν. ὅτῳ δὲ αἰσθάνεται, οὐχ ὁμοίως
φανερόν. διὸ κἂν ἀπορήσειέ τις τίνι αἰσθάνεται²
τῆς ὀσμῆς, εἴπερ ἀναπνέουσι μὲν γίνεται τὸ ὀσμᾶσθαι μοναχῶς (τοῦτο γὰρ φαίνεται ἐπὶ τῶν ἀναπνεόντων συμβαῖνον πάντων), ἐκείνων δ' οὐθὲν
ἀναπνεῖ αἰσθάνεται μέντοι, εἰ μή τις παρὰ τὰς
20 πέντε αἰσθήσεις ἑτέρα. τοῦτο δ' ἀδύνατον· τοῦ
γὰρ ὀσφραντοῦ ὄσφρησις, ἐκεῖνα δὲ τούτου αἰσθάνεται,³ ἀλλ' οὐ τὸν αὐτὸν ἴσως τρόπον, ἀλλὰ τοῖς
μὲν ἀναπνέουσι τὸ πνεῦμα ἀφαιρεῖ τὸ ἐπικείμενον
ὥσπερ πῶμά τι (διὸ οὐκ αἰσθάνεται μὴ ἀναπνέοντα),
τοῖς δὲ μὴ ἀναπνέουσιν ἀφῄρηται τοῦτο, καθάπερ
25 ἐπὶ τῶν ὀφθαλμῶν τὰ μὲν ἔχει βλέφαρα τῶν ζῴων,
ὧν μὴ ἀνακαλυφθέντων οὐ δύναται ὁρᾶν, τὰ δὲ
σκληρόφθαλμα οὐκ ἔχει, διόπερ οὐ προσδεῖται
οὐδενὸς τοῦ ἀνακαλύψοντος, ἀλλ' ὁρᾷ ἐκ τοῦ δυνα-
τοῦ ὄντος⁴ αὐτοῦ⁵ εὐθύς. ὁμοίως δὲ καὶ τῶν ἄλλων
ζῴων ὁτιοῦν οὐδὲν δυσχεραίνει τῶν καθ' αὑτὰ δυσω-
30 δῶν τὴν ὀσμήν, ἂν μή τι τύχῃ φθαρτικὸν ὄν. ὑπὸ
τούτων δ' ὁμοίως φθείρεται καθάπερ καὶ οἱ ἄνθρωποι ὑπὸ τῆς τῶν ἀνθράκων ἀτμίδος καρηβαροῦσι
καὶ φθείρονται πολλάκις· οὕτως ὑπὸ τῆς τοῦ θείου

____

¹ om. EMY, Biehl.　　　² αἰσθάνονται LSP.
³ αἰσθάνονται LSUP.　　⁴ ὄντος] ὁρᾶν LSU.　　⁵ αὐτῷ LSUP.

it is clear that animals which do not breathe have perception of a scented object; for fishes and the whole class of insects, owing to the nutritive kind of smell, have a very accurate perception, even at a distance, of their proper food, although they may be far away from it. *E.g.*, bees [show it about honey], and the class of small ants which some call *knipes*, and among marine animals the murex, and many other similar creatures, have a keen perception of their food by its smell. But the organ whereby they have this perception is not so certain. So one might be in considerable doubt with what organ they appre- hend smell, if smelling only occurs when they breathe (which is clearly the case with all animals that do breathe), and all the above animals possess this sense although none of them breathes; unless there is another sense besides the accepted five. But this is impossible; for a perception of odour is a sense of smell, and these animals do perceive it, but perhaps not in the same way; but, in the case of animals which breathe, the breath removes something which lies on the organ like a kind of lid (and so they do not perceive smell unless they breathe), but in the case of non-breathing animals this is removed; just as some animals have eyelids on their eyes, and cannot see when these are shut, but the hard-eyed animals have no eyelids, and so do not need anything to uncover the eyes, but can see directly as soon as the object is within visible distance. Similarly none of the lower animals objects to the smell of things which are *per se* malodorous, except such as are actually destructive. By these they are destroyed exactly in the same way as men get headaches from, and are often killed by, the fumes of charcoal; so the other

Smelling without breathing.

445 a δυνάμεως καὶ τῶν ἀσφαλτωδῶν φθείρεται τἄλλα
ζῷα, καὶ φεύγει διὰ τὸ πάθος. αὑτῆς δὲ καθ᾽
αὑτὴν τῆς δυσωδίας οὐδὲν φροντίζουσιν, καίτοι
πολλὰ τῶν φυομένων δυσώδεις ἔχει τὰς ὀσμάς, ἐὰν
5 μή τι συμβάλληται πρὸς τὴν γεῦσιν ἢ τὴν ἐδωδὴν
αὐτοῖς. ἔοικε δ᾽ ἡ αἴσθησις ἡ τοῦ ὀσφραίνεσθαι
περιττῶν οὐσῶν τῶν αἰσθήσεων καὶ τοῦ ἀριθμοῦ
ἔχοντος μέσον τοῦ περιττοῦ καὶ αὐτὴ μέση εἶναι
τῶν τε ἁπτικῶν, οἷον ἁφῆς καὶ γεύσεως καὶ τῶν δι᾽
ἄλλου αἰσθητικῶν, οἷον ὄψεως καὶ ἀκοῆς. διὸ καὶ
10 τὸ ὀσφραντὸν τῶν θρεπτικῶν ἐστι πάθος τι (ταῦτα
δ᾽ ἐν τῷ ἁπτῷ γένει), καὶ τοῦ ἀκουστοῦ δὲ καὶ τοῦ
ὁρατοῦ· διὸ καὶ ἐν ἀέρι καὶ ἐν ὕδατι ὀσμῶνται.
ὥστ᾽ ἐστὶ τὸ ὀσφραντὸν κοινόν τι τούτων ἀμφοτέρων, ὃ καὶ τῷ ἁπτῷ ὑπάρχει καὶ τῷ ἀκουστῷ καὶ
τῷ διαφανεῖ· διὸ εὐλόγως παρείκασται ξηρότητος
15 ἐν ὑγρῷ καὶ χυτῷ οἷον βαφή τις εἶναι καὶ πλύσις.
πῶς μὲν οὖν εἴδη δεῖ λέγειν καὶ πῶς οὐ δεῖ τοῦ
ὀσφραντοῦ, ἐπὶ τοσοῦτον εἰρήσθω.

Ὃ δὲ λέγουσί τινες τῶν Πυθαγορείων, οὐκ ἔστιν
εὔλογον· τρέφεσθαι γάρ φασιν ἔνια ζῷα ταῖς ὀσμαῖς.
πρῶτον μὲν γὰρ ὁρῶμεν ὅτι τὴν τροφὴν δεῖ εἶναι
20 σύνθετον· καὶ γὰρ τὰ τρεφόμενα οὐχ ἁπλᾶ ἐστιν,
διὸ καὶ περιττώματα γίνεται τῆς τροφῆς, ἢ ἐν
αὐτοῖς ἢ ἔξω, ὥσπερ τοῖς φυτοῖς. ἔτι δ᾽ οὐδὲ τὸ
ὕδωρ ἐθέλει αὐτὸ μόνον ἄμικτον ὂν τρέφειν· σωματῶδες γάρ τι δεῖ εἶναι τὸ συστησόμενον. ἔτι πολὺ
ἧττον εὔλογον τὸν ἀέρα σωματοῦσθαι. πρὸς δὲ τούτοις, ὅτι πᾶσίν ἐστι τοῖς ζῴοις τόπος δεκτικὸς τῆς
25 τροφῆς, ἐξ οὗ ὅταν εἰσέλθῃ[1] λαμβάνει τὸ σῶμα· τοῦ
δ᾽ ὀσφραντοῦ ἐν τῇ κεφαλῇ τὸ αἰσθητήριον, καὶ μετὰ

___
[1] ὅταν εἰσέλθῃ EMYP : ἕλκον LSU.

animals are destroyed by the action of sulphur and bituminous substances, and avoid them because of their effect. But they disregard the unpleasant smell in itself (and yet many plants have offensive smells), unless it affects the taste or edibility of food. As the number of the senses is odd, and an odd number always has a middle unit, smell would seem to be a middle term between the tactual senses—*viz.*, touch and taste—on the one hand, and on the other the senses which perceive through a medium, *viz.*, sight and hearing. So the object of smell is an affection of substances used for food (for these belong to the class of tangible objects), and also of the audible and visible. Hence creatures smell in both air and water. So the object of smell is common to both spheres ; it belongs to the tangible and also to the audible and the transparent ; so it was reasonably described as a dipping or washing of the dry in the wet and fluid. So much for the question how far we should speak of species of smellable objects.

But the theory advanced by some of the Pythagoreans is not reasonable ; for they say that some animals are nourished by smells. For first of all we see that food must be composite ; for the animals nourished are not simple, and for this reason there is waste matter left over from food either in the bodies themselves or outside as in plants ; not even water will serve for food, if it is unmixed ; for to cohere a thing must be corporeal. It is even less probable that air can be made corporeal. Besides, it is evident that all animals possess a region which receives the food, and from which, when it has entered, the body draws it. Now the organ of smell lies in the head,

The supposed nutritive function of smell.

259

445 a

πνευματώδους εἰσέρχεται ἀναθυμιάσεως, ὥστ᾽ εἰς
τὸν ἀναπνευστικὸν βαδίζοι ἂν τόπον. ὅτι μὲν οὖν
οὐ συμβάλλεται εἰς τροφὴν τὸ ὀσφραντόν, ἢ ὀσφραντόν, δῆλον· ὅτι μέντοι εἰς ὑγίειαν, καὶ ἐκ τῆς
30 αἰσθήσεως καὶ ἐκ τῶν εἰρημένων φανερόν, ὥστε
ὅπερ ὁ χυμὸς ἐν τῷ θρεπτικῷ καὶ πρὸς τὰ τρεφό-
445 b μενα, τοῦτ᾽ ἐστὶ πρὸς ὑγίειαν τὸ ὀσφραντόν. καθ᾽
ἕκαστον μὲν οὖν αἰσθητήριον διωρίσθω τὸν τρόπον
τοῦτον.

VI. Ἀπορήσειε δ᾽ ἄν τις, εἰ πᾶν σῶμα εἰς ἄπειρον διαιρεῖται, ἆρα καὶ τὰ παθήματα τὰ αἰσθητά,
5 οἷον χρῶμα καὶ χυμὸς καὶ ὀσμὴ καὶ βάρος καὶ
ψόφος καὶ ψυχρὸν καὶ θερμὸν καὶ κοῦφον καὶ
σκληρὸν καὶ μαλακόν, ἢ ἀδύνατον; ποιητικὸν γάρ
ἐστιν ἕκαστον αὐτῶν τῆς αἰσθήσεως· τῷ δύνασθαι
γὰρ κινεῖν αὐτὴν λέγεται πάντα. ὥστ᾽ ἀνάγκη τήν
τε αἴσθησιν εἰς ἄπειρα διαιρεῖσθαι καὶ πᾶν εἶναι
10 μέγεθος αἰσθητόν· ἀδύνατον γὰρ λευκὸν μὲν ὁρᾶν,
μὴ ποσὸν δέ. εἰ γὰρ μὴ οὕτως, ἐνδέχοιτ᾽ ἂν εἶναι
τι σῶμα μηδὲν ἔχον χρῶμα μηδὲ βάρος μηδ᾽ ἄλλο
τι τοιοῦτον πάθος· ὥστ᾽ οὐδ᾽ αἰσθητὸν ὅλως,
ταῦτα γὰρ τὰ αἰσθητά. τὸ ἄρ᾽ αἰσθητὸν ἔσται
συγκείμενον οὐκ ἐξ αἰσθητῶν. ἀλλ᾽ ἀναγκαῖον· οὐ
15 γὰρ δὴ ἔκ γε τῶν μαθηματικῶν. ἔτι τίνι κρινοῦμεν
ταῦτα ἢ γνωσόμεθα; ἢ τῷ νῷ; ἀλλ᾽ οὐ νοητά,
οὐδὲ νοεῖ ὁ νοῦς τὰ ἐκτὸς μὴ μετ᾽ αἰσθήσεως ὄντα.
ἅμα δ᾽ εἰ ταῦτ᾽ ἔχει οὕτως, ἔοικε μαρτυρεῖν τοῖς τὰ

and smell enters at the same time as a vapour inhaled with the breath, so that it must go to the respiratory region. So it is obvious that smell *qua* smell does not contribute to nutrition ; but it is equally clear, both from our own sensation and from what we have already said, that it does contribute to health ; so that smell is in relation to health in general what flavour is in nutrition relatively to what is nourished. Let this then suffice for an account of the several sense organs.

VI. A difficulty might arise as to whether, if every body is susceptible of infinite division, its sensible qualities, *e.g.*, colour, flavour, smell, weight, sound, cold and heat, lightness, hardness and softness, are also so susceptible, or is this impossible ? For each of these produces sensation ; in fact they are all called " sensible " from their capacity to excite sensation. Then on the above assumption sensation must be capable of infinite division, and every magnitude must be perceptible ; for it is impossible to see a white thing which is not a magnitude. Otherwise it would be possible for a body to exist which had neither colour nor weight, nor any other attribute ; so that it would not be perceptible at all, for these are the perceptible attributes. In this case every perceptible body will consist of imperceptible parts. But its parts must be perceptible ; for they cannot consist of mathematical abstractions. Again, by what faculty should we discern or cognize these ? By the mind ? But they are not apprehended by the mind, nor does the mind cognize any external objects which are unconnected with sensation. At the same time, if our supposition is true, it would seem to support the

Can sensation be infinitely divided ?

If so, what is the meaning of an imperceptible part ?

261

445 b

ἄτομα ποιοῦσι μεγέθη· οὕτω γὰρ ἂν λύοιτο ὁ λόγος.
20 ἀλλ' ἀδύνατα· εἴρηται δὲ περὶ αὐτῶν ἐν τοῖς λόγοις
τοῖς περὶ κινήσεως.ᵃ περὶ δὲ τῆς λύσεως αὐτῶν
ἅμα δῆλον ἔσται καὶ διὰ τί πεπέρανται τὰ εἴδη καὶ
χρώματος καὶ χυμοῦ καὶ φθόγγων καὶ τῶν ἄλλων
αἰσθητῶν. ὧν μὲν γάρ ἐστιν ἔσχατα, ἀνάγκη
πεπεράνθαι τὰ ἐντός· τὰ δ' ἐναντία ἔσχατα. πᾶν
25 δὲ τὸ αἰσθητὸν ἔχει ἐναντίωσιν, οἷον ἐν χρώματι τὸ
λευκὸν καὶ τὸ μέλαν, ἐν χυμῷ γλυκὺ καὶ πικρόν·
καὶ ἐν τοῖς ἄλλοις δὴ πᾶσίν ἐστιν ἔσχατα τὰ ἐναντία. τὸ μὲν οὖν συνεχὲς εἰς ἄπειρα τέμνεται
ἄνισα, εἰς δ' ἴσα πεπερασμένα· τὸ δὲ μὴ καθ'
αὑτὸ συνεχὲς εἰς πεπερασμένα εἴδη. ἐπεὶ οὖν τὰ
30 μὲν πάθη ὡς εἴδη λεκτέον, ὑπάρχει δὲ συνέχεια
ἀεὶ ἐν τούτοις, ληπτέον ὅτι τὸ δυνάμει καὶ τὸ
ἐνεργείᾳ ἕτερον· καὶ διὰ τοῦτο τὸ μυριοστημόριον
446 a λανθάνει τῆς κέγχρου ὁρωμένης, καίτοι ἡ ὄψις
ἐπελήλυθεν, καὶ ὁ ἐν τῇ διέσει φθόγγος λανθάνει,
καίτοι συνεχοῦς ὄντος ἀκούει τοῦ μέλους παντός.
τὸ δὲ διάστημα τὸ τοῦ μεταξὺ πρὸς τοὺς ἐσχάτους
λανθάνει. ὁμοίως δὲ καὶ ἐν τοῖς ἄλλοις αἰσθητοῖς
5 τὰ μικρὰ πάμπαν· δυνάμει γὰρ ὁρατά, ἐνεργείᾳ δ'
οὔ, ὅταν μὴ χωρὶς ᾖ¹· καὶ γὰρ ἐνυπάρχει δυνάμει
ἡ ποδιαία τῇ δίποδι, ἐνεργείᾳ δ' ἤδη διαιρεθείσῃ.²
χωριζόμεναι δ' αἱ τηλικαῦται ὑπεροχαὶ εὐλόγως
μὲν ἂν καὶ διαλύοιντο εἰς τὰ περιέχοντα, ὥσπερ
10 καὶ ἀκαριαῖος χυμὸς εἰς τὴν θάλατταν ἐκχυθείς.
οὐ μὴν ἀλλ' ἐπειδὴ οὐδ' ἡ τῆς αἰσθήσεως ὑπεροχὴ

---

¹ χωρισθῇ LSU, Bekker.
² Bywater : διαιρεθεῖσα codd.

ᵃ *Physics* vi. 1, 2.

atomic theory; for our difficulty might be solved on those lines. But the theory is impossible; it has been discussed in our treatise on movement.[a] The solution of these questions will make it clear why the kinds of colour, flavour, sounds, and other sensible objects are limited; for where there are extremes, the intermediate stages must be limited; and the contraries are extremes. Every sensible object involves contrariety, for instance in colour white and black, in flavour sweet and bitter; and in all other sensibles the contraries are extremes. Now what is continuous can be divided into an infinite number of unequal parts, but into a finite number of equal parts; while that which is not in itself continuous can be divided into a finite number of species. Since then the attributes in question may be spoken of as species, and continuity is always inherent in them, we must realize the difference between the potential and the actual; it is because of this that, when a grain of millet is looked at, the ten-thousandth part of it cannot be seen, although vision has covered it all, and the sound of quarter-tone escapes us, although one can hear the whole continuous scale; but the interval between the extremes escapes us. The same thing is true of all very small quantities in the other sensible objects; potentially they are visible but not actually, unless they are isolated from the whole. (Potentially the one-foot length exists in the two-foot length, but actually only after its division.) Such small increments, when isolated, might well be merged in their environment, like a flavoured droplet when poured into the sea. But it is important to realize that as the increment of sense is not perceptible by itself,

Very small parts must be potentially if not actually perceptible.

446 a

καθ' αὑτὴν αἰσθητὴ οὐδὲ χωριστή (δυνάμει γὰρ
ἐνυπάρχει ἐν τῇ ἀκριβεστέρᾳ ἡ ὑπεροχή), οὐδὲ τὸ
τηλικοῦτον αἰσθητὸν χωριστὸν ἔσται ἐνεργείᾳ
αἰσθέσθαι, ἀλλ' ὅμως ἔσται αἰσθητόν· δυνάμει τε
15 γάρ ἐστιν ἤδη, καὶ ἐνεργείᾳ ἔσται προσγενόμενον.
ὅτι μὲν οὖν ἔνια μεγέθη καὶ πάθη λανθάνει, καὶ
διὰ τίν' αἰτίαν, καὶ πῶς αἰσθητὰ καὶ πῶς οὔ,
εἴρηται. ὅταν δὲ δὴ ἐνυπάρχοντα οὕτω ἤδη πρὸς
αὑτὰ¹ ᾖ ὥστε καὶ ἐνεργείᾳ αἰσθητὰ εἶναι, καὶ μὴ
μόνον ἐν τῷ ὅλῳ ἀλλὰ καὶ χωρὶς ᾖ,² πεπερασμένα
20 ἀνάγκη εἶναι τὸν ἀριθμὸν καὶ χρώματα καὶ χυμοὺς
καὶ φθόγγους.

Ἀπορήσειε δ' ἄν τις, ἆρ' ἀφικνοῦνται ἢ τὰ
αἰσθητὰ ἢ αἱ κινήσεις αἱ ἀπὸ τῶν αἰσθητῶν, ὁποτέρως ποτὲ γίνεται ἡ αἴσθησις, ὅταν ἐνεργῶσιν, εἰς
τὸ μέσον πρῶτον, οἷον ἥ τε ὀσμὴ φαίνεται ποιοῦσα
καὶ ὁ ψόφος· πρότερον γὰρ ὁ ἐγγὺς αἰσθάνεται τῆς
25 ὀσμῆς, καὶ ὁ ψόφος ὕστερον ἀφικνεῖται τῆς πληγῆς.
ἆρ' οὖν οὕτω καὶ τὸ ὁρώμενον καὶ τὸ φῶς; καθάπερ καὶ Ἐμπεδοκλῆς φησιν ἀφικνεῖσθαι πρότερον
τὸ ἀπὸ τοῦ ἡλίου φῶς εἰς τὸ μεταξὺ πρὶν πρὸς
τὴν ὄψιν ἢ ἐπὶ τὴν γῆν. δόξειε δ' ἂν εὐλόγως
τοῦτο συμβαίνειν· τὸ γὰρ κινούμενον κινεῖταί ποθέν
30 ποι, ὥστ' ἀνάγκη εἶναί τινα καὶ χρόνον ἐν ᾧ κινεῖται
446 b ἐκ θατέρου πρὸς θάτερον· ὁ δὲ χρόνος πᾶς δι-
αιρετός, ὥστε ἦν ὅτε οὔπω ἑωρᾶτο ἀλλ' ἔτ' ἐφέρετο
ἡ ἀκτὶς ἐν τῷ μεταξύ. καὶ εἰ ἅπαν ἅμα ἀκούει

¹ ἤδη πρὸς αὑτὰ EMY : πως ἄττα Bekker.

nor isolable (for it exists only potentially in a more distinctly perceptible whole), so neither will it be possible actually to perceive its equally small object when separated from the whole, yet it will be perceptible ; for it is so already potentially, and will become so actually by aggregation. We have shown, then, that some magnitudes and some qualities escape us ; we have explained the reason for this, and the sense in which they are perceptible, and the sense in which they are not. Thus when they are so inter-related in an aggregate as to be perceptible actually, not merely in the whole, but even in isolation, it follows that their colours, flavours and sounds must be limited in number.

There is a further question to be considered : do these sensible objects, or the movements arising from them (in whichever way sensation arises), when they are actualized, arrive first to a mid-point, as smell and sound seem to do ? For the man next to the smell perceives it sooner, and the sound of the blow reaches us after the blow has been struck. Is the same thing true of the object seen and light ? Empedocles for instance states that the light from the sun reaches an intermediate point, before it reaches the vision, or the earth. This sounds a probable account of what happens ; for that which is moved is moved from some source, and in some direction, so that some interval of time must elapse in which the movement takes place from the one point to the other. But all time is divisible, so that an interval existed during which the light was not yet seen, but the ray was still moving in the intermediate space. And even supposing that " hearing " and " having heard," " per-

Does a perceptible object become first potentially and then actually perceptible?

---

² virgulam hic habuit Alexander, post πεπερασμένα codd.

eiving " and " having perceived " are simultaneous,
and involve no process of becoming, but exist without
any such process, nevertheless the interval still
exists, just as sound has not yet reached our ear,
although the blow causing it has been struck. (The
alteration in the letters of a word as heard proves that
some movement takes place in the intervening space[a];
for the reason why the listener does not hear what is
said, is that air in moving towards him has undergone
some change.) Is the same thing true of colour and
light ? For it is not true that the one sees, and the
other is seen, just because the two are in a certain
relation, e.g., that of equality ; for in that case there
would be no need for each of them to be in some
particular place ; for when things are equal it makes
no difference whether they are near to or far from
one another. Now it is reasonable to suppose that
the same thing happens with sound and smell ; for
just as their media, air and water, are continuous,
so are they, and yet the movement of both is divided
into parts. And so there is a sense in which the first
and last hear and smell the same thing, and also a
sense in which they do not. But some find a further
difficulty in this ; for they say that it is impossible
for one person to hear, or see, or smell the same thing
as another ; for they argue that it is impossible for
several separate persons to hear or smell the same
thing ; for in that case a single thing would be
separate from itself. The original cause of the movement, e.g., the bell, or the incense, or the fire, which
all perceive, is the same and numerically one, but
the subjective perceptions, though specifically the
same, are numerically different, for many see, smell,

In what
sense can
two persons
perceive the
same thing ?

446 b

25 καὶ ἀκούουσιν. ἔστι δ᾽ οὔτε σώματα ταῦτα, ἀλλὰ
πάθος καὶ κίνησίς τις (οὐ γὰρ ἂν τοῦτο συνέβαινεν),
οὔτ᾽ ἄνευ σώματος.

Περὶ δὲ τοῦ φωτὸς ἄλλος λόγος· τῷ εἶναι γάρ
τι φῶς ἐστίν, ἀλλ᾽ οὐ κίνησίς τις. ὅλως δὲ οὐδὲ
ὁμοίως ἐπί τε ἀλλοιώσεως ἔχει καὶ φορᾶς· αἱ μὲν
30 γὰρ φοραὶ εὐλόγως εἰς τὸ μεταξὺ πρῶτον ἀφ-
447 a ικνοῦνται (δοκεῖ δ᾽ ὁ ψόφος εἶναι φερομένου τινὸς
κίνησις), ὅσα δ᾽ ἀλλοιοῦται, οὐκέτι ὁμοίως· ἐνδέχεται γὰρ ἀθρόον ἀλλοιοῦσθαι, καὶ μὴ τὸ ἥμισυ
πρότερον, οἷον τὸ ὕδωρ ἅμα πᾶν πήγνυσθαι. οὐ
5 μὴν ἀλλ᾽ ἂν ᾖ πολὺ τὸ θερμαινόμενον ἢ πηγνύ-
μενον, τὸ ἐχόμενον ὑπὸ τοῦ ἐχομένου πάσχει, τὸ
δὲ πρῶτον ὑπ᾽ αὐτοῦ τοῦ ἀλλοιοῦντος μεταβάλλει,
καὶ οὐκ ἀνάγκη ἅμα ἀλλοιοῦσθαι καὶ ἀθρόον. ἦν
δ᾽ ἂν καὶ τὸ γενέσθαι ὥσπερ ἡ ὀσμή, εἰ ἐν ὑγρῷ
ἦμεν καὶ πορρωτέρω ἔτι πρὶν θιγεῖν αὐτοῦ ᾐσθανόμεθα. εὐλόγως δ᾽ ἂν ἐστὶ μεταξὺ τοῦ αἰσθητηρίου,
10 οὐχ ἅμα πάντα πάσχει, πλὴν ἐπὶ τοῦ φωτὸς διὰ
τὸ εἰρημένον. διὰ τὸ αὐτὸ δὲ καὶ ἐπὶ τοῦ ὁρᾶν·
τὸ γὰρ φῶς ποιεῖ τὸ ὁρᾶν.

VII. Ἔστι δέ τις ἀπορία καὶ ἄλλη τοιάδε περὶ
αἰσθήσεως, πότερον ἐνδέχεται δυεῖν ἅμα αἰσθάνεσθαι ἐν τῷ αὐτῷ καὶ ἀτόμῳ χρόνῳ, ἢ οὔ, εἰ δὴ
15 ἀεὶ ἡ μείζων κίνησις τὴν ἐλάττω ἐκκρούει· διὸ
ἐπιφερομένων ἐπὶ τὰ ὄμματα οὐκ αἰσθάνονται, ἐὰν
τύχωσι σφόδρα τι ἐννοοῦντες ἢ φοβούμενοι ἢ
ἀκούοντες πολὺν ψόφον. τοῦτο δὴ ὑποκείσθω,
καὶ ὅτι ἑκάστου μᾶλλον ἔστιν αἰσθάνεσθαι ἁπλοῦ

¹ δυεῖν B.

or hear them at the same time. These are not bodies, but are an affection or movement of some kind (for otherwise the effect would not be what it is), though they imply body.

With light it is different; for light is due to the existence of something, but is not a movement. Light is a special case. Generally speaking, change of state and travel in space are different; for spatial movements naturally first reach the intervening space (and sound is held to be the movement of something travelling), but with things which change their state the position is not the same; for it is possible that such change of state should occur in a thing all at once, and not in half first; for instance, water may all freeze at one time. It must, however, be admitted that when a large body grows hot or freezes, each part is affected by the next, while the first part owes its change to the actual cause of the change; the whole need not change all together at the same time. Tasting would be like smelling, if we lived in water, and perceived at a distance before making contact. Naturally where there is a medium between sense-organ and object, the parts are not all affected simultaneously, except in the case of light, for the reason given, and of vision too for the same reason; for light causes vision.

VII. There is a further question about sensation, whether it is possible to perceive two things in one and the same indivisible time or not, if we assume that the stronger always overrides the weaker stimulus; which is why we do not see things presented to our eyes, if we happen to be engrossed in thought, or in a state of fear, or listening to a loud noise. Let this be granted, and also that it is more possible to perceive each individual thing when Can two things be perceived simultaneously?

447 a

ὄντος ἢ κεκραμένου, οἷον οἴνου ἀκράτου ἢ κεκρα-
20 μένου, καὶ μέλιτος, καὶ χρόας, καὶ τῆς νήτης μόνης
ἢ ἐν τῷ διὰ πασῶν, διὰ τὸ ἀφανίζειν ἄλληλα.
τοῦτο δὲ ποιεῖ ἐξ ὧν ἕν τι γίνεται. εἰ δὴ ἡ μείζων
τὴν ἐλάττω κίνησιν ἐκκρούει, ἀνάγκη, ἂν ἅμα
ὦσι, καὶ αὐτὴν ἧττον αἰσθητὴν εἶναι ἢ εἰ μόνη
ἦν· ἀφῄρηται[1] γάρ τι ἡ ἐλάττων μιγνυμένη, εἴπερ
25 ἅπαντα τὰ ἁπλᾶ μᾶλλον αἰσθητά ἐστιν. ἐὰν ἄρα
ἴσαι ὦσιν ἕτεραι οὖσαι, οὐδετέρας ἔσται αἴσθησις·
ἀφανιεῖ γὰρ ἡ ἑτέρα ὁμοίως τὴν ἑτέραν. ἁπλῆς
δ' οὐκ ἔστιν αἰσθάνεσθαι. ὥστε ἢ οὐδεμία ἔσται
αἴσθησις ἢ ἄλλη ἐξ ἀμφοῖν. ὅπερ καὶ γίνεσθαι
δοκεῖ ἐκ τῶν κεραννυμένων ἐν ᾧ ἂν μιχθῶσιν·
ἐπεὶ οὖν ἐκ μὲν ἐνίων γίνεταί τι, ἐκ δ' ἐνίων οὐ
30 γίνεται, τοιαῦτα δὲ τὰ ὑφ' ἑτέραν αἴσθησιν
447 b (μίγνυνται γὰρ ὧν τὰ ἔσχατα ἐναντία[2]· οὐκ ἔστι δ'
ἐκ λευκοῦ καὶ ὀξέος ἓν γενέσθαι ἀλλ' ἢ κατὰ
συμβεβηκός, ἀλλ' οὐχ ὡς ἐξ ὀξέος καὶ βαρέος
συμφωνία)· οὐκ ἄρα οὐδ' αἰσθάνεσθαι ἐνδέχεται
αὐτῶν ἅμα. ἴσαι μὲν γὰρ οὖσαι αἱ κινήσεις ἀφα-
5 νιοῦσιν ἀλλήλας, ἐπεὶ μία οὐ γίνεται ἐξ αὐτῶν·
ἐὰν δ' ἄνισοι, ἡ κρείττων αἴσθησιν ἐμποιήσει.[3] ἔτι[4]
μᾶλλον ἅμα δυεῖν αἴσθοιτ' ἂν ἡ ψυχὴ τῇ μιᾷ
αἰσθήσει ὧν μία αἴσθησις, οἷον ὀξέος καὶ βαρέος

---

[1] ἀφαιρεῖται LSUP.
[2] ἐναντία] ἐν EMY, Bekker.
[3] punctum pro commate ponendum cens. Thurot.
[4] ἔτι LSU, Alexander : ἐπεί.

simple, than when mixed with another. For instance it is easier to taste pure wine than wine mixed with water, and so also with honey, or with colour ; and the tonic is easier to hear by itself than when sounded with the octave, because they tend to obscure one another. This always happens with individual things out of which one whole is formed. If then the stronger overrides the weaker stimulus, it follows that, if they occur together, the stronger must be less perceptible than if it occurred alone ; for the weaker by mixing with it has subtracted something, since all simple things are more perceptible. If then the stimuli are equal and different, there can be no perception of either ; for each will similarly efface the other. But it is impossible in this case to perceive either in its simple form. Hence there will be either no perception at all, or a complex perception different from that of either. The latter seems actually to result when two things are blended, in whatever whole they are combined. Some stimuli, then, effect a complex sensation, and others do not ; of the latter class are such as fall under different senses (for mixture can only take place between things whose extremes are contraries ; so no single whole can be formed from white and high, except incidentally, not as a concord is formed from high and low pitch) ; and so it is impossible to perceive them both together. The stimuli, if equal, will cancel each other, since no one impulse arises from them ; but if they are not equal the greater will induce sensation. Again, the soul would be more likely to perceive two things at the same time with one sense, if both were objects of the same sense, such as high and low ; for the simul-

μᾶλλον γὰρ ἅμα ἡ κίνησις τῆς μιᾶς ταύτης ἢ
τοῖν δυοῖν, οἷον ὄψεως καὶ ἀκοῆς.

10 Τῇ μιᾷ δὲ ἅμα δυοῖν οὐκ ἔστιν αἰσθάνεσθαι ἂν
μὴ μιχθῇ· τὸ γὰρ μῖγμα ἓν βούλεται εἶναι, τοῦ
δ' ἑνὸς μία αἴσθησις, ἡ δὲ μία ἅμα αὐτή. ὥστ'
ἐξ ἀνάγκης τῶν μεμιγμένων ἅμα αἰσθάνεται, ὅτι
μιᾷ αἰσθήσει κατ' ἐνέργειαν αἰσθάνεται· ἑνὸς μὲν
γὰρ ἀριθμῷ ἡ κατ' ἐνέργειαν μία, εἴδει δὲ ἡ κατὰ
15 δύναμιν μία. καὶ εἰ μία τοίνυν ἡ αἴσθησις ἡ κατ'
ἐνέργειαν, ἓν ἐκεῖνα ἐρεῖ. μεμῖχθαι ἄρα ἀνάγκη
αὐτά. ὅταν ἄρα μὴ ᾖ μεμιγμένα, δύο ἔσονται
αἰσθήσεις αἱ κατ' ἐνέργειαν. ἀλλὰ κατὰ μίαν
δύναμιν καὶ ἄτομον χρόνον μίαν ἀνάγκη εἶναι τὴν
ἐνέργειαν· μιᾶς γὰρ εἰσάπαξ μία κίνησις καὶ
20 χρῆσις, μία δὲ ἡ δύναμις. οὐκ ἄρα ἐνδέχεται δυεῖν
ἅμα αἰσθάνεσθαι τῇ μιᾷ αἰσθήσει. ἀλλὰ μὴν εἰ
τὰ ὑπὸ τὴν αὐτὴν αἴσθησιν ἅμα ἀδύνατον, ἐὰν ᾖ
δύο, δῆλον ὅτι ἧττον ἔτι τὰ κατὰ δύο αἰσθήσεις
ἐνδέχεται ἅμα αἰσθάνεσθαι, οἷον λευκὸν καὶ γλυκύ.
φαίνεται γὰρ τὸ μὲν τῷ ἀριθμῷ ἓν ἡ ψυχὴ οὐδενὶ
25 ἑτέρῳ λέγειν ἀλλ' ἢ τῷ ἅμα, τὸ δὲ τῷ εἴδει ἓν
τῇ κρινούσῃ αἰσθήσει καὶ τῷ τρόπῳ. λέγω δὲ
τοῦτο, ὅτι ἴσως τὸ λευκὸν καὶ τὸ μέλαν, ἕτερον
τῷ εἴδει ὄν, ἡ αὐτὴ κρίνει, καὶ τὸ γλυκὺ καὶ τὸ

ᵃ A. seems to mean this. If you look at (e.g.) two white
objects and are conscious of them as two white objects you
are receiving not one impression, but two, and therefore
cannot receive both with one sense faculty simultaneously.
If the two white objects give only a general impression of
white (i.e., specifically or as a class), then your perception may
be one, but it is only potential and not actualized.

taneous movement of this one sensation is greater than it would be in the case of two, such as vision and hearing.

But it is not possible to perceive two objects with one sense, unless they are combined ; for the combination tends to be one object, and the perception of one object is one, and as one is coinstantaneous with itself. Hence the perception of combined objects is necessarily simultaneous, because we perceive them with a perception actually one ; because the perception of that which is numerically one is actually one, while that of an object specifically one is only potentially one.[a] Now if the actual perception is one, it will assert its objects to be one. Hence they must be combined. Thus when they are not combined, the actual sensations will be two. But in the case of a single faculty in an indivisible moment of time, the activity must be single ; for the stimulation and exercise of one faculty at a single instant must be one, and the faculty in this case is only one. Thus it is not possible to perceive two objects at once with a single sense. Moreover if it is impossible to perceive two things at the same moment when they fall under the same sense, clearly it is still more impossible to perceive simultaneously two things which fall under two senses, such as white and sweet. For apparently the soul asserts numerical unity solely in virtue of simultaneous perception, while it asserts specific unity in virtue of the discriminating sense and its method. I mean that presumably the same sense discerns white and black, which are specifically different ; and a sense self-identical but different

447 b

πικρόν, ἡ αὐτὴ μὲν ἑαυτῇ, ἐκείνης δ' ἄλλη, ἀλλ'
ἑτέρως ἑκάτερον τῶν ἐναντίων, ὡς δ' αὔτως ἑαυταῖς

30 τὰ σύστοιχα, οἷον ὡς ἡ γεῦσις τὸ γλυκύ, οὕτως

448 a ἡ ὄψις τὸ λευκόν· ὡς δ' αὕτη τὸ μέλαν, οὕτως
ἐκείνη τὸ πικρόν.

Ἔτι εἰ αἱ τῶν ἐναντίων κινήσεις ἐναντίαι, ἅμα
δὲ τὰ ἐναντία ἐν τῷ αὐτῷ καὶ ἀτόμῳ οὐκ ἐνδέχεται
ὑπάρχειν, ὑπὸ δὲ τὴν αἴσθησιν τὴν μίαν ἐναν-
5 τία ἐστίν, οἷον γλυκὺ πικρῷ, οὐκ¹ ἂν ἐνδέχοιτο
αἰσθάνεσθαι ἅμα. ὁμοίως δὲ δῆλον ὅτι οὐδὲ τὰ
μὴ ἐναντία· τὰ μὲν γὰρ τοῦ λευκοῦ τὰ δὲ τοῦ
μέλανός ἐστιν, καὶ ἐν τοῖς ἄλλοις ὁμοίως, οἷον
τῶν χυμῶν οἱ μὲν τοῦ γλυκέος οἱ δὲ τοῦ πικροῦ.
οὐδὲ τὰ μεμιγμένα ἅμα (λόγοι γάρ εἰσιν ἀντι-
10 κειμένων, οἷον τὸ διὰ πασῶν καὶ τὸ διὰ πέντε), ἂν
μὴ ὡς ἓν αἰσθάνηται. οὕτως δ' εἷς λόγος ὁ τῶν
ἄκρων γίνεται, ἄλλως δ' οὔ· ἔσται γὰρ ἅμα ὁ
μὲν πολλοῦ πρὸς ὀλίγον ἢ περιττοῦ πρὸς ἄρτιον,
ὁ δ' ὀλίγου πρὸς πολὺ ἢ ἀρτίου πρὸς περιττόν.
εἰ οὖν πλεῖον ἔτι ἀπέχει ἀλλήλων καὶ διαφέρει τὰ
15 συστοίχως μὲν λεγόμενα ἐν ἄλλῳ δὲ γένει τῶν ἐν
τῷ αὐτῷ γένει λεγομένων² (οἷον τὸ γλυκὺ καὶ τὸ
λευκὸν καλῶ σύστοιχα, γένει δ' ἕτερα), τὸ γλυκὺ
δὲ τοῦ μέλανος πλεῖον ἔτι τῷ εἴδει διαφέρει ἢ τοῦ
λευκοῦ,³ ἔτι ἂν ἧττον ἅμα ἐνδέχοιτο αὐτὰ αἰσθά-

---

¹ τούτων οὐκ LSU, Bekker.
² λεγομένων] λέγω δ' LSUP, Bekker.
³ τοῦ λευκοῦ LSU, Alexander : τὸ λευκόν.

from the former discerns sweet and bitter; but while these senses differ in their method of perceiving their respective contraries, they are similar in their method of perceiving corresponding qualities; for instance vision apprehends white in the same way as taste apprehends sweet; and as the former apprehends black, so the latter apprehends bitter.

Moreover, if the stimuli aroused by contraries are also contrary, and if contraries cannot both reside at once in what is the same and indivisible, and if contraries, *e.g.*, sweet and bitter, fall under one sense, it cannot be possible to perceive these contraries simultaneously. In the same way clearly things which are not contraries cannot be simultaneously perceived; for some colours are akin to white and some to black, and similarly with other sensibles; *e.g.*, some flavours are akin to sweet and some to bitter. Nor can combined objects be simultaneously perceived (for they are ratios of opposites, *e.g.*, the chord of the octave or the fifth), unless they are perceived as one. For in this way the ratio of the extremes becomes one, but in no other way; for we shall have the ratio of many to few, or odd to even, and also that of few to many or even to odd. If then the objects which, though in different genera, are called corresponding (*e.g.*, I call sweet and white corresponding, though different in genus) are remoter and more different from one another than members of the same genus, and in form sweet differs from black even more than from white, it would be even less possible to perceive these ⟨*viz.*, sweet and black⟩ simultane-

275

448 a

νεσθαι ἢ τὰ τῷ γένει ταὐτά. ὥστ᾿ εἰ μὴ ταῦτα,
οὐδ᾿ ἐκεῖνα.

20 Ὃ δὲ λέγουσί τινες τῶν περὶ τὰς συμφωνίας,
ὅτι οὐχ ἅμα μὲν ἀφικνοῦνται οἱ ψόφοι, φαίνονται
δέ, καὶ λανθάνει, ὅταν ὁ χρόνος ᾖ ἀναίσθητος,
πότερον ὀρθῶς λέγεται ἢ οὔ; τάχα γὰρ ἂν φαίη
τις καὶ νῦν παρὰ τοῦτο δοκεῖν ἅμα ὁρᾶν καὶ
ἀκούειν, ὅτι οἱ μεταξὺ χρόνοι λανθάνουσιν. ἢ
25 τοῦτ᾿ οὐκ ἀληθές, οὐδ᾿ ἐνδέχεται χρόνον εἶναι
ἀναίσθητον οὐδένα οὐδὲ λανθάνειν, ἀλλὰ παντὸς
ἐνδέχεται αἰσθάνεσθαι. εἰ γὰρ ὅτε αὐτὸς αὑτοῦ
τις αἰσθάνεται ἢ ἄλλου ἐν συνεχεῖ χρόνῳ, μὴ
ἐνδέχεται τότε λανθάνειν ὅτι ἐστίν, ἔστι δέ τις ἐν
τῷ συνεχεῖ καὶ τοσοῦτος ὅσος ὅλως ἀναίσθητός
ἐστι, δῆλον ὅτι τότε λανθάνοι ἂν εἰ ἔστιν αὐτὸς
30 αὑτόν, καὶ εἰ ὁρᾷ καὶ[1] αἰσθάνεται· καὶ εἰ αἰσθάνεται
448 b ἔτι, οὐκ ἂν εἴη οὔτε χρόνος οὔτε πρᾶγμα οὐδὲν
ὃ αἰσθάνεται ἢ ἐν ᾧ, εἰ μὴ οὕτως, ὅτι ἐν τούτου
τινὶ ἢ ὅτι τούτου τι ὁρᾷ, εἴπερ ἔστι τι μέγεθος
καὶ χρόνου καὶ πράγματος ἀναίσθητον ὅλως διὰ
μικρότητα· εἰ γὰρ τὴν ὅλην ὁρᾷ καὶ αἰσθάνεται τὸν
5 αὐτὸν συνεχῶς χρόνον οὕτω τῷ ἐν τούτου[2] τινί,
ἀφῃρήσθω ἢ [τὸ][3] ΓΒ, ἐν ᾗ οὐκ ᾐσθάνετο. οὐκοῦν
ἐν ταύτης τινὶ ἢ ταύτης τι, ὥσπερ τὴν γῆν ὁρᾷ
ὅλην, ὅτι τοδὶ αὐτῆς, καὶ ἐν τῷ ἐνιαυτῷ βαδίζει,
ὅτι ἐν τῷδε τῷ μέρει αὐτοῦ. ἀλλὰ μὴν ἐν τῷ
10 ΒΓ οὐδὲν αἰσθάνεται. τῷ ἄρα ἐν τούτου τινὶ τοῦ

---

[1] καί] καὶ οὐκ LSU, Bekker.
[2] οὕτω τῷ ἐν τούτου Alexander, Beare : οὐ τῶν νῦν τούτων.
[3] Biehl.

ously than things belonging to the same class ⟨*e.g.*,
white and black⟩. So that if the latter is impossible,
the former is certainly so.

Now for the point which some writers about concords raise ; they say that sounds do not arrive
simultaneously, but only seem to do so, and deceive
us, when the interval of time is imperceptible. Are
they right or not ? If so, one might say at once that
this is why we seem to see and hear simultaneously,
because the intervening time is not noticeable. Probably this is untrue, and it cannot be that any time
is imperceptible and unnoticed ; it is possible to perceive every moment of it. For if, when a man perceives himself or anything else in continuous time,
it is impossible for him to be unaware of his existence,
and if again in continuous time there can be a time
so short as to be quite imperceptible, it is clear that,
during that time, he would be unaware of his existence, and of the fact of his seeing or perceiving ; and,
if he still perceives, there could be no object which,
or time in which, he perceives, except in the sense
that he sees in some part of the time, or some part
of the object ; if, that is, there exists any magnitude,
whether time or object, which is quite imperceptible
owing to its smallness ; for if he sees the whole line
and perceives it for a time continuously the same, in
the sense that he does so in some part of this time ;
from the whole line AB let a part CB be cut off, being
a time in which he perceives nothing. Then he perceives in a part of the time, or a part of the line, in
the sense that one sees the whole world by seeing
part of it, or walks in a year by walking in a certain
part of it. But by our assumption during the part
BC he sees nothing. In that case he is said to see

ΑΒ αἰσθάνεσθαι λέγεται τοῦ ὅλου αἰσθάνεσθαι καὶ
τὴν ὅλην. ὁ δ' αὐτὸς λόγος καὶ ἐπὶ τῆς ΑΓ· ἀεὶ
γὰρ ἐν τινὶ καὶ τινός, ὅλου δ' οὐκ ἔστιν αἰσθάνεσθαι. ἅπαντα μὲν οὖν αἰσθητά ἐστιν, ἀλλ' οὐ
15 φαίνεται ὅσα ἐστίν· τοῦ γὰρ ἡλίου τὸ μέγεθος ὁρᾷ
καὶ τὸ τετράπηχυ πόρρωθεν, ἀλλ' οὐ φαίνεται ὅσον,
ἀλλ' ἐνίοτε ἀδιαίρετον, ὁρᾷ δ' οὐκ ἀδιαίρετον. ἡ
δ' αἰτία εἴρηται ἐν τοῖς ἔμπροσθεν περὶ τούτου. ὅτι
μὲν οὖν οὐθείς ἐστι χρόνος ἀναίσθητος, ἐκ τούτων
φανερόν.

Περὶ δὲ τῆς πρότερον λεχθείσης ἀπορίας σκεπτέον, πότερον ἐνδέχεται ἅμα πλειόνων αἰσθάνεσθαι
ἢ οὐκ ἐνδέχεται. τὸ δ' ἅμα λέγω ἐν ἑνὶ καὶ
20 ἀτόμῳ χρόνῳ πρὸς ἄλληλα. πρῶτον μὲν οὖν ἆρ'
ὧδ' ἐνδέχεται, ἅμα μέν, ἑτέρῳ δὲ τῆς ψυχῆς
αἰσθάνεσθαι, καὶ οὐ τῷ ἀτόμῳ, οὕτω δ' ἀτόμῳ
ὡς παντὶ ὄντι συνεχεῖ; ἢ ὅτι πρῶτον μὲν τὰ
κατὰ τὴν μίαν αἴσθησιν, οἷον λέγω ὄψιν, εἰ ἔσται
ἄλλῳ αἰσθανομένη ἄλλου καὶ ἄλλου χρώματος,
25 πλείω τε μέρη ἕξει εἴδει ταὐτά; καὶ γὰρ ἃ
αἰσθάνεται, ἐν τῷ αὐτῷ γένει ἐστίν. εἰ δὲ ὅτι
ὡς δύο ὄμματα φαίη τις, οὐδὲν κωλύει οὕτω καὶ
ἐν τῇ ψυχῇ, ὅτι ἴσως ἐκ μὲν τούτων ἕν τι γίνεται
καὶ μία ἡ ἐνέργεια αὐτῶν· εἰ δέ, ᾗ[1] μὲν ἓν τὸ
ἐξ ἀμφοῖν, ἓν καὶ[2] τὸ αἰσθανόμενον ἔσται, εἰ δὲ

---

[1] εἰ δὲ ᾗ Ε, εἰ δὲ ἡ MY : ἐκεῖ δὲ εἰ vulgo.
[2] ἓν καὶ ΕΜΥ : ἐκεῖνο.

---

[a] *Cf.* 445 b 11.

the whole line AB in the whole time, because he sees a part of it in a part of the time. The same argument applies to the part AC ; for it will be found that one always perceives a part in a part of the time, and that one can never perceive the whole. Therefore all magnitudes can be perceived, but their size is not apparent to our senses ; for instance one sees a magnitude such as that of the sun or a six-foot rod at a distance, but its actual size is not apparent ; indeed it sometimes seems indivisible, but what we see is not really indivisible. The reason for this has been given in our previous account of the subject.[a] It is clear then from this that there is no such thing as imperceptible time.

But in regard to the difficulty mentioned before we must consider whether it is possible or impossible to perceive more than one thing at once. By " at once " I mean in a time which is one and indivisible for the several things relative to one another. In the first place is it possible to perceive two things at once, but with a different part of the soul, and not with the indivisible part, but with one indivisible in the sense of being all continuous ? Or does this imply in the first place that in respect of a single sense, e.g., sight, if the soul is to perceive one colour with one part and another with another, it will possess a number of parts the same in species ? for the objects which it perceives are in the same class. If one were to urge, that just as there are two eyes, so there is nothing to prevent there being two identical parts in the soul, we should answer that probably one unit is made up of the two eyes, and in actual operation they are one ; but if so, in so far as they constitute one organ, that which perceives will also be one ;

A further discussion of simultaneous perception.

Has the soul separate parts ?

448 b
30 χωρίς, οὐχ ὁμοίως ἕξει. ἔτι αἰσθήσεις αἱ αὐταὶ
449 a πλείους ἔσονται, ὥσπερ εἴ τις ἐπιστήμας διαφόρους
φαίη· οὔτε γὰρ ἡ ἐνέργεια ἄνευ τῆς καθ᾽ αὑτὴν
ἔσται δυνάμεως, οὔτ᾽ ἄνευ ταύτης ἔσται αἴσθησις.

Εἰ δὲ τούτων ἐν ἑνὶ καὶ ἀτόμῳ ⟨μὴ⟩[1] αἰσθάνεται,
δῆλον ὅτι καὶ τῶν ἄλλων· μᾶλλον γὰρ ἐνεδέχετο
5 τούτων ἅμα πλειόνων ἢ τῶν τῷ γένει ἑτέρων.
εἰ δὲ δὴ ἄλλῳ μὲν γλυκέος ἄλλῳ δὲ λευκοῦ αἰσθάνεται ἡ ψυχὴ μέρει, ἤτοι τὸ ἐκ τούτων ἕν τί
ἐστιν ἢ οὐχ ἕν. ἀλλ᾽ ἀνάγκη ἕν· ἐν γάρ τι τὸ
αἰσθητικόν ἐστι μέρος. τίνος οὖν ἐκεῖνο ἑνός;
οὐδὲν γὰρ ἐκ τούτων ἕν. ἀνάγκη ἄρα ἕν τι εἶναι
τῆς ψυχῆς, ᾧ ἅπαντα αἰσθάνεται, καθάπερ εἴρη-
10 ται πρότερον, ἄλλο δὲ γένος δι᾽ ἄλλου. ἆρ᾽ οὖν
ᾗ μὲν ἀδιαίρετόν ἐστι κατ᾽ ἐνέργειαν, ἕν τί ἐστι
τὸ αἰσθητικὸν γλυκέος καὶ λευκοῦ, ὅταν δὲ δι-
αιρετὸν γένηται κατ᾽ ἐνέργειαν, ἕτερον; ἢ ὥσπερ
ἐπὶ τῶν πραγμάτων αὐτῶν ἐνδέχεται, οὕτω καὶ
ἐπὶ τῆς ψυχῆς. τὸ γὰρ αὐτὸ καὶ ἓν ἀριθμῷ
15 λευκὸν καὶ γλυκύ ἐστι, καὶ ἄλλα πολλά, εἰ μὴ
χωριστὰ τὰ πάθη ἀλλήλων, ἀλλὰ τὸ εἶναι ἕτερον
ἑκάστῳ. ὁμοίως τοίνυν θετέον καὶ ἐπὶ τῆς ψυχῆς
τὸ αὐτὸ καὶ ἓν εἶναι ἀριθμῷ τὸ αἰσθητικὸν πάντων,
τῷ μέντοι εἶναι ἕτερον καὶ ἕτερον τῶν μὲν γένει

---

[1] μὴ supplendum censuit Alexander.

while if they act separately the analogy will break down. Moreover the same senses will then become plural, just as one might speak of different branches of knowledge. For there will be no activity without its proper faculty, nor without activity will there be any perception.

If the soul does not perceive these things in one indivisible time, clearly the same is true in all other cases; for it would be more possible for it to perceive several of these simultaneously than things different in genus. If then it is true that the soul perceives sweet with one part and white with another part, then either the compound of these is one, or it is not. But it must be one; for the perceptive faculty is one. What one object, then, does that one faculty perceive? For surely no one object can be composed of these.[a] There must then, as has been said before, be one part of the soul with which it perceives everything, although it perceives different objects with different parts. Then is the faculty which perceives sweet and white one *qua* indivisible in operation, but different when it has become divisible in operation? More likely what is possible in objects of sense is also possible in the soul. For that which is numerically one and the same may be white and sweet and have many other qualities as well, if the attributes are not separable from one another, but differ individually in their being. So we must suppose the same thing to be true of the soul, and that the general faculty of perception is one and the same numerically, but different in its being; different relatively to its objects sometimes in genus and sometimes in species.

This theory is unsound.

---

[a] *Viz.*, heterogeneous objects like white and sweet.

449 a

τῶν δὲ εἴδει. ὥστε καὶ αἰσθάνοιτ' ἂν ἅμα τῷ
20 αὐτῷ καὶ ἑνί, λόγῳ δ' οὐ τῷ αὐτῷ.

Ὅτι δὲ τὸ αἰσθητὸν πᾶν ἐστι μέγεθος καὶ οὐ
ἔστιν ἀδιαίρετον αἰσθητόν, δῆλον. ἔστι γὰρ ὅθε
μὲν οὐκ ἂν ὀφθείη, ἄπειρον τὸ ἀπόστημα, ὅθε
δὲ ὁρᾶται, πεπερασμένον. ὁμοίως δὲ καὶ τ
ὀσφραντὸν καὶ ἀκουστὸν καὶ ὅσων μὴ αὐτῶν ἁπτό
25 μενοι αἰσθάνονται. ἔστι δέ τι ἔσχατον τοῦ ἀποστή
ματος ὅθεν οὐχ ὁρᾶται, καὶ πρῶτον ὅθεν ὁρᾶται
τοῦτο δὴ ἀνάγκη ἀδιαίρετον εἶναι, οὗ ἐν μὲν τῷ
ἐπέκεινα οὐκ ἐνδέχεται αἰσθάνεσθαι ὄντος, ἐν δ
τῷ ἐπὶ ταδὶ ἀνάγκη αἰσθάνεσθαι. εἰ δή τί ἐστι
ἀδιαίρετον αἰσθητόν, ὅταν τεθῇ ἐπὶ τῷ ἐσχάτῳ
30 ὅθεν ἐστὶν ὕστατον μὲν οὐκ αἰσθητὸν πρῶτον δ
αἰσθητόν, ἅμα συμβήσεται ὁρατὸν εἶναι καὶ ἀόρατον
τοῦτο δ' ἀδύνατον.

449 b

Περὶ μὲν οὖν τῶν αἰσθητηρίων καὶ τῶν αἰσθητῶ
τίνα τρόπον ἔχει καὶ κοινῇ καὶ καθ' ἕκαστο
αἰσθητήριον εἴρηται· τῶν δὲ λοιπῶν πρῶτο
σκεπτέον περὶ μνήμης καὶ τοῦ μνημονεύειν.

So that simultaneous perception would be possible with a part which is one and the same, but not the same in conceptual relation.

It is of course clear that every sensible object is a magnitude, and that no sensible object is indivisible. The distance from which an object cannot be seen is indeterminate, but the distance from which it can be seen is determinate. This is also true of the objects of smell and hearing, and all the other objects which we perceive without contact. But there is in interval of space a certain point, the last from which the object cannot be seen, and the first from which it is seen. This point, beyond which if an object is, one cannot perceive it, and short of which the object must be perceptible, is surely of necessity indivisible. If, then, any sensible object is indivisible, when it is placed at the limiting point, that is the last point at which it cannot be seen and the first at which it can be seen, it will be both visible and invisible at the same time ; which is impossible.

Every sensible object is a magnitude.

Concerning the sense organs and objects of sense, their character in general and in relation to each sense organ, we have concluded our discussion. Of the remaining subjects we must first consider memory and remembering.

ON MEMORY AND
RECOLLECTION

INTRODUCTION

THIS section displays Aristotle's psychology at its best, because he is here able to argue logically from facts of common experience, and is not unduly handicapped by the limitations of his physical theories. Moreover, his powers of analysis give him for once (in examining the process of recollection) a distinct advantage over Plato.

Memory is an affection of the common sense-faculty, which (being discriminative of time) can distinguish between fresh images of sensation or thought and the impressed images persisting from previous experiences. It can also (though it does not always) relate these impressed images to the experiences which produced them; its capacity for doing this depends upon the "depth" of the original impression, which varies with the circumstances and age of the percipient subject.

The theory of *Recollection* is chiefly notable for the clear grasp which it shows (1) of the general principle of the association of ideas, and of the distinction between natural and habitual association, and (2) of the twofold nature of the process as something set up by a deliberate mental act in a corporeal substrate, so that the train of thought, once started, may continue automatically without any further conscious effort.

ΠΕΡΙ ΜΝΗΜΗΣ ΚΑΙ ΑΝΑΜΝΗΣΕΩΣ

I. Περὶ μνήμης καὶ τοῦ μνημονεύειν λεκτέον
τί ἐστι, καὶ διὰ τίν' αἰτίαν γίγνεται, καὶ τίνι τῶν
τῆς ψυχῆς μορίων συμβαίνει τοῦτο τὸ πάθος καὶ
τὸ ἀναμιμνήσκεσθαι· οὐ γὰρ οἱ αὐτοί εἰσι μνημονικοὶ καὶ ἀναμνηστικοί, ἀλλ' ὡς ἐπὶ τὸ πολὺ
μνημονικώτεροι μὲν οἱ βραδεῖς, ἀναμνηστικώτεροι
δὲ οἱ ταχεῖς καὶ εὐμαθεῖς.

Πρῶτον μὲν οὖν ληπτέον ποῖά ἐστι τὰ μνημο-
νευτά· πολλάκις γὰρ ἐξαπατᾷ τοῦτο. οὔτε γὰρ
τὸ μέλλον ἐνδέχεται μνημονεύειν, ἀλλ' ἔστι δοξα-
στὸν καὶ ἐλπιστόν (εἴη δ' ἂν καὶ ἐπιστήμη τις
ἐλπιστική, καθάπερ τινές φασι τὴν μαντικήν),
οὔτε τοῦ παρόντος, ἀλλ' αἴσθησις· ταύτῃ γὰρ οὔτε
τὸ μέλλον οὔτε τὸ γενόμενον γνωρίζομεν, ἀλλὰ τὸ
παρὸν μόνον. ἡ δὲ μνήμη τοῦ γενομένου· τὸ δὲ
παρὸν ὅτε πάρεστιν, οἷον τοδὶ τὸ λευκὸν ὅτε ὁρᾷ,
οὐδεὶς ἂν φαίη μνημονεύειν, οὐδὲ τὸ θεωρούμενον,
ὅτε θεωρῶν τυγχάνει καὶ ἐννοῶν· ἀλλὰ τὸ μὲν
αἰσθάνεσθαί φησι, τὸ δ' ἐπίστασθαι μόνον· ὅταν

288

ON MEMORY AND RECOLLECTION

I. Our task is now to discuss memory and remembering : what it is, why it occurs and to what part of the soul this affection and that of recollection belongs. Men who have good memories are not the same as those who are good at recollecting, in fact generally speaking the slow-witted have better memories, but the quick-witted and those who learn easily are better at recollecting.

First, then, we must comprehend what sort of things are objects of memory ; for mistakes are frequent on this point. It is impossible to remember the future, which is an object of conjecture or expectation (there might even be a science of expectation as some say there is of divination); nor is there memory of the present, but only perception ; for it is neither the future nor the past that we cognize by perception, but only the present. But memory is of the past ; no one could claim to remember the present while it is present. For instance one cannot remember a particular white object while one is looking at it, nor can one remember a subject of theoretical speculation while one is actually speculating and thinking about it. One merely claims to perceive the former, and to know the latter. But when one

What can be remembered ?

289

449 b

δ' ἄνευ τῶν ἐνεργειῶν ἔχῃ τὴν ἐπιστήμην καὶ
20 τὴν αἴσθησιν, οὕτω μέμνηται [τὰς τοῦ τριγώνου
ὅτι δύο ὀρθαῖς ἴσαι],¹ τὸ μὲν ὅτι ἔμαθεν ἢ ἐθεώρησεν,
τὸ δὲ ὅτι ἤκουσεν ἢ εἶδεν ἢ ὅ τι τοιοῦτον· ἀεὶ
γὰρ ὅταν ἐνεργῇ κατὰ τὸ μνημονεύειν, οὕτως ἐν
τῇ ψυχῇ λέγει ὅτι πρότερον τοῦτο ἤκουσεν ἢ
ᾔσθετο ἢ ἐνόησεν.

Ἔστι μὲν οὖν ἡ μνήμη οὔτε αἴσθησις οὔτε
25 ὑπόληψις, ἀλλὰ τούτων τινὸς ἕξις ἢ πάθος, ὅταν
γένηται χρόνος. τοῦ δὲ νῦν ἐν τῷ νῦν οὐκ ἔστι
μνήμη, καθάπερ εἴρηται καὶ πρότερον, ἀλλὰ τοῦ
μὲν παρόντος αἴσθησις, τοῦ δὲ μέλλοντος ἐλπίς,
τοῦ δὲ γενομένου μνήμη. διὸ μετὰ χρόνου πᾶσα
μνήμη. ὥσθ' ὅσα χρόνου αἰσθάνεται, ταῦτα μόνα
30 τῶν ζῴων μνημονεύει, καὶ τούτῳ ᾧ αἰσθάνεται.

Ἐπεὶ δὲ περὶ φαντασίας εἴρηται πρότερον ἐν
τοῖς περὶ ψυχῆς, καὶ νοεῖν οὐκ ἔστιν ἄνευ φαν-
450 a τάσματος· συμβαίνει γὰρ τὸ αὐτὸ πάθος ἐν τῷ
νοεῖν ὅπερ καὶ ἐν τῷ διαγράφειν· ἐκεῖ τε γὰρ οὐθὲν
προσχρώμενοι τῷ τὸ ποσὸν ὡρισμένον εἶναι τὸ
τριγώνου, ὅμως γράφομεν ὡρισμένον κατὰ τὸ
ποσόν· καὶ ὁ νοῶν ὡσαύτως, κἂν μὴ ποσὸν νοῇ,
τίθεται πρὸ ὀμμάτων ποσόν, νοεῖ δ' οὐχ ᾗ ποσόν.
5 ἂν δ' ἡ φύσις ᾖ τῶν ποσῶν, ἀόριστον δέ, τίθεται
μὲν ποσὸν ὡρισμένον, νοεῖ δ' ᾗ ποσὸν μόνον. διὰ
τίνα μὲν οὖν αἰτίαν οὐκ ἐνδέχεται νοεῖν οὐδὲν ἄνευ
τοῦ συνεχοῦς, οὐδ' ἄνευ χρόνου τὰ μὴ ἐν χρόνῳ

¹ τὰς . . . ἴσαι damnavit Freudenthal.

has knowledge or sensation without the actualization of these faculties, then one remembers : in the former case that he learned or thought out the fact, and in the latter that he heard or saw it or perceived it in some other such way ; for when a man is exercising his memory he always says in his mind that he has heard, or felt, or thought this before.

Memory, then, is neither sensation nor judgement, but is a state or affection of one of these, when time has elapsed. There can be no memory of something now present at the present time, as has been said, but sensation refers to what is present, expectation to what is future, and memory to what is past. All memory, then, implies lapse of time. Hence only those living creatures which are conscious of time can be said to remember, and they do so with that part which is conscious of time. *What is memory ?*

We have already dealt with imagination in the treatise *On the Soul.*[a] It is impossible even to think without a mental picture. The same affection is involved in thinking as in drawing a diagram ; for in this case although we make no use of the fact that the magnitude of a triangle is a finite quantity, yet we draw it as having a finite magnitude. In the same way the man who is thinking, though he may not be thinking of a finite magnitude, still puts a finite magnitude before his eyes, though he does not think of it as such. And even if the nature of the object is quantitative, but indeterminate, he still puts before him a finite magnitude, although he thinks of it as merely quantitative. Why it is impossible to think of anything without continuity, or to think of things which are timeless except in terms of time, is another *The part played by imagina- tion.*

---

[a] *De An.* iii. 7, etc.

450 a

ὄντα, ἄλλος λόγος. μέγεθος δ' ἀναγκαῖον γνωρί-
10 ζειν καὶ κίνησιν ᾧ καὶ χρόνον, καὶ τὸ φάντασμα
τῆς κοινῆς αἰσθήσεως πάθος ἐστίν. ὥστε φανερὸν
ὅτι τῷ πρώτῳ αἰσθητικῷ τούτων ἡ γνῶσίς ἐστιν.
ἡ δὲ μνήμη καὶ ἡ τῶν νοητῶν οὐκ ἄνευ φαντά-
σματός ἐστιν. ὥστε τοῦ διανοουμένου[1] κατὰ συμβεβηκὸς ἂν εἴη, καθ' αὑτὸ δὲ τοῦ πρώτου αἰσθη-
15 τικοῦ. διὸ καὶ ἑτέροις τισὶν ὑπάρχει τῶν ζῴων,
καὶ οὐ μόνον ἀνθρώποις καὶ τοῖς ἔχουσι δόξαν ἢ
φρόνησιν. εἰ δὲ τῶν νοητικῶν τι μορίων ἦν, οὐκ
ἂν ὑπῆρχε πολλοῖς τῶν ἄλλων ζῴων, ἴσως δ'
οὐδενὶ τῶν θνητῶν, ἐπεὶ οὐδὲ νῦν πᾶσι διὰ τὸ μὴ
πάντα χρόνου αἴσθησιν ἔχειν· ἀεὶ γὰρ ὅταν ἐνεργῇ
20 τῇ μνήμῃ, καθάπερ καὶ πρότερον εἴπομεν, ὅτι εἶδε
τοῦτο ἢ ἤκουσεν ἢ ἔμαθε, προσαισθάνεται ὅτι
πρότερον· τὸ δὲ πρότερον καὶ ὕστερον ἐν χρόνῳ
ἐστίν.

Τίνος μὲν οὖν τῶν τῆς ψυχῆς ἐστιν ἡ μνήμη,
φανερόν, ὅτι οὗπερ καὶ ἡ φαντασία· καὶ ἔστι
μνημονευτὰ καθ' αὑτὰ μὲν ὅσα ἐστὶ φανταστά,
25 κατὰ συμβεβηκὸς δὲ ὅσα μὴ ἄνευ φαντασίας.
ἀπορήσειε δ' ἄν τις πῶς ποτὲ τοῦ μὲν πάθους
παρόντος τοῦ δὲ πράγματος ἀπόντος μνημονεύεται
τὸ μὴ παρόν. δῆλον γὰρ ὅτι δεῖ νοῆσαι τοιοῦτον
τὸ γινόμενον διὰ τῆς αἰσθήσεως ἐν τῇ ψυχῇ καὶ
τῷ μορίῳ τοῦ σώματος τῷ ἔχοντι αὐτήν, οἷον
30 ζωγράφημά τι τὸ πάθος, οὗ φαμὲν τὴν ἕξιν μνήμην

---

[1] διανοουμένου Bywater : νοουμένου.

292

question. Now we must cognize magnitude and motion with the faculty with which we cognize time, and the image is an affection of the common sense-faculty. Thus it is clear that the cognition of these things belongs to the primary sense-faculty. But memory, even of the objects of thought, implies a mental picture. Hence it would seem to belong incidentally to the thinking faculty, but essentially to the primary sense-faculty. Hence memory is found not only in man and beings which are capable of opinion and thought, but also in some other animals. If it formed part of the intellectual faculty, it would not belong, as it does, to many other animals; probably not to any mortal being, since even as it is it does not belong to all, because they have not all a consciousness of time; for, as we said before, whenever a man actively remembers that he has seen, heard or learned something, he always has the additional consciousness that he did so *before*; now "before" and "after" relate to time.

It is obvious, then, that memory belongs to that part of the soul to which imagination belongs; all things which are imaginable are essentially objects of memory, and those which necessarily involve imagination are objects of memory only incidentally. The question might be asked how one can remember something which is not present, since it is only the affection that is present, and the fact is not. For it is obvious that one must consider the affection which is produced by sensation in the soul, and in that part of the body which contains the soul—the affection, the lasting state of which we call memory

Memory and imagination.

293

450 a

εἶναι· ἡ γὰρ γινομένη κίνησις ἐνσημαίνεται οἷον
τύπον τινὰ τοῦ αἰσθήματος, καθάπερ οἱ σφραγιζό-
450 b μενοι τοῖς δακτυλίοις. διὸ καὶ τοῖς μὲν ἐν κινήσει
πολλῇ διὰ πάθος ἢ δι᾽ ἡλικίαν οὖσιν οὐ γίνεται
μνήμη, καθάπερ ἂν εἰς ὕδωρ ῥέον ἐμπιπτούσης
τῆς κινήσεως καὶ τῆς σφραγῖδος· τοῖς δὲ διὰ τὸ
ψήχεσθαι, καθάπερ τὰ παλαιὰ τῶν οἰκοδομημάτων,
5 καὶ διὰ σκληρότητα τοῦ δεχομένου τὸ πάθος οὐκ
ἐγγίνεται ὁ τύπος. διόπερ οἵ τε σφόδρα νέοι καὶ
οἱ γέροντες ἀμνήμονές εἰσιν· ῥέουσι γὰρ οἱ μὲν
διὰ τὴν αὔξησιν, οἱ δὲ διὰ τὴν φθίσιν. ὁμοίως
δὲ καὶ οἱ λίαν ταχεῖς καὶ οἱ λίαν βραδεῖς οὐδέτεροι
φαίνονται μνήμονες· οἱ μὲν γάρ εἰσιν ὑγρότεροι τοῦ
10 δέοντος, οἱ δὲ σκληρότεροι· τοῖς μὲν οὖν οὐ μένει
τὸ φάντασμα ἐν τῇ ψυχῇ, τῶν δ᾽ οὐχ ἅπτεται.

Ἀλλ᾽ εἰ δὴ τοιοῦτόν ἐστι τὸ συμβαῖνον περὶ
τὴν μνήμην, πότερον τοῦτο μνημονεύει τὸ πάθος,
ἢ ἐκεῖνο ἀφ᾽ οὗ ἐγένετο; εἰ μὲν γὰρ τοῦτο, τῶν
ἀπόντων οὐδὲν ἂν μνημονεύοιμεν· εἰ δ᾽ ἐκεῖνο,
15 πῶς αἰσθανόμενοι τούτου μνημονεύομεν, οὗ μὴ
αἰσθανόμεθα, τὸ ἀπόν; εἴ τ᾽ ἐστὶν ὅμοιον ὥσπερ
τύπος ἢ γραφὴ ἐν ἡμῖν, τούτου αὐτοῦ ἡ αἴσθησις
διὰ τί ἂν εἴη μνήμη ἑτέρου, ἀλλ᾽ οὐκ αὐτοῦ
τούτου; ὁ γὰρ ἐνεργῶν τῇ μνήμῃ θεωρεῖ τὸ πάθος
τοῦτο καὶ αἰσθάνεται τούτου. πῶς οὖν τὸ μὴ παρὸν
μνημονεύει; εἴη γὰρ ἂν καὶ ὁρᾶν τὸ μὴ παρὸν
20 καὶ ἀκούειν. ἢ ἔστιν ὡς ἐνδέχεται καὶ συμβαίνει
τοῦτο; οἷον γὰρ τὸ ἐν τῷ πίνακι γεγραμμένον
καὶ ζῷόν ἐστι καὶ εἰκών, καὶ τὸ αὐτὸ καὶ ἓν
τοῦτ᾽ ἐστὶν ἄμφω, τὸ μέντοι εἶναι οὐ ταὐτὸν ἀμ-

—as a kind of picture ; for the stimulus produced
impresses a sort of likeness of the percept, just as
when men seal with signet rings. Hence in some
people, through disability or age, memory does not
occur even under a strong stimulus, as though the
stimulus or seal were applied to running water ; while
in others owing to detrition like that of old walls in
buildings, or to the hardness of the receiving surface,
the impression does not penetrate. For this reason
the very young and the old have poor memories ;
they are in a state of flux, the young because of
their growth, the old because of their decay. For a
similar reason neither the very quick nor the very
slow appear to have good memories ; the former are
moister than they should be, and the latter harder ;
with the former the picture does not remain in the
soul, with the latter it makes no impression.

Now if memory really occurs in this way, is what
one remembers the present affection, or the original
from which it arose ? If the former, then we could
not remember anything in its absence ; if the latter,
how can we, by perceiving the affection, remember
the absent fact which we do not perceive ? If there
is in us something like an impression or picture, why
should the perception of just this be memory of something else and not of itself ? For when one exercises
his memory this affection is what he considers and
perceives. How, then, does he remember what is
not present ? This would imply that one can also
see and hear what is not present. But surely in a
sense this can and does occur. Just as the picture
painted on the panel is at once a picture and a portrait, and though one and the same, is both, yet the
essence of the two is not the same, and it is possible

What does
one actually
remember ?

to think of it both as a picture and as a portrait, so in the same way we must regard the mental picture within us both as an object of contemplation in itself and as a mental picture of something else. In so far as we consider it in itself, it is an object of contemplation or a mental picture, but in so far as we consider it in relation to something else, *e.g.*, as a likeness, it is also an aid to memory. Hence when the stimulus of it is operative, if the soul perceives the impression as independent, it appears to occur as a thought, or a mental picture ; but if it is considered in relation to something else, it is as though one contemplated a figure in a picture as a portrait, *e.g.*, of Coriscus, although he has not just seen Coriscus. As in this case the affection caused by the contemplation differs from that which is caused when one contemplates the object merely as a painted picture, so in the soul the one object appears as a mere thought, but the other, being (as in the former case) a likeness, is an aid to memory. And for this reason sometimes we do not know, when such stimuli occur in our soul from an earlier sensation, whether the phenomenon is due to sensation, and we are in doubt whether it is memory or not. But sometimes it happens that we reflect and remember that we have heard or seen this something before. Now this occurs whenever we first think of it as itself, and then change and think of it as referring to something else. The opposite also occurs, as happened to Antipheron of Oreus, and other lunatics ; for they spoke of their mental pictures as if they had actually taken place, and as if they actually remembered them. This happens when one regards as a likeness what is not a likeness. Memorizing preserves the memory of something by constant reminding. This is nothing but the repeated contemplation of an object as a likeness, and not independently.

Thus we have explained (*a*) what memory or remembering is : that it is a state induced by a mental image, related as a likeness to that of which it is an image ; and (*b*) to what part of us it pertains : that it pertains to the primary sense-faculty, *i.e.*, that with which we perceive time.

II. It remains to speak about recollecting. Now Recollection first of all we must assume as our basis the truths tion and established in our preliminary discussion.[a] For recollection is neither the recovery nor the acquisition of memory ; for when one first learns or receives a sense impression, one does not recover any memory [b] (for none has gone before), nor does one acquire it for the first time ; it is only at the moment when the state or affection has been induced that there is memory ; so that memory is not induced at the same time as the original affection. Moreover, as soon as the affection is completely induced in the individual and ultimate sense organ, the affection—or knowledge, if one can call the state or affection knowledge (and there is nothing to prevent our remembering incidentally some objects of our knowledge)—is already present in the affected subject ; but memory proper is not established until time has elapsed ; for one remembers in the present what one saw or suffered in the past ; one does not remember in the present what one experiences in the present. Moreover it is evident that it is possible to remember things which are not recalled at the moment, but which one has

[b] A. here rejects Plato's identification of " learning " with " recollection."

451 b

αἰσθόμενον ἢ παθόντα· ἀλλ' ὅταν ἀναλαμβάνῃ ἣν
πρότερον εἶχεν ἐπιστήμην ἢ αἴσθησιν ἢ οὗ ποτὲ τὴν
ἕξιν ἐλέγομεν μνήμην, τοῦτ' ἐστὶ καὶ τότε τὸ
5 ἀναμιμνήσκεσθαι τῶν εἰρημένων τι. τὸ δὲ μνημονεύειν συμβαίνει, καὶ μνήμη ἀκολουθεῖ. οὐδὲ δὴ
ταῦτα ἁπλῶς, ἐὰν ἔμπροσθεν ὑπάρξαντα πάλιν
ἐγγένηται, ἀλλ' ἔστιν ὡς, ἔστι δ' ὡς οὔ. δὶς γὰρ
μαθεῖν καὶ εὑρεῖν ἐνδέχεται τὸν αὐτὸν τὸ αὐτό.
10 δεῖ οὖν διαφέρειν τὸ ἀναμιμνήσκεσθαι τούτων, καὶ
ἐνούσης πλείονος ἀρχῆς ἢ ἐξ ἧς μανθάνουσιν
ἀναμιμνήσκεσθαι.

Συμβαίνουσι δ' αἱ ἀναμνήσεις ἐπειδὴ πέφυκεν
ἡ κίνησις ἥδε γενέσθαι μετὰ τήνδε· εἰ μὲν ἐξ ἀνάγκης, δῆλον ὡς ὅταν ἐκείνην[1] κινηθῇ, τήνδε κινη-
θήσεται· εἰ δὲ μὴ ἐξ ἀνάγκης ἀλλ' ἔθει, ὡς ἐπὶ
τὸ πολὺ κινηθήσεται. συμβαίνει δ' ἐνίας[2] ἅπαξ
15 ἐθισθῆναι μᾶλλον ἢ ἄλλας[2] πολλάκις κινουμένους·
διὸ ἔνια ἅπαξ ἰδόντες μᾶλλον μνημονεύομεν ἢ
ἕτερα πολλάκις. ὅταν οὖν ἀναμιμνησκώμεθα,
κινούμεθα τῶν προτέρων τινὰ κινήσεων, ἕως ἂν
κινηθῶμεν μεθ' ἣν ἐκείνη εἴωθεν. διὸ καὶ τὸ
ἐφεξῆς θηρεύομεν νοήσαντες ἀπὸ τοῦ νῦν ἢ ἄλλου
20 τινός, καὶ ἀφ' ὁμοίου ἢ ἐναντίου ἢ τοῦ σύνεγγυς.
διὰ τοῦτο γίνεται ἡ ἀνάμνησις· αἱ γὰρ κινήσεις
τούτων τῶν μὲν αἱ αὐταί, τῶν δ' ἅμα, τῶν δὲ

---

[1] ἐκείνην Beare : ἐκείνῃ.
[2] ἐνίας . . . ἄλλας Freudenthal : ἐνίους . . . ἄλλους.

---

[a] *i.e.*, although memory does not imply recollection.

300

perceived or suffered all along. But when one re-
covers some previous knowledge or sensation or experience, the continuing state of which we described
before as memory, then this process is recollection
of one of the aforesaid objects. Yet [a] the process of
recollection implies memory, and is followed by
memory. Nor is it true to say without qualification
that recollection is the re-introduction of something
which existed in us before ; in one sense this is true,
in another not ; for it is possible for the same man
to learn or discover the same thing twice. Hence
recollection must differ from these acts ; it must
imply some originative principle beyond that from
which we learn in the first instance.

Acts of recollection occur when one impulse naturally succeeds another : now if the sequence is necessary, it is plain that whoever experiences one impulse
will also experience the next ; but if the sequence
is not necessary, but customary, the second experience will normally follow. But it happens that some
impulses become habitual to us more readily from
a single experience than others do from many ; and
so we remember some things that we have seen once
better than others that we have seen many times.
When we recollect, then, we re-experience one of our
former impulses, until at last we experience that
which customarily precedes the one which we require.
This is why we follow the trail in order, starting in
thought from the present, or some other concept, and
from something similar or contrary to, or closely connected with, what we seek. This is how recollection
takes place ; for the impulses from these experiences
are sometimes identical and sometimes simultaneous
with those of what we seek, and sometimes form a

301

451 b

μέρος ἔχουσιν, ὥστε τὸ λοιπὸν μικρὸν ὃ ἐκινήθη
μετ' ἐκεῖνο.

Ζητοῦσι μὲν οὖν οὕτω, καὶ μὴ ζητοῦντες δ'
οὕτως ἀναμιμνήσκονται, ὅταν μεθ' ἑτέραν κίνησιν
25 ἐκείνη γένηται· ὡς δὲ τὰ πολλὰ ἑτέρων γενομένων
κινήσεων οἵων εἴπομεν, ἐγένετο ἐκείνη. οὐδὲν δὲ
δεῖ σκοπεῖν τὰ πόρρω, πῶς μεμνήμεθα, ἀλλὰ τὰ
σύνεγγυς· δῆλον γὰρ ὅτι ὁ αὐτός ἐστι τρόπος,
λέγω δὲ τὸ ἐφεξῆς, οὐ προζητήσας οὐδ' ἀναμνη-
σθείς. τῷ γὰρ ἔθει ἀκολουθοῦσιν αἱ κινήσεις ἀλλή-
30 λαις, ἥδε μετὰ τήνδε. καὶ ὅταν τοίνυν ἀναμιμνήσκεσθαι βούληται, τοῦτο ποιήσει· ζητήσει λαβεῖν
ἀρχὴν κινήσεως, μεθ' ἣν ἐκείνη ἔσται. διὸ τάχιστα
452 a καὶ κάλλιστα γίνονται ἀπ' ἀρχῆς αἱ ἀναμνήσεις·
ὡς γὰρ ἔχουσι τὰ πράγματα πρὸς ἄλληλα τῷ
ἐφεξῆς, οὕτω καὶ αἱ κινήσεις. καὶ ἔστιν εὐμνη-
μόνευτα ὅσα τάξιν τινὰ ἔχει, ὥσπερ τὰ μαθήματα·
τὰ δὲ φαῦλα[1] χαλεπῶς. καὶ τούτῳ διαφέρει τὸ
5 ἀναμιμνήσκεσθαι τοῦ πάλιν μανθάνειν, ὅτι δυνή-
σεταί πως δι' αὑτοῦ κινηθῆναι ἐπὶ τὸ μετὰ τὴν
ἀρχήν. ὅταν δὲ μή, ἀλλὰ δι' ἄλλου, οὐκέτι
μέμνηται.

Πολλάκις δ' ἤδη μὲν ἀδυνατεῖ ἀναμνησθῆναι,
ζητῶν[2] δὲ δύναται καὶ εὑρίσκει. τοῦτο δὲ γίνεται
κινοῦντι πολλά, ἕως ἂν τοιαύτην κινήσῃ κίνησιν ᾗ
10 ἀκολουθήσει τὸ πρᾶγμα. τὸ γὰρ μεμνῆσθαί ἐστι τὸ

---

[1] φαύλως καὶ LSU, Bekker.   [2] ζητεῖν LSU, Bekker.

---

[a] Sc., which is different from the natural order.

part of them ; so that the remaining portion which we experienced after that is relatively small.

This is the way in which men try to recollect, and the way in which they recollect, even if they do not try to : *viz.*, when one impulse follows upon another. Generally speaking, it is when other impulses, such as we have mentioned, have first been aroused that the particular impulse follows. We need not inquire how we remember when the extremes of the series are far apart, but only when they are near together ; for it is clear that the method is the same in both cases, I mean by following a chain of succession, without previous search or recollection. For custom makes the impulses follow one another in a certain order.[a] Thus when a man wishes to recall anything, this will be his method ; he will try to find a starting-point for an impulse which will lead to the one he seeks. This is why acts of recollection are achieved soonest and most successfully when they start from the beginning of a series ; for just as the objects are related to each other in an order of succession, so are the impulses. Those subjects which possess an orderly arrangement, like mathematical problems, are the easiest to recollect ; ill-arranged subjects are recovered with difficulty. It is in this that the difference between recollecting and learning afresh lies, that in the former one will be able in some way to move on by his own effort to the term next after the starting-point. When he cannot do this himself, but only through another agency, he no longer remembers.

It often happens that one cannot recollect at the moment, but can do so by searching, and finds what he wants. This occurs by his initiating many impulses, until at last he initiates one such that it will lead to the object of his search. For remembering

The method.

Why it may fail.

452 a

ἐνεῖναι δυνάμει τὴν κινοῦσαν· τοῦτο δέ, ὥστ' ἐ
αὐτοῦ καὶ ὧν ἔχει κινήσεων κινηθῆναι, ὥσπε
εἴρηται. δεῖ δὲ λαβέσθαι ἀρχῆς. διὸ ἀπὸ τόπω
δοκοῦσιν ἀναμιμνήσκεσθαι ἐνίοτε. τὸ δ' αἴτιον ὅτ
ταχὺ ἀπ' ἄλλου ἐπ' ἄλλο ἔρχονται, οἷον ἀπὸ γάλα
κτος ἐπὶ λευκόν, ἀπὸ λευκοῦ δ' ἐπ' ἀέρα, καὶ ἀπ
15 τούτου ἐφ' ὑγρόν, ἀφ' οὗ ἐμνήσθη μετοπώρου, ταύ
την ἐπιζητῶν τὴν ὥραν. ἔοικε δὴ καθόλου ἀρχ
καὶ τὸ μέσον πάντων· εἰ γὰρ μὴ πρότερον, ὅτα
ἐπὶ τοῦτο ἔλθῃ, μνησθήσεται, ἢ οὐκέτ' οὐδὲ ἄλλο
θεν, οἷον εἴ τις νοήσειεν ἐφ' ὧν ΑΒΓΔΕΖΗΘ
20 εἰ γὰρ μὴ ἐπὶ τοῦ Α[1] μέμνηται, ἐπὶ τοῦ Ε[2] ἐμνήσθη
ἐντεῦθεν γὰρ ἐπ' ἄμφω κινηθῆναι ἐνδέχεται, κα
ἐπὶ τὸ Δ καὶ ἐπὶ τὸ Ζ.[3] εἰ δὲ μὴ τούτων τι ἐπιζη
τεῖ, ἐπὶ τὸ Ζ[4] ἐλθὼν μνησθήσεται, εἰ τὸ Η ἢ τὸ Θ
ἐπιζητεῖ. εἰ δὲ μή, ἐπὶ τὸ Δ[6]· καὶ οὕτως ἀεί. το
δ' ἀπὸ τοῦ αὐτοῦ ἐνίοτε μὲν μνησθῆναι ἐνίοτε δ
25 μή, αἴτιον ὅτι ἐπὶ πλείω ἐνδέχεται κινηθῆναι ἀπ
τῆς αὐτῆς ἀρχῆς, οἷον ἀπὸ τοῦ Γ ἐπὶ τὸ Ζ ἢ τὸ Δ

Ἐὰν οὖν μὴ διὰ παλαιοῦ κινῆται, ἐπὶ τὸ συνηθέ
στερον κινεῖται· ὥσπερ γὰρ φύσις ἤδη τὸ ἔθος
διὸ ἃ πολλάκις ἐννοοῦμεν, ταχὺ ἀναμιμνησκόμεθα

---

[1] A Ross : E.   [2] E Ross : EΘ.   [3] Z vet. trans. : E
[4] Z Ross : Γ.   [5] Θ EY : Z.   [6] Δ Y : A.

---

[a] The given text and translation follow the suggestion o
W. D. Ross, quoted by G. R. T. Ross in his commentary t
the *De Sensu* and *De Memoria*. For another interpretatio
see J. R. Smyly in *C.R.* xx. 248-249.

[b] It may be possible to travel from A to H, when it i

consists in the potential existence in the mind of the effective stimulus ; and this, as has been said, in such a way that the subject is stimulated from himself, and from the stimuli which he contains within him. But one must secure a starting-point. This is why some people seem, in recollecting, to proceed from *loci*. The reason for this is that they pass rapidly from one step to the next ; for instance from milk to white, from white to air, from air to damp ; from which one remembers autumn, if this is the season that he is trying to recall. Generally speaking the middle point seems to be a good point to start from ; for one will recollect when one comes to this point, if not before, or else one will not recollect from any other. For instance,[a] suppose one were thinking of a series, which may be represented by the letters ABCDEFGH ; if one does not recall what is wanted at A, yet one does at E ; for from that point it is possible to travel in either direction, that is either towards D or towards F. If one does not want one of these, he will remember by passing on to F, if he wants G or H. If not, he passes on to D.[b] Success is always achieved in this way. The reason why we sometimes recollect and sometimes do not, although starting from the same point, is that it is possible to travel from the same starting-point to more than one destination ; for instance from C we may go direct to F or only to D.

If one is not moving along an old path, one's movement tends towards the more customary ; for custom now takes the place of nature. Hence we remember quickly things which are often in our thoughts ; for

not possible to travel from C to H, because, as is said above, the essential to success is the seizing on the right point from which to start.

452 a

30 ὥσπερ γὰρ φύσει τόδε μετὰ τόδε ἐστίν, οὕτω κα
ἐνεργείᾳ· τὸ δὲ πολλάκις φύσιν ποιεῖ. ἐπεὶ δ᾿

452 b ἐν τοῖς φύσει γίνεται καὶ παρὰ φύσιν καὶ ἀπ
τύχης, ἔτι μᾶλλον ἐν τοῖς δι᾿ ἔθος, οἷς ἡ φύσις γ
μὴ ὁμοίως ὑπάρχει· ὥστε κινηθῆναι ἐνίοτε κἀκε
καὶ ἄλλως, ἄλλως τε καὶ ὅταν ἀφέλκῃ ἐκεῖθε

5 αὐτόσε πη. διὰ τοῦτο καὶ ὅταν δέῃ ὄνομα μνη
μονεῦσαι, παρόμοιον μέν, εἰς δ᾿ ἐκεῖνο σολοικίζομεν
τὸ μὲν οὖν ἀναμιμνήσκεσθαι τοῦτον συμβαίνει τὸ
τρόπον.

Τὸ δὲ μέγιστον, γνωρίζειν δεῖ τὸν χρόνον,
μέτρῳ ἢ ἀορίστως. ἔστω δέ τι ᾧ κρίνει τὸν πλείω
καὶ ἐλάττω· εὔλογον δ᾿ ὥσπερ τὰ μεγέθη· νοεῖ γὰ

10 τὰ μεγάλα καὶ πόρρω οὐ τῷ ἀποτείνειν ἐκεῖ τὴ
διάνοιαν, ὥσπερ τὴν ὄψιν φασί τινες (καὶ γὰρ μ
ὄντων ὁμοίως νοήσει), ἀλλὰ τῇ ἀνάλογον κινήσει
ἔστι γὰρ ἐν αὐτῇ τὰ ὅμοια σχήματα καὶ κινήσεις
τίνι οὖν διοίσει, ὅταν τὰ μείζω νοῇ,[2] ὅτι ἐκεῖνα νοεῖ
ἢ τὰ ἐλάττω; πάντα γὰρ τὰ ἐντὸς ἐλάττω, ὥσπερ

15 ἀνάλογον καὶ τὰ ἐκτός. ἔστι δ᾿ ἴσως ὥσπερ κα
τοῖς εἴδεσιν ἀνάλογον λαβεῖν ἄλλο ἐν αὐτῷ, οὕτ
καὶ τοῖς ἀποστήμασιν. ὥσπερ οὖν εἰ τὴν ΑΒ ΒΓ
κινεῖται, ποιεῖ τὴν Γ[3]Δ· ἀνάλογον γὰρ ἡ ΑΓ κα
ἡ ΓΔ. τί οὖν μᾶλλον τὴν ΓΔ ἢ τὴν ΖΗ ποιεῖ

---

[1] ἐπεὶ δ᾿ ὥσπερ LSUM.
[2] νοῇ ὅτι EMY : νοῇ ; ἢ ὅτι vulgo.
[3] ΓΔ M : ΑΔ.

as in nature one thing follows another, so also in the actualization of these stimuli ; and the frequency has the effect of nature. But since in purely natural phenomena some things occur contrary to nature, and owing to chance, so still more in matters of habit, to which the term " natural " does not belong in the same sense ; so that the mind is sometimes impelled not only in the required direction but also otherwise, especially when something diverts it from that direction, and turns it towards itself. This is why when we want to remember a name, we remember one rather like it, but fail to enunciate the one we want. Recollection, then, occurs in this way.

But the most important point is to cognize the time, either exactly or indeterminately. Let it be granted that one possesses a faculty by which to distinguish greater and lesser time ; it is natural to suppose that we can distinguish these as we distinguish magnitudes. For the mind does not think of large things at a distance by stretching out to them, as some think that vision operates (for the mind will think of them equally if they are not there), but one thinks of them by a proportionate mental impulse ; for there are similar figures and movements in the mind. How then, when the mind thinks of bigger things, will its thinking of them differ from its thinking of smaller things ? For all internal things are smaller, and as it were proportionate to those outside. Perhaps, just as we may suppose that there is something in man proportionate to the forms,[a] we may assume that there is something similarly proportionate to their distances. *E.g.*, if one experiences the impulses AB, BE, he can imagine CD ; for AC and CD are in the same ratio as AB : BE. Why then does he imagine CD rather than FG ? Surely because

Time is an important factor.

---

[a] *Sc.* of external objects.

452 b

20 ἢ ὡς ἡ ΑΓ πρὸς τὴν ΑΒ ἔχει, οὕτως ἡ τὸ Θ πρὸς τὴν Ι¹ ἔχει. ταύτας οὖν ἅμα κινεῖται. ἂν δὲ τὴν ΖΗ βούληται νοῆσαι, τὴν μὲν ΒΕ ὁμοίως νοεῖ, ἀντὶ δὲ τῶν ΘΙ τὰς ΚΛ νοεῖ. αὗται γὰρ ἔχουσιν ὡς ΖΑ πρὸς ΒΑ.

Ὅταν οὖν ἅμα ἥ τε τοῦ πράγματος γίνεται κίνησις καὶ ἡ τοῦ χρόνου, τότε τῇ μνήμῃ ἐνεργεῖ.
25 ἂν δ' οἴηται μὴ ποιῶν, οἴεται μνημονεύειν· οὐθὲν γὰρ κωλύει διαψευσθῆναί τινα καὶ δοκεῖν μνημονεύειν μὴ μνημονεύοντα. ἐνεργοῦντα δὲ τῇ μνήμῃ μὴ οἴεσθαι ἀλλὰ λανθάνειν μεμνημένον οὐκ ἔστιν· τοῦτο γὰρ ἦν αὐτὸ τὸ μεμνῆσθαι. ἀλλ' ἐὰν ἡ τοῦ πράγματος γένηται χωρὶς τῆς τοῦ χρόνου ἢ αὕτη ἐκείνης, οὐ μέμνηται.

30 Ἡ δὲ τοῦ χρόνου διττή ἐστιν· ὁτὲ μὲν γὰρ μέτρῳ
453 a οὐ μέμνηται αὐτό, οἷον ὅτι τρίτην ἡμέραν ὁδήποτε ἐποίησεν, ὁτὲ δὲ καὶ μέτρῳ· ἀλλὰ μέμνηται καὶ ἐὰν μὴ μέτρῳ. εἰώθασι δὲ λέγειν ὅτι μέμνηνται μέν, πότε μέντοι οὐκ ἴσασιν, ὅταν τοῦ πότε μὴ γνωρίζωσι τὸ ποσὸν μέτρῳ.

¹ Ι ΕΜΥ : Μ plerique et codd. et edd.

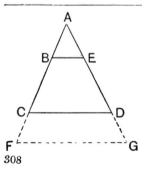

^a If this figure represents A.'s train of thought his argument is as follows. If a man thinks of the ratio AB : BE, he can pass directly to the ratio AC : CD, for by the figure given he knows the ratio AC : AB, which we may call H : I. But to proceed to AF : FG he would first have to determine the ratio K : L, that is the ratio FA : BA, which he does not know, because it is " outside."

AC has the same ratio to AB as H to I. Thus he has these impulses simultaneously. But if he wants to imagine FG, he keeps BE in mind, but instead of H : I, he thinks K : L ; for these are in the same relation as FA to BA.[a]

Thus it is when the impulse relating to the fact and that relating to its time occur together that one actually remembers. If one thinks that he experiences these impulses without doing so, he thinks that he remembers ; for there is nothing to prevent a man from being deceived about it, and from supposing that he remembers when he does not. But when a man actually remembers he cannot suppose that he does not, and remember without being aware of it ; for remembering, as we have seen, essentially involves awareness. But if the impulse relating to the fact takes place apart from that relating to the time, or *vice versa*, one does not remember.

The impulse relating to the time is of two kinds. Sometimes one remembers a fact without an exact estimate of time, such as that one did so and so *the day before yesterday*, and sometimes with an exact estimate ; but it is still an act of memory, even if there is no exact estimate of time. Men are accustomed to say that they remember an occurrence, but that they do not know when it occurred, when they do not know the length of the period exactly.

Recollection may be accurate except for time.

453 a

5 Ὅτι μὲν οὖν οὐχ οἱ αὐτοὶ μνημονικοὶ καὶ ἀναμνηστικοί, ἐν τοῖς πρότερον εἴρηται. διαφέρει δὲ
τοῦ μνημονεύειν τὸ ἀναμιμνήσκεσθαι οὐ μόνον κατὰ
τὸν χρόνον, ἀλλ᾽ ὅτι τοῦ μὲν μνημονεύειν καὶ τῶν
ἄλλων ζῴων μετέχει πολλά, τοῦ δ᾽ ἀναμιμνήσκεσθαι οὐδὲν ὡς εἰπεῖν τῶν γνωριζομένων ζῴων, πλὴν
10 ἄνθρωπος. αἴτιον δ᾽ ὅτι τὸ ἀναμιμνήσκεσθαί ἐστιν
οἷον συλλογισμός τις· ὅτι γὰρ πρότερον εἶδεν ἢ
ἤκουσεν ἤ τι τοιοῦτον ἔπαθε, συλλογίζεται ὁ ἀναμιμνησκόμενος, καὶ ἔστιν οἷον ζήτησίς τις. τοῦτο
δ᾽ οἷς καὶ τὸ βουλευτικὸν ὑπάρχει, φύσει μόνοις
συμβέβηκεν· καὶ γὰρ τὸ βουλεύεσθαι συλλογισμός
τίς ἐστιν.

15 Ὅτι δὲ σωματικόν τι τὸ πάθος καὶ ἡ ἀνάμνησις ζήτησις ἐν τοιούτῳ φαντάσματος, σημεῖον τὸ
παρενοχλεῖν ἐνίους ἐπειδὰν μὴ δύνωνται ἀναμνησθῆναι καὶ πάνυ ἐπέχοντες τὴν διάνοιαν, καὶ οὐκέτ᾽
ἐπιχειροῦντας ἀναμιμνήσκεσθαι οὐδὲν ἧττον, καὶ
μάλιστα τοὺς μελαγχολικούς· τούτους γὰρ φαντά-
20 σματα κινεῖ μάλιστα. αἴτιον δὲ τοῦ μὴ ἐπ᾽ αὐτοῖς
εἶναι τὸ ἀναμιμνήσκεσθαι, ὅτι καθάπερ τοῖς βάλ-
λουσιν οὐκέτι ἐπ᾽ αὐτοῖς τὸ στῆσαι, οὕτω καὶ
ὁ ἀναμιμνησκόμενος καὶ θηρεύων σωματικόν τι
κινεῖ, ἐν ᾧ τὸ πάθος. μάλιστα δ᾽ ἐνοχλοῦνται οἷς
25 ἂν ὑγρότης τύχῃ ὑπάρχουσα περὶ τὸν αἰσθητικὸν
τόπον· οὐ γὰρ ῥᾳδίως παύεται κινηθεῖσα, ἕως ἂν
ἐπέλθῃ τὸ ζητούμενον καὶ εὐθυπορήσῃ ἡ κίνησις.
διὸ καὶ ὀργαὶ καὶ φόβοι, ὅταν τι κινήσωσιν, ἀντι-
κινούντων πάλιν τούτων οὐ καθίστανται, ἀλλ᾽ ἐπὶ
τὸ αὐτὸ ἀντικινοῦσιν. καὶ ἔοικε τὸ πάθος τοῖς

We have said before that those who have good Memory and
memories are not the same as those who recollect recollection
quickly. Recollecting differs from remembering not further
merely in the matter of time, but also because, while guished.
many other animals share in memory, one may say
that none of the known animals can recollect except
man. This is because recollecting is, as it were, a
kind of inference ; for when a man is recollecting
he infers that he has seen or heard or experienced
something of the sort before, and the process is a kind
of search. This power can only belong by nature to
such animals as have the faculty of deliberation ; for
deliberation too is a kind of inference.

That the experience is in some sense physical, and Recollec-
that recollection is the search for a mental picture in tion has
the physical sphere, is proved by the annoyance which actions.
some men show when in spite of great concentration
they cannot remember, and which persists even when
they have abandoned the attempt to recollect,
especially in the case of the melancholic ; for these
are especially affected by mental pictures. The
reason why the recollecting does not lie in their power
is that, just as when men throw stones it no longer
lies in their power to stop them, so the man who is
employed in recollecting and search sets in motion
a bodily part in which the affection resides. And
most disturbed are those who have moisture about
the sensitive region ; for the moisture, once set in
motion, does not readily stop until the object sought
comes round again, and the impulse follows a straight
course. For this reason too outbursts of temper or
fear, when they have once produced an impulse, do
not cease even when the subjects of them set up
counter movements, but continue their original
activity in spite of these. This affection is like that

311

which occurs in the case of names, tunes and sayings, when any of them has been very much on our lips ; for even though we give up the habit and do not mean to yield to it, we find ourselves continually singing or saying the familiar sounds.

Dwarfish people and those who have large upper extremities have poorer memories than their opposites, because they carry a great weight on their organ of perception, and their impulses cannot, from the first, keep their direction, but are scattered, and do not easily travel in a straight course in their recollecting. The very young and the very old have inferior memories because of the forces at work in them ; for the latter are in a state of rapid decay, and the former in a state of rapid growth ; small children moreover are dwarfish, until they are advanced in age.

This concludes our account of memory and remembering, what their nature is, and with what part of the soul animals remember ; also of recollecting, what it is, and how it occurs, and for what reason.

Reasons for good and bad memories.

INTRODUCTION

THIS section is so closely connected with the next two (*On Dreams* and *On Prophecy in Sleep*) that the three may be conveniently considered together. The discussion is eminently reasonable throughout, and is based on careful observation of actual facts ; error is almost confined to the physiological theories, which nevertheless contain enough general resemblance to the truth to earn a qualified admiration.

Sleep is the natural intermission of consciousness in the common sense-organ (*i.e.*, the heart). Its object is to give rest to the senses (and any higher faculty which implies consciousness) ; hence it is confined to animals and man. It is caused in the following way. When food is taken in, its heat causes evaporation, which carries certain liquid and solid elements of the food upwards through the veins to the head. Here these elements are cooled and condensed, producing a heaviness of the upper extremity ; and when thoroughly cooled they descend again in a mass towards the centre of the body, where they compress the " hot substance " (? the non-evaporable part of the food), and thereby so act upon the common sense-faculty that it ceases to function until the process of digestion is complete, *i.e.*, until the thinner and thicker constituents of the blood have been separated in the central chamber of the heart and dispatched to the upper and lower extremities respectively. The details of the theory are unimportant; what matters is the recognition of a relation between sleep and digestion, and of the withdrawal of blood from the brain during sleep.

Dreams are due to the stimulation of the imaginative faculty by persistent sensory impressions. These impressions are always setting up stimulatory movements, which are obscured during consciousness by the activity of the senses and intellect, but in sleep are unimpeded except by such disturbances as are due to the process of digestion in its early stages, to personal idiosyncrasy, or to disease. Dreams must be distinguished from the faint sense-perceptions which sometimes occur in sleep.

Towards the question whether dreams can be significant Aristotle's attitude is cautious. Most cases of the fulfilment of dreams are probably pure coincidences; there may, however, be exceptions. Slight physical disturbances can affect the imagination more easily in sleep, and therefore dreams may have a medical significance. Again, the experiences of a dream may modify subsequent waking conduct. Finally, certain mental or emotional conditions, such as vacancy of mind, excitability or strong affection, make some people peculiarly susceptible to stimuli proceeding, in some way difficult to explain, from external events. There is no evidence that dreams have a divine origin.

ΠΕΡΙ ΥΠΝΟΥ ΚΑΙ
ΕΓΡΗΓΟΡΣΕΩΣ

I. Περὶ δὲ ὕπνου καὶ ἐγρηγόρσεως σκεπτέον, τίνα
τε τυγχάνει ὄντα, καὶ πότερον ἴδια τῆς ψυχῆς ἢ τοῦ
σώματος ἢ κοινά, κἂν ᾖ κοινά, τίνος μορίου τῆς
ψυχῆς ἢ τοῦ σώματος· καὶ διὰ τίν' αἰτίαν ὑπάρχει
15 τοῖς ζῴοις· καὶ πότερον ἅπαντα κεκοινώνηκεν
αὐτῶν ἀμφοτέρων, ἢ τὰ μὲν ὕπνου τὰ δὲ θατέρου
μόνον, ἢ τὰ μὲν οὐδετέρου τὰ δ' ἀμφοτέρων· πρὸς
δὲ τούτοις τί ἐστι τὸ ἐνύπνιον, καὶ διὰ τίν' αἰτίαν
οἱ καθεύδοντες ὁτὲ μὲν ὀνειρώττουσιν ὁτὲ δ' οὔ·
ἢ συμβαίνει μὲν ἀεὶ τοῖς καθεύδουσιν ἐνυπνιάζειν,
20 ἀλλ' οὐ μνημονεύουσιν· καὶ εἰ τοῦτο γίνεται, διὰ
τίνα αἰτίαν γίνεται· καὶ πότερον ἐνδέχεται τὰ
μέλλοντα προορᾶν ἢ οὐκ ἐνδέχεται, καὶ τίνα τρόπον,
εἰ ἐνδέχεται· καὶ πότερον τὰ μέλλοντα ὑπ' ἀνθρώπου πράσσεσθαι μόνον, ἢ καὶ ὧν τὸ δαιμόνιον ἔχει
τὴν αἰτίαν καὶ φύσει γίνεται ἢ ἀπὸ ταὐτομάτου.
25   Πρῶτον μὲν οὖν τοῦτό γε φανερόν, ὅτι τῷ αὐτῷ
τοῦ ζῴου ἥ τε ἐγρήγορσις ὑπάρχει καὶ ὁ ὕπνος·
ἀντίκεινται γάρ, καὶ φαίνεται στέρησίς τις ὁ ὕπνος
τῆς ἐγρηγόρσεως· ἀεὶ γὰρ τὰ ἐναντία καὶ ἐπὶ τῶν
ἄλλων καὶ ἐν τοῖς φυσικοῖς ἐν τῷ αὐτῷ δεκτικῷ

ON SLEEP AND WAKING

I. WE have now to turn our attention to sleep and Problems arising from the subject. waking. What are they? Are they peculiar to the soul or to the body, or do they belong to both? If they belong to both, to what part of the soul or body do they belong? Why are they characteristic of animals? Do all animals share in both, or do some share in sleep only, and others in waking only, or some in neither and others in both? Furthermore what is a dream, and why do men when asleep sometimes dream and sometimes not? Or do sleepers always dream, but sometimes fail to remember their dreams? If the latter is true, why does it occur? Is it possible or impossible to foresee the future? If it is possible, in what way? Again, does the possibility cover only actions to be performed by man, or those also which are due to superhuman agency and are brought about naturally or spontaneously?

To begin with, this at any rate is obvious, that Sleep and wakefulness are complementary. waking and sleep belong to the same part of the animal; for they are opposites, and sleep is apparently a privation of waking; for in natural objects, as in all other cases, contraries evidently occur in the same receptive material, and are affections of the

319

453 b

φαίνεται γινόμενα καὶ τοῦ αὐτοῦ ὄντα πάθη, λέγω
30 δ' οἷον ὑγίεια καὶ νόσος, καὶ κάλλος καὶ αἶσχος, καὶ
ἰσχὺς καὶ ἀσθένεια, καὶ ὄψις καὶ τυφλότης, καὶ
454 a ἀκοὴ καὶ κωφότης. ἔτι δὲ καὶ ἐκ τῶνδε δῆλον.
ᾧ γὰρ τὸν ἐγρηγορότα γνωρίζομεν, τούτῳ καὶ
τὸν ὑπνοῦντα· τὸν γὰρ αἰσθανόμενον ἐγρηγορέναι
νομίζομεν καὶ τὸν ἐγρηγορότα πάντα ἢ τῶν ἔξωθέν
τινος αἰσθάνεσθαι ἢ τῶν ἐν αὑτῷ κινήσεων. εἰ
5 τοίνυν τὸ ἐγρηγορέναι ἐν μηδενὶ ἄλλῳ ἐστὶν ἢ
τῷ αἰσθάνεσθαι, δῆλον ὅτι ᾧπερ αἰσθάνεται, τούτῳ
καὶ ἐγρήγορε τὰ ἐγρηγορότα καὶ καθεύδει τὰ καθεύδοντα.

Ἐπεὶ δ' οὔτε τῆς ψυχῆς ἴδιον τὸ αἰσθάνεσθαι οὔτε
τοῦ σώματος (οὗ γὰρ ἡ δύναμις, τούτου καὶ ἡ
ἐνέργεια· ἡ δὲ λεγομένη αἴσθησις, ὡς ἐνέργεια,
10 κίνησίς τις διὰ τοῦ σώματος τῆς ψυχῆς ἐστί),
φανερὸν ὡς οὔτε τῆς ψυχῆς τὸ πάθος ἴδιον, οὔτ'
ἄψυχον σῶμα δυνατὸν αἰσθάνεσθαι.

Διωρισμένων δὲ πρότερον ἐν ἑτέροις περὶ τῶν
λεγομένων ὡς μορίων τῆς ψυχῆς, καὶ τοῦ μὲν
θρεπτικοῦ χωριζομένου τῶν ἄλλων ἐν τοῖς ἔχουσι
σώμασι ζωήν, τῶν δ' ἄλλων οὐδενὸς ἄνευ τούτου,
15 δῆλον ὡς ὅσα μὲν αὐξήσεως καὶ φθίσεως μετέχει
μόνον τῶν ζώντων, ὅτι τούτοις οὐχ ὑπάρχει ὕπνος
οὐδ' ἐγρήγορσις, οἷον τοῖς φυτοῖς· οὐ γὰρ ἔχουσι τὸ
αἰσθητικὸν μόριον, οὔτ' εἰ χωριστόν ἐστιν οὔτ' εἰ
μὴ χωριστόν· τῇ γὰρ δυνάμει καὶ τῷ εἶναι χωριστόν
ἐστιν.
20 Ὁμοίως δὲ καὶ ὅτι οὐδέν ἐστιν ὃ ἀεὶ ἐγρήγορεν
ἢ ἀεὶ καθεύδει, ἀλλὰ τοῖς αὐτοῖς ὑπάρχει τῶν ζώων

---

^a i.e., sleep.   ^b 432 a 15, etc.

same subject; I mean, *e.g.*, health and disease, beauty and ugliness, strength and weakness, sight and blindness, hearing and deafness. And the same point is clear from the following facts. The criterion by which we know that a man is awake is the same as that by which we know that a man is asleep; for we reckon that the man who is conscious is awake, and that anyone who is awake is conscious of some stimulus, either external or internal. If, then, waking consists in nothing else than the exercise of consciousness, clearly it is in virtue of the part by which they perceive that animals wake when they are awake and sleep when they are asleep.

Since the exercise of sense-perception does not belong exclusively either to soul or to body (for a potentiality and its actuality reside in the same subject; and what we call sensation, as actuality, is a movement of the soul through the agency of the body), it is clear that the affection *a* is not peculiar to the soul, nor is a body without soul capable of sensation.

Elsewhere *b* we have previously distinguished the " parts " of the soul, the nutritive part being found without the others in bodies which have life, whereas none of the others is found without the nutritive part. It is clear that such living things as partake only of growth and decay, *e.g.*, plants, are not subject to sleep, nor to waking; because they do not possess the perceptive part—whether this can exist separately from the others or not. It is, of course, separable as a faculty and as an entity.

Similarly it is obvious that there is no animal which is always awake or always asleep, but both these

Both belong to soul and body.

454 a

ἀμφότερα τὰ πάθη ταῦτα· οὐδὲ¹ γὰρ εἴ τί ἐστι
ζῷον ἔχον² αἴσθησιν, τοῦτ᾽ ἐνδέχεται οὔτε καθεύδειν οὔτ᾽ ἐγρηγορέναι· ἄμφω γάρ ἐστι τὰ πάθη
ταῦτα περὶ αἴσθησιν τοῦ πρώτου αἰσθητικοῦ. οὐκ
ἐνδέχεται δὲ οὐδὲ θάτερον τούτων ἀεὶ ὑπάρχειν

25 τῷ αὐτῷ, οἷον ἀεί τι γένος ζῴων καθεύδειν ἢ ἀεί
τι ἐγρηγορέναι, ὅτι³ ὅσων ἐστί τι ἔργον κατὰ
φύσιν, ὅταν ὑπερβάλλῃ τὸν χρόνον ᾧ δύναται χρόνῳ
τι ποιεῖν, ἀνάγκη ἀδυνατεῖν, οἷον τὰ ὄμματα ὁρῶντα
καὶ παύεσθαι τοῦτο ποιοῦντα, ὁμοίως δὲ καὶ χεῖρα

30 καὶ ἄλλο πᾶν οὗ ἐστί τι ἔργον. εἰ δή τινός ἐστιν
ἔργον τὸ αἰσθάνεσθαι, καὶ τοῦτο ἂν ὑπερβάλλῃ
ὅσον ἦν χρόνον δυνάμενον αἰσθάνεσθαι συνεχῶς,
ἀδυνατήσει καὶ οὐκέτι τοῦτο ποιήσει. εἰ τοίνυν τὸ

454 b ἐγρηγορέναι τούτῳ⁴ ὥρισται τῷ λελύσθαι τὴν
αἴσθησιν, τῶν δ᾽ ἐναντίων τὸ μὲν ἀνάγκη παρεῖναι
τὸ δ᾽ οὔ, τὸ δ᾽ ἐγρηγορέναι τῷ καθεύδειν ἐναντίον,
καὶ ἀναγκαῖον παντὶ θάτερον ὑπάρχειν, ἀναγκαῖον
ἂν εἴη καθεύδειν. εἰ οὖν τὸ τοιοῦτον πάθος ὕπνος,

5 τοῦτο δ᾽ ἐστὶν ἀδυναμία δι᾽ ὑπερβολὴν τοῦ ἐγρηγορέναι, ἡ δὲ τοῦ ἐγρηγορέναι ὑπερβολὴ ὁτὲ μὲν
νοσώδης ὁτὲ δ᾽ ἄνευ νόσου γίνεται, ὥστε καὶ ἡ
ἀδυναμία καὶ ἡ διάλυσις ὡσαύτως ἔσται, ἀνάγκη
πᾶν τὸ ἐγρηγορὸς ἐνδέχεσθαι καθεύδειν· ἀδύνατον
γὰρ ἀεὶ ἐνεργεῖν. ὁμοίως δὲ οὐδὲ καθεύδειν οὐδὲν

10 ἀεὶ ἐνδέχεται. ὁ γὰρ ὕπνος πάθος⁵ τι τοῦ αἰσθητικοῦ μορίου ἐστίν, οἷον δεσμὸς καὶ ἀκινησία τις, ὥστ᾽

___

¹ οὐδὲ L : οὐ EVY : οὔτε.   ² ἔχον] μὴ ἔχον ci. Beare.
³ ὅτι Becker : ἔτι.   ⁴ τούτῳ EVM : om. LSUY.
⁵ om. EVY, Bekker.

affections belong to the same animals. For no animal which has sense-perception can be neither asleep nor awake ; for both these affections are of the sensitivity of the primary sense-faculty. Nor can either sleep or waking be a permanent attribute of the same animal : for instance no species of animal can be permanently asleep or permanently awake. For all parts which have a natural function, when they overpass the period for which they are able to perform their function, must become exhausted ; *e.g.*, the eyes must become exhausted with seeing, and cease to see, and similarly the hand, and any other part which has a function. If then some part has sensation as its function, this too, if it exceeds the due period of its capacity for continuous sensation, will become exhausted, and cease to perform its function. Therefore if waking is defined by liberty of sensation, and if of two contraries one must be present and the other not, and further if waking is the contrary of sleeping, and if one or the other must be present in every animal, then sleeping must be a necessity. Thus if this sort of affection is sleep, and this is a state of incapacity due to excess of waking, and excess of waking occurs sometimes as the result of disease, and sometimes when there is no disease, so that the same will be true of the state of incapacity or dissolution of activity, everything which wakes must be capable of sleeping ; for it cannot always be active. Similarly nothing can be always asleep. For sleep is an affection of the sensitive part of us—a kind of fetter or immobilization ; whence it follows that everything

Neither condition can be permanent.

Sleep is logically necessary.

454 b

ἀνάγκη πᾶν τὸ καθεῦδον ἔχειν τὸ αἰσθητικὸν μόριον. αἰσθητικὸν δὲ τὸ δυνατὸν αἰσθάνεσθαι κατ
ἐνέργειαν· ἐνεργεῖν δὲ τῇ αἰσθήσει κυρίως κα
ἁπλῶς ἀδύνατον καθεῦδον ἅμα· διὸ ἀναγκαῖο
ὕπνον πάντα ἐγερτὸν εἶναι.

15 Τὰ μὲν οὖν ἄλλα σχεδὸν πάντα δῆλα κοινωνοῦνθ'
ὕπνου, καὶ πλωτὰ καὶ πτηνὰ καὶ πεζά (καὶ γὰρ τὸ
τῶν ἰχθύων γένη πάντα καὶ τὰ τῶν μαλακίω
ὦπται καθεύδοντα, καὶ τἆλλα πάνθ' ὅσαπερ ἔχε
ὀφθαλμούς· καὶ γὰρ τὰ σκληρόφθαλμα φανερὰ κα
τὰ ἔντομα κοιμώμενα· βραχύυπνα δὲ τὰ τοιαῦτα
20 πάντα, διὸ καὶ λάθοι ἄν τινα πολλάκις πότεροι
μετέχουσι τοῦ καθεύδειν ἢ οὔ), τῶν δ' ὀστρακο-
δέρμων κατὰ μὲν τὴν αἴσθησιν οὐδέ πω γέγονε
φανερὸν εἰ καθεύδουσιν· εἰ δέ τῳ πιθανὸς ὁ λεχθεὶς
λόγος, τοῦτο[1] πεισθήσεται.

Ὅτι μὲν οὖν ὕπνου κοινωνεῖ τὰ ζῷα πάντα,
25 φανερὸν ἐκ τούτων· τῷ γὰρ αἴσθησιν ἔχειν ὥρισται
τὸ ζῷον, τῆς δ' αἰσθήσεως τρόπον τινὰ τὴν μὲ
ἀκινησίαν καὶ οἷον δεσμὸν τὸν ὕπνον εἶναί φαμεν,
τὴν δὲ λύσιν καὶ τὴν ἄνεσιν ἐγρήγορσιν. τῶν δὲ
φυτῶν οὐδὲν οἷόν τε κοινωνεῖν οὐδετέρου τούτων
τῶν παθημάτων. ἄνευ μὲν γὰρ αἰσθήσεως οὐχ
30 ὑπάρχει οὔθ' ὕπνος οὔτ' ἐγρήγορσις· οἷς δ' αἴσθησις
ὑπάρχει, καὶ τὸ λυπεῖσθαι καὶ τὸ χαίρειν· οἷς δὲ
ταῦτα, καὶ ἐπιθυμία. τοῖς δὲ φυτοῖς οὐδὲν ὑπάρχε
τούτων. σημεῖον δ' ὅτι καὶ τὸ ἔργον τὸ αὑτοῦ
455 a ποιεῖ τὸ θρεπτικὸν μόριον ἐν τῷ καθεύδειν μᾶλλον
ἢ ἐν τῷ ἐγρηγορέναι· τρέφεται γὰρ καὶ αὐξάνεται
τότε μᾶλλον, ὡς οὐδὲν προσδεόμενα πρὸς ταῦτα
τῆς αἰσθήσεως.

[1] τοῦτο Bywater : τούτῳ.

which sleeps must possess this sensitive part. But only that is sensitive which is capable of actual sensation ; and active exercise of sensation in the strict and unqualified sense is impossible while the subject is asleep. Hence all sleep must admit of being awakened.

Practically all other animals, aquatic, winged and terrestrial, partake in sleep (for all kinds of fishes and molluscs have been observed asleep, and all other animals which have eyes ; for clearly even hard-eyed animals and insects repose, but these creatures only sleep for a short time, so that one might often doubt whether they partake in sleep or not), but in the case of the testacea direct observation has not yet proved whether they sleep or not. But if the foregoing argument appeals to anyone, he will be satisfied that they do. Living things which sleep.

Therefore that all animals partake in sleep is obvious from the following considerations. The animal is defined by the possession of sensation, and we hold that sleep is in some way the immobilization or fettering of sensation, and that the release or relaxation of this is waking. But none of the plants can share in either of these affections, for neither sleep nor waking is possible without sensation. But creatures which have sensation have also pain and pleasure, and those which have these must also have desire ; but none of these are possessed by plants. Evidence of this is afforded by the fact that the nutritive faculty ⟨in animals⟩ exercises its function more in sleep than in waking ; for in sleep they absorb more nourishment and grow more rapidly, which implies that they do not require sense-perception for these two purposes.

325

II. Διὰ τί δὲ καθεύδει καὶ ἐγρήγορε, καὶ διὰ
5 ποίαν τιν' αἴσθησιν ἢ ποίας, εἰ διὰ πλείους, σκεπτέον. ἐπεὶ δ' ἔνια μὲν τῶν ζῴων ἔχει τὰς αἰσθήσεις
πάσας, ἔνια δ' οὐκ ἔχει,[1] οἷον ὄψιν, τὴν δ' ἀφὴν καὶ
τὴν γεῦσιν ἅπαντ' ἔχει, πλὴν εἴ τι τῶν ζῴων ἀτελές
(εἴρηται δὲ περὶ αὐτῶν ἐν τοῖς περὶ ψυχῆς),[a]
ἀδύνατον δ' ἐστὶν ἁπλῶς ὁποιανοῦν αἴσθησιν αἰ-
10 σθάνεσθαι τὸ καθεῦδον ζῷον, φανερὸν ὅτι πᾶσιν
ἀναγκαῖον ὑπάρχειν ταὐτὸ πάθος ἐν τῷ καλουμένῳ
ὕπνῳ· εἰ γὰρ τῇ μὲν τῇ δὲ μή, ταύτῃ καθεῦδον
αἰσθήσεται, τοῦτο δ' ἀδύνατον.

Ἐπεὶ δ' ὑπάρχει καθ' ἑκάστην αἴσθησιν τὸ μέν
τι ἴδιον τὸ δέ τι κοινόν, ἴδιον μὲν οἷον τῇ ὄψει τὸ
15 ὁρᾶν, τῇ δ' ἀκοῇ τὸ ἀκούειν, ταῖς δ' ἄλλαις κατὰ
τὸν αὐτὸν τρόπον· ἔστι δέ τις καὶ κοινὴ δύναμις
ἀκολουθοῦσα πάσαις, ᾗ καὶ ὅτι ὁρᾷ καὶ ἀκούει[2]
αἰσθάνεται (οὐ γὰρ δὴ τῇ γε ὄψει ὁρᾷ ὅτι ὁρᾷ,
καὶ κρίνει δὴ καὶ δύναται κρίνειν ὅτι ἕτερα τὰ
γλυκέα τῶν λευκῶν οὔτε γεύσει οὔτε ὄψει οὔτ'
20 ἀμφοῖν, ἀλλά τινι κοινῷ μορίῳ τῶν αἰσθητηρίων
ἁπάντων· ἔστι μὲν γὰρ μία αἴσθησις, καὶ τὸ
κύριον αἰσθητήριον ἕν, τὸ δ' εἶναι αἰσθήσει τοῦ
γένους ἑκάστου ἕτερον, οἷον ψόφου καὶ χρώματος),
τοῦτο δ' ἅμα τῷ ἁπτικῷ μάλισθ' ὑπάρχει (τοῦτο
μὲν γὰρ χωρίζεται τῶν ἄλλων αἰσθητηρίων, τὰ δ'
25 ἄλλα τούτου ἀχώριστα· εἴρηται δὲ περὶ αὐτῶν ἐν
τοῖς περὶ ψυχῆς θεωρήμασιν)·[b] φανερὸν τοίνυν ὅτι

---

[1] ἔχει VYE₂ : ἔχουσιν. [2] ἀκούει καὶ EMPSY.

[a] De An. 414 b 3. [b] De An. iii. 13.

II. We must now consider why a creature sleeps *The purpose* *of sleep.* or wakes, and to what sort of sense or senses (if there be more than one) these affections are due. Since some animals have all the senses and some have not (*e.g.*, some have not sight), but all have touch and taste, unless they are imperfect (we have touched upon this subject in the treatise *On the Soul* *a*), and since it is impossible, strictly speaking, for the sleeping animal to have any sensation at all, it is evident that in the state called sleep all experience the same affection ; for if one sense is affected and another not, then the animal will perceive with the latter sense while it is asleep, and this is impossible.

Now every sense has both a special function of its *Special and* own and something shared with the rest. The special *"common"* *sense.* function, *e.g.*, of the visual sense is seeing, that of the auditory, hearing, and similarly with the rest ; but there is also a common faculty associated with them all, whereby one is conscious that one sees and hears (for it is not by *sight* that one is aware that one sees ; and one judges and is capable of judging that sweet is different from white not by taste, nor by sight, nor by a combination of the two, but by some part which is common to all the sense organs ; for there is one sense-faculty, and one paramount sense organ, but the mode of its sensitivity varies with each class of sensible objects, *e.g.*, sound and colour) ; and this is closely connected with the sense of touch (for this is separable from the other sense organs, but the others are inseparable from it. We have discussed this in our speculations *On the Soul* *b*). It is clear then that

waking and sleep are an affection of this common sense-organ. This explains why they are attributes of all animals; for touch alone is common to all animals. For if sleep were due to an affection of all the senses, it would be strange that these senses, which need not and in a way cannot act simultaneously, should necessarily be inoperative and immobile simultaneously; the direct opposite would be more probable—that they should not all rest simultaneously. But our present theory gives a rational account of them also. For when that sense organ which is master of all the rest, and which all the rest subserve, is affected in any way, all the others must be affected too, but it is not necessary that when one of them is incapacitated this master sense should also be incapacitated.

Sleep is an affection of the "common sense."

It is clear for many reasons that sleep does not consist simply in the fact that the senses are inactive or out of use, nor in the inability to feel any sensation. For this happens in fainting fits; fainting is an incapacitation of the senses, and some forms of derangement also are of this kind. Again those who have the veins in the neck compressed become unconscious. But sleep results when the incapacity for use occurs not in any chance sense organ, nor for any haphazard cause, but when, as we have just explained, it resides in the primary organ by which one perceives all things; when this is incapacitated, all the sense organs must lose their capacity for sensation; but when one of the latter is incapacitated, the former need not be so affected.

We have next to consider the cause of sleep, and what sort of affection it is. Now there are several causes—we recognize as such the final, the efficient,

The cause of sleep.

ARISTOTLE

ὅθεν ἡ ἀρχὴ τῆς κινήσεως, καὶ τὴν ὕλην καὶ τὸν
λόγον αἴτιον εἶναί φαμεν), πρῶτον μὲν οὖν ἐπειδὴ
λέγομεν τὴν φύσιν ἕνεκά του ποιεῖν, τοῦτο δ' ἀγαθόν τι, τὴν δ' ἀνάπαυσιν παντὶ τῷ πεφυκότι κινεῖσθαι, μὴ δυναμένῳ δ' ἀεὶ καὶ συνεχῶς κινεῖσθαι
20 μεθ' ἡδονῆς ἀναγκαῖον εἶναι καὶ ὠφέλιμον, τῷ δ'
ὕπνῳ αὐτῇ τῇ ἀληθείᾳ[1] προσάπτουσι τὴν μεταφορὰν
ταύτην ὡς ἀναπαύσει ὄντι· ὥστε σωτηρίας ἕνεκα
τῶν ζῴων ὑπάρχει. ἡ δ' ἐγρήγορσις τέλος· τὸ γὰρ
αἰσθάνεσθαι καὶ τὸ φρονεῖν πᾶσι τέλος οἷς ὑπάρχει
25 θάτερον αὐτῶν· βέλτιστα γὰρ ταῦτα, τὸ δὲ τέλος
βέλτιστον. ὥστε ἀναγκαῖον ἑκάστῳ τῶν ζῴων
ὑπάρχειν τὸν ὕπνον. λέγω δ' ἐξ ὑποθέσεως τὴν
ἀνάγκην, ὅτι εἰ ζῷον ἔσται ἔχον τὴν αὐτοῦ φύσιν,
ἐξ ἀνάγκης τιν' ὑπάρχειν αὐτῷ δεῖ, καὶ τούτων
ὑπαρχόντων ἕτερα ὑπάρχειν.

Ἔτι δὲ ποίας κινήσεως καὶ πράξεως ἐν τοῖς
30 σώμασι γιγνομένης συμβαίνει τό τε ἐγρηγορέναι
καὶ τὸ καθεύδειν τοῖς ζῴοις, μετὰ ταῦτα λεκτέον.
τοῖς μὲν οὖν ἄλλοις ζῴοις καθάπερ τοῖς ἐναίμοις
ὑποληπτέον εἶναι τὰ αἴτια τοῦ πάθους, ἢ ταὐτὰ ἢ
τὰ ἀνάλογον, τοῖς δ' ἐναίμοις ἅπερ τοῖς ἀνθρώποις·
ὥστ' ἐκ τούτων πάντα θεωρητέον.

Ὅτι μὲν οὖν ἡ τῆς αἰσθήσεως ἀρχὴ γίνεται ἀπὸ
τοῦ αὐτοῦ μέρους τοῖς ζῴοις ἀφ' οὗπερ καὶ ἡ τῆς
κινήσεως, διώρισται πρότερον ἐν ἑτέροις. αὕτη δ'
ἐστὶ τριῶν διωρισμένων τόπων ὁ μέσος κεφαλῆς
καὶ τῆς κάτω κοιλίας. τοῖς μὲν οὖν ἐναίμοις τοῦτ'
5 ἐστὶ τὸ περὶ τὴν καρδίαν μέρος· πάντα γὰρ τὰ

---

[1] δι' αὐτὴν τὴν ἀλήθειαν LMSU.

[a] The three positions are the head, the heart and the
stomach. *Cf. De Part. An.* 656 b 5.

330

the material and the formal cause. First of all then, since we hold that nature acts with some end in view, and that this end is a good, and that to everything which naturally moves, but cannot with pleasure move always and continuously, rest is necessary and beneficial ; and since in the light of the facts sleep is metaphorically called " rest " : it follows that the object of sleep is to preserve animal life. But the *end* of an animal is the state of waking ; for perception or thinking is the proper end of all creatures which have either of these capacities, since they represent what is best, and the end is what is best. Hence sleep belongs necessarily to every animal. I use the word " necessarily " on the assumption that if an animal is to exist and to realize its own nature certain characteristics must necessarily belong to it, and that if these belong to it others must belong also.

Next we have to say from what kind of movement and action, taking place in the body, sleep and waking arise in animals. We must assume that the causes of the affection in other animals are the same as, or analogous to, those operative in sanguineous animals, and the same for sanguineous animals as for men. Thus we must approach the whole subject in the light of these known facts.

The nature of sleep and waking.

In another place it has been laid down that sense-perception originates in the same part of an animal's body as movement does. This spot is of three definite positions, that which lies between the head and the lower part of the abdomen.[a] In sanguineous animals this is the region about the heart ; for all sanguineous

331

456 a

ἔναιμα καρδίαν ἔχει, καὶ ἡ ἀρχὴ τῆς κινήσεως καὶ
τῆς αἰσθήσεως τῆς κυρίας ἐντεῦθέν ἐστιν. τῆς μὲν
οὖν κινήσεως φανερὸν ὅτι καὶ ἡ τοῦ πνεύματος
ἀρχὴ καὶ ὅλως ἡ τῆς καταψύξεώς ἐστιν ἐνταῦθα,
καὶ τὸ ἀναπνεῖν τε καὶ τὸ ὑγρῷ καταψύχεσθαι πρός
10 γε σωτηρίαν τοῦ ἐν τούτῳ τῷ μορίῳ θερμοῦ ἡ
φύσις πεπόρικεν. ῥηθήσεται δὲ περὶ αὐτῆς ὕστερον
καθ᾽ αὑτήν. τοῖς δ᾽ ἀναίμοις καὶ τοῖς ἐντόμοις καὶ
μὴ δεχομένοις πνεῦμα ἐν τῷ ἀνάλογον τὸ σύμφυτον
πνεῦμα ἀναφυσώμενον καὶ συνιζάνον φαίνεται.
δῆλον δὲ τοῦτο ἐπὶ τῶν ὁλοπτέρων, οἷον σφηκῶν
15 καὶ μελισσῶν, καὶ ἐν ταῖς μυίαις καὶ ὅσα τοιαῦτα.
ἐπεὶ δὲ κινεῖν μέν τι ἢ ποιεῖν ἄνευ ἰσχύος ἀδύνατον,
ἰσχὺν δὲ ποιεῖ ἡ τοῦ πνεύματος κάθεξις, τοῖς μὲν
εἰσφερομένοις ἡ θύραθεν, τοῖς δὲ μὴ ἀναπνέουσιν
ἡ σύμφυτος (διὸ καὶ βομβοῦντα φαίνεται τὰ πτε-
ρωτά, ὅταν κινῆται, τῇ τρίψει τοῦ πνεύματος
20 προσπίπτοντος πρὸς τὸ ὑπόζωμα τῶν ὁλοπτέρων),
κινεῖται δὲ πᾶν αἰσθήσεώς τινος γινομένης, ἢ οἰκείας ἢ ἀλλοτρίας, ἐν τῷ πρώτῳ αἰσθητηρίῳ· εἰ
δὴ[1] ἐστὶν ὁ ὕπνος καὶ ἡ ἐγρήγορσις πάθη τοῦ μορίου
τούτου, ἐν ᾧ μὲν τόπῳ καὶ ἐν ᾧ μορίῳ πρώτῳ
γίνεται ὁ ὕπνος καὶ ἡ ἐγρήγορσις, φανερόν.

25 Κινοῦνται δ᾽ ἔνιοι καθεύδοντες καὶ ποιοῦσι πολλὰ
ἐγρηγορικά, οὐ μέντοι ἄνευ φαντάσματος καὶ αἰσθήσεώς τινος· τὸ γὰρ ἐνύπνιόν ἐστιν αἴσθημα
τρόπον τινά. λεκτέον δὲ περὶ αὐτῶν ὕστερον.
διότι δὲ τὰ μὲν ἐνύπνια μνημονεύουσιν ἐγερθέντες,

----

[1] δὴ Bonitz : δ᾽.

animals possess a heart, and both movement and the dominant sense-perception originate there. As for movement, it is clear that breathing and in general the process of cooling takes its rise here, and that nature has supplied both breathing and the power of cooling by moisture with a view to the conservation of the heat in that part. We shall discuss this separately later on. In bloodless animals and insects and creatures which do not respire, the naturally inherent breath is seen expanding and contracting in the part which corresponds to the heart in other animals. This is obvious in the case of the holoptera, such as wasps and bees, and among flies and similar creatures. Since it is impossible to make any movement, or do any action, without strength, and the holding of the breath produces strength—breath from outside in the case of animals which inhale, and inherent breath in the case of those which do not (which is why winged insects of the class holoptera are observed to buzz when they move, through the friction of the breath pulsating against the diaphragm) ; and since all sensual movement is associated with some sensation, either internal or external, in the primary sense organ : then if sleep and waking are affections of this part of the body, it is clear in what place and in what part of the body sleep and waking originate.

Some people move when they are asleep, and perform various waking acts, but not without some mental image and sense-perception ; for a dream is in a way a sense-impression. We must discuss this subject later. Why men remember their dreams after they have been awakened, and yet fail to

Dreams in sleep.

τὰς δ' ἐγρηγορικὰς πράξεις οὐ μνημονεύουσιν, ἐν
τοῖς προβληματικοῖς εἴρηται.

30 III. Ἐχόμενον δὲ τῶν εἰρημένων ἐστὶν ἐπελθεῖν
τίνων γινομένων καὶ πόθεν ἡ ἀρχὴ τοῦ πάθους
γίγνεται, τοῦ τ' ἐγρηγορέναι καὶ τοῦ καθεύδειν.
φανερὸν δὴ ὅτι ἐπεὶ ἀναγκαῖον τῷ ζώῳ, ὅταν
αἴσθησιν ἔχῃ, τότε πρῶτον τροφήν τε λαμβάνειν καὶ
αὔξησιν, τροφὴ δ' ἐστὶ πᾶσιν ἡ ἐσχάτη τοῖς μὲν
35 ἐναίμοις ἡ τοῦ αἵματος φύσις τοῖς δ' ἀναίμοις τὸ
456 b ἀνάλογον, τόπος δὲ τοῦ αἵματος αἱ φλέβες, τούτων
δ' ἀρχὴ ἡ καρδία (φανερὸν δὲ τὸ λεχθὲν ἐκ τῶν
ἀνατομῶν)· τῆς μὲν οὖν θύραθεν τροφῆς εἰσιούσης
εἰς τοὺς δεκτικοὺς τόπους γίνεται ἡ ἀναθυμίασις εἰς
5 τὰς φλέβας, ἐκεῖ δὲ μεταβάλλουσα ἐξαιματοῦται
καὶ πορεύεται ἐπὶ τὴν ἀρχήν. εἴρηται δὲ περὶ
τούτων ἐν τοῖς περὶ τροφῆς· νῦν δ' ἀναληπτέον
ὑπὲρ αὐτῶν τούτου χάριν, ὅπως τὰς ἀρχὰς τῆς
κινήσεως θεωρήσωμεν, καὶ τί πάσχοντος τοῦ
μορίου τοῦ αἰσθητικοῦ συμβαίνει ἡ ἐγρήγορσις καὶ
ὁ ὕπνος. οὐ γάρ ἐστιν ὁ ὕπνος ἡτισοῦν ἀδυναμία
10 τοῦ αἰσθητικοῦ, καθάπερ εἴρηται· καὶ γὰρ ἔκνοια
καὶ πνιγμός τις καὶ λιποψυχία ποιεῖ τὴν τοιαύτην
ἀδυναμίαν. ἤδη δὲ γεγένηταί τισι καὶ φαντασία
λιποψυχήσασιν ἰσχυρῶς. τοῦτο μὲν οὖν ἔχει τινὰ
ἀπορίαν· εἰ γὰρ ἐνδέχεται καταδαρθεῖν τὸν λιπο-
ψυχήσαντα, ἐνδέχοιτ' ἂν ἐνύπνιον εἶναι καὶ τὸ
15 φάντασμα. πολλὰ δ' ἐστιν ἃ λέγουσιν οἱ σφόδρα
λιποψυχήσαντες καὶ δόξαντες τεθνάναι· περὶ ὧν τὸν
αὐτὸν λόγον ὑποληπτέον εἶναι πάντων.

Ἀλλὰ γὰρ ὥσπερ εἴπομεν, οὐκ ἔστιν ὁ ὕπνος
ἀδυναμία πᾶσα τοῦ αἰσθητικοῦ, ἀλλ' ἐκ τῆς περὶ
τὴν τροφὴν ἀναθυμιάσεως γίνεται τὸ πάθος τοῦτο·

remember the waking acts which they have done, has been discussed in our treatise on *Problems*.

III. Next in order we must consider in what cir- The physical explanation of sleep. cumstances and whence the affection of waking and sleeping takes its rise. It must be when an animal has sensation that it first takes in food and exhibits growth. In all cases food in its final form is, for sanguineous animals, the natural substance blood, and for bloodless animals something analogous to this. The blood has its place in the veins, and the starting-point of these is the heart. This is clear from the study of anatomy. When food from without enters the places designed to receive it, the evaporation from it passes into the veins, and changing there becomes blood and makes its way to their starting-point. This subject has been treated in our discussion of nutrition, but we must recapitulate here with a view to investigating the first steps of the process, and in what way the sensitive part is affected to make waking and sleep occur. For sleep, as has been said, is not any and every incapacity of the sensitive faculty; for such incapacity is produced by unconsciousness, throttling and faintness. Also imagination has been known to occur even in a severe faint. This fact involves a difficulty; if it is possible for one who has fainted to fall asleep, his imaginary vision might be a dream. Again, words are often spoken by people who are in so deep a trance that they seem to be dead. The same explanation must be assumed for all these cases.

But as we have said, sleep is not every incapacity How food acts. of the sensitive faculty. This affection arises from the evaporation due to food; for that which is vaporized

456 b

20 ἀναγκαῖον γὰρ τὸ ἀναθυμιώμενον μέχρι του ὠθεῖ-
σθαι, εἶτ' ἀντιστρέφειν καὶ μεταβάλλειν καθάπερ
εὔριπον. τὸ δὲ θερμὸν ἑκάστου τῶν ζῴων πρὸς τὸ
ἄνω πέφυκε φέρεσθαι· ὅταν δ' ἐν τοῖς ἄνω τόποις
γένηται, ἀθρόον πάλιν ἀντιστρέφει καὶ καταφέρεται.
25 διὸ μάλιστα γίνονται ὕπνοι ἀπὸ τῆς τροφῆς· ἀθρόον
γὰρ πολὺ τό τε ὑγρὸν καὶ τὸ σωματῶδες ἀνα-
φέρεται. ἱστάμενον μὲν οὖν βαρύνει καὶ ποιεῖ
νυστάζειν· ὅταν δὲ ῥέψῃ κάτω καὶ ἀντιστρέψαι
ἀπώσῃ τὸ θερμόν, τότε γίνεται ὁ ὕπνος καὶ τὸ ζῷον
καθεύδει. σημεῖον δὲ τούτων καὶ τὰ ὑπνωτικά·
30 πάντα γὰρ καρηβαρίαν ποιεῖ, καὶ τὰ ποτὰ καὶ
τὰ βρωτά, μήκων, μανδραγόρας, οἶνος, αἷραι. καὶ
καταφερόμενοι καὶ νυστάζοντες τοῦτο δοκοῦσι
πάσχειν, καὶ ἀδυνατοῦσιν αἴρειν τὴν κεφαλὴν καὶ
τὰ βλέφαρα. καὶ μετὰ τὰ σῖτα μάλιστα τοιοῦτος
ὁ ὕπνος· πολλὴ γὰρ ἡ ἀπὸ τῶν σιτίων ἀναθυμίασις.
35 ἔτι δ' ἐκ κόπων ἐνίων· ὁ μὲν γὰρ κόπος συντη-
κτικόν, τὸ δὲ σύντηγμα γίνεται ὥσπερ τροφὴ ἄ-
457 a πεπτος, ἂν μὴ ψυχρὸν ᾖ. καὶ νόσοι δέ τινες ταὐτὸ
τοῦτο ποιοῦσιν, ὅσαι ἀπὸ περιττώματος ὑγροῦ καὶ
θερμοῦ, οἷον συμβαίνει τοῖς πυρέττουσι καὶ ἐν τοῖς
ληθάργοις. ἔτι δ' ἡ πρώτη ἡλικία· τὰ γὰρ παιδία
5 καθεύδει σφόδρα διὰ τὸ τὴν τροφὴν ἄνω φέρεσθαι
πᾶσαν. σημεῖον δὲ τὸ ὑπερβάλλειν τὸ μέγεθος τῶν
ἄνω πρὸς τὰ κάτω κατὰ τὴν πρώτην ἡλικίαν, διὰ
τὸ ἐπὶ ταῦτα γίνεσθαι τὴν αὔξησιν. διὰ ταύτην δὲ
τὴν αἰτίαν καὶ ἐπιληπτικὰ γίνεται· ὅμοιον γὰρ ὁ
ὕπνος ἐπιλήψει, καὶ ἔστι τρόπον τινὰ ὁ ὕπνος
10 ἐπίληψις. διὸ καὶ συμβαίνει πολλοῖς ἡ ἀρχὴ τούτου
τοῦ πάθους καθεύδουσιν, καὶ καθεύδοντες μὲν
ἁλίσκονται, ἐγρηγορότες δ' οὔ· ὅταν γὰρ πολὺ

must be driven forward for a space, and then turn and change its course, like the tide in a narrow strait. Now in every animal the hot tends to rise ; when it reaches the upper parts, it turns back and descends in a dense mass. So sleepiness mostly occurs after food, for then both liquid and solid matter are carried up in considerable bulk. As this becomes stationary it weighs one down and makes him nod ; when it has shifted downwards, and by its return has driven back the hot, then sleep occurs, and the animal falls asleep. Narcotics prove this ; for they all, both liquid and solid (*e.g.*, poppy, mandragora, wine and darnel), cause head-heaviness. When men are heavy and nod, they seem to suffer this affection, and cannot raise their heads or eyelids. Sleep of this kind occurs most often after meals, for evaporation from the food is considerable ; but it sometimes follows fatigue, for fatigue acts as a solvent, and the dissolved matter, unless it is cold, has the effect of undigested food. Certain diseases produce the same result, such as arise from an excess of moisture and heat, as is the case with the feverish and comatose. The same thing is true of early childhood ; for children sleep a great deal, because all the food is borne upwards. The greater size of the upper parts in comparison with the lower in early youth proves this, and is due to the fact that growth takes place in the upward direction. Hence too they are liable to epilepsy, for sleep is like epilepsy ; indeed in a sense sleep *is* an epileptic fit. Consequently for many people epilepsy begins in sleep, and they are regularly seized with it when asleep, but not when awake. For when a large volume of vapour is carried up, as it descends again it swells the veins and chokes the passage through which respiration passes. This is why wines are not good for infants, nor for wet nurses (probably it makes no difference whether the infants or their nurses drink them); they should drink wine diluted, and in small quantities; for wine is spirituous, especially dark wine. The upper parts of infants are so full of food that for five months they do not even turn their necks; for as in the case of the very drunk, much moisture is carried upwards. Probably this is why the embryo lies quiet in the womb at first. Generally speaking, people with inconspicuous veins, and dwarfish or big-headed types, are addicted to sleep; for the veins of the former are narrow, so that the descending moisture cannot easily flow through, and with dwarfs and the big-headed the upward surge of the evaporation is considerable. But those with marked veins are not much given to sleep owing to the easy flow in the veins, unless of course they have any counteracting affection. Nor are the melancholic inclined to sleep much; for the region within is chilled, so that there is not much evaporation in their case. For this reason also they are inclined to eat much though they are spare; for their condition of body is as if they did not profit by their food. Black bile also being by nature cold cools the nutritive region and other parts, wherever there is potentially a secretion of this kind.

Thus it is obvious from what has been said that sleep is a sort of concentration or natural recession inwards of the hot matter, due to the cause described above. Hence the movement of a sleepy person is considerable. But where the heat fails he grows

Heat as a factor

457 b

ψύχεται, καὶ διὰ ψύξιν καταπίπτει τὰ βλέφαρα.
5 καὶ τὰ μὲν ἄνω κατέψυκται καὶ τὰ ἔξω, τὰ δ᾽ ἐντὸς
καὶ τὰ κάτω θερμά, οἷον τὰ περὶ τοὺς πόδας καὶ
τὰ εἴσω.

Καίτοι τις ἀπορήσειεν ἄν, ὅτι μετὰ τὰ σιτία
ἰσχυρότατος ὁ ὕπνος γίνεται, καὶ ἔστιν ὑπνωτικὰ
οἶνος καὶ ἄλλα θερμότητα[1] ἔχοντα τοιαῦτα.[2] ἔστι
10 δ᾽ οὐκ εὔλογον τὸν μὲν ὕπνον εἶναι κατάψυξιν, τὰ
δ᾽ αἴτια τοῦ καθεύδειν θερμά. πότερον οὖν τοῦτο
συμβαίνει ὅτι ὥσπερ ἡ κοιλία κενὴ μὲν οὖσα θερμή
ἐστιν, ἡ δὲ πλήρωσις αὐτὴν καταψύχει διὰ τὴν
κίνησιν, οὕτω καὶ οἱ ἐν τῇ κεφαλῇ πόροι καὶ τόποι
καταψύχονται ἀναφερομένης τῆς ἀναθυμιάσεως; ἢ
15 ὥσπερ τοῖς προσχεομένοις τὸ θερμὸν ἐξαίφνης
φρίκη γίνεται, κἀκεῖ ἀνιόντος τοῦ θερμοῦ ἀθροι-
ζόμενον τὸ ψυχρὸν καταψύχει, καὶ τὸ κατὰ φύσιν
θερμὸν ποιεῖ ἐξαδυνατεῖν καὶ ὑποχωρεῖν; ἔτι δὲ
πολλῆς ἐμπιπτούσης τροφῆς, ἣν ἀνάγει τὸ θερμόν,
ὥσπερ τὸ πῦρ ἐπιτιθεμένων τῶν ξύλων, καταψύχεται, ἕως ἂν καταπεφθῇ.
20 Γίνεται γὰρ ὁ ὕπνος, ὥσπερ εἴρηται, τοῦ σωμα-
τώδους ἀναφερομένου ὑπὸ τοῦ θερμοῦ διὰ τῶν φλεβῶν πρὸς τὴν κεφαλήν. ὅταν δὲ μηκέτι δύνηται,
ἀλλὰ τῷ πλήθει ὑπερβάλλῃ τὸ ἀναχθέν, πάλιν ἀνταπ-
ωθεῖ[3] καὶ κάτω ῥεῖ. διὸ καταπίπτουσί τε ὑπο-
25 σπωμένου τοῦ θερμοῦ τοῦ ἀνάγοντος οἱ ἄνθρωποι
(μόνον γὰρ ὀρθὸν τῶν ζῴων), καὶ ἐπιπεσὸν μὲν
ἔκνοιαν ποιεῖ, ὕστερον δὲ φαντασίαν. ἢ αἱ μὲν νῦν
λεγόμεναι λύσεις ἐνδεχόμεναι μέν εἰσι τοῦ γίνεσθαι
τὴν κατάψυξιν· οὐ μὴν ἀλλὰ κύριός γ᾽ ἐστὶν ὁ τόπος
ὁ περὶ τὸν ἐγκέφαλον, ὥσπερ ἐν ἄλλοις εἴρηται.

---

[1] θερμότητα] θερμότητας LSUM. [2] τοιαῦτα] τοιαύτας LU.

cold, and owing to this chilling his eyelids fall. The parts above and outside are now cold, but the parts below and within, *e.g.*, those about the feet and the interior of the body, are hot.

Yet one might see some difficulty in the fact that sleep is deepest after food, and that wine and other things which are naturally heating tend to produce sleep. It does not seem logical that sleep should be a chilling, while the causes of sleep are hot. Is it possible that just as the stomach is hot when empty, but the filling of it chills it because of the movement, so also the passages and regions of the head grow cold as the evaporation rises? Or is the explanation that, just as those over whom hot water is poured suddenly shiver. so in this case as the hot rises the gathering cold chills it, and makes what is naturally hot lose its power and withdraw? Besides, when much food is introduced, the hot, carrying it up, is chilled just like fire when logs are put on it, until the food is digested. *A difficulty in this explanation.*

For sleep comes, as has been said, when the solid part [a] is carried upwards by the hot through the veins to the head. But when that which is carried upwards becomes excessive in amount and can no longer ascend, it forces the hot back again and flows downwards. And so when the heat with its raising force is withdrawn, men sink down (for they alone of living creatures are erect), and the process produces loss of consciousness, and afterwards imagination. Or perhaps, while these are possible explanations of the chilling, yet the controlling factor is the region about the brain, as has been said elsewhere. The brain, or *Various suggestions.*

----

[a] *Sc.*, of the evaporation.

[3] ἀνταπωθεῖται plerique edd.

457 b

πάντων δ' ἐστὶ τῶν ἐν τῷ σώματι ψυχρότατον ὁ
ἐγκέφαλος, τοῖς δὲ μὴ ἔχουσι τὸ ἀνάλογον τούτῳ
μόριον. ὥσπερ οὖν τὸ ἀπατμίζον ὑγρὸν ὑπὸ τῆς
τοῦ ἡλίου θερμότητος, ὅταν ἔλθῃ εἰς τὸν ἄνω τόπον,
διὰ τὴν ψυχρότητα αὐτοῦ καταψύχεται καὶ συστὰν

458 a

καταφέρεται γενόμενον πάλιν ὕδωρ, οὕτως ἐν τῇ
ἀναφορᾷ τοῦ θερμοῦ τῇ πρὸς τὸν ἐγκέφαλον ἡ μὲν
περιττωματικὴ ἀναθυμίασις εἰς φλέγμα συνέρχεται
(διὸ καὶ οἱ κατάρροι φαίνονται γιγνόμενοι ἐκ τῆς
κεφαλῆς), ἡ δὲ τρόφιμος καὶ μὴ νοσώδης καταφέρεται συνισταμένη καὶ καταψύχει τὸ θερμόν.
πρὸς δὲ τὸ καταψύχεσθαι καὶ μὴ δέχεσθαι ῥᾳδίως
τὴν ἀναθυμίασιν συμβάλλεται καὶ ἡ λεπτότης καὶ
ἡ στενότης τῶν περὶ τὸν ἐγκέφαλον φλεβῶν. τῆς
μὲν οὖν καταψύξεως τοῦτ' ἐστὶν αἴτιον, καίπερ
τῆς ἀναθυμιάσεως ὑπερβαλλούσης τῇ θερμότητι.

Ἐγείρεται δ', ὅταν πεφθῇ καὶ κρατήσῃ ἡ συν-
εωσμένη θερμότης ἐν ὀλίγῳ πολλὴ ἐκ τοῦ περιεστῶτος, καὶ διακριθῇ τό τε σωματωδέστερον αἷμα
καὶ τὸ καθαρώτατον. ἔστι δὲ λεπτότατον μὲν αἷμα
καὶ καθαρώτατον τὸ ἐν τῇ κεφαλῇ, παχύτατον δὲ
καὶ θολερώτατον τὸ ἐν τοῖς κάτω μέρεσιν. παντὸς
δὲ τοῦ αἵματος ἀρχή, ὥσπερ εἴρηται καὶ ἐνταῦθα καὶ
ἐν ἄλλοις, ἡ καρδία. τῶν δ' ἐν τῇ καρδίᾳ ἑκατέρας
τῆς θαλάμης κοινὴ ἡ μέση· ἐκείνων δ' ἑκατέρα
δέχεται ἐξ ἑκατέρας τῆς φλεβός, τῆς τε μεγάλης
καλουμένης καὶ τῆς ἀορτῆς· ἐν δὲ τῇ μέσῃ γίνεται
ἡ διάκρισις. ἀλλὰ τὸ μὲν διορίζειν περὶ τούτων
ἑτέρων ἐστὶ λόγων οἰκειότερον. διὰ δὲ τὸ γίνεσθαι
ἀδιακριτώτερον τὸ αἷμα μετὰ τὴν τῆς τροφῆς
προσφορὰν ὁ ὕπνος γίνεται, ἕως ἂν διακριθῇ
τοῦ αἵματος τὸ μὲν καθαρώτερον εἰς τὰ ἄνω,
τὸ δὲ θολερώτερον εἰς τὰ κάτω· ὅταν δὲ τοῦτο

in animals which have none, whatever part corresponds to the brain, is the coldest of all parts of the body. Just as moisture vaporized by the heat of the sun, when it reaches the upper region, is chilled by the coldness of it, and after condensing becomes water again, and is carried down, so in the rising of hot matter towards the brain, the excrementitious vapour collects into phlegm (which is why catarrhs are observed to arise from the head), while the nutritive and wholesome evaporation is condensed and carried down and chills the hot. The fineness and narrowness of the veins about the brain contribute to the chilling of it and to the difficulty with which it admits the evaporation. This, then, is the cause of the chilling, although the evaporation is exceedingly hot.

Awakening occurs when digestion is complete ; Awakening. when the heat, withdrawn from the surrounding parts and concentrated in large quantity in a small space, prevails, and the more corporeal is separated from the pure blood. The blood in the head is the rarest and purest, that in the lower parts of the body is thickest and most turbid. But the source of all the blood, as has been said in this treatise and in others, is the heart ; and of the compartments of the heart the middle communicates with both the others ; and each of these receives blood from one of the blood-vessels, *viz.*, the so-called " great vein " and the aorta. The separation takes place in the middle compartment. A detailed treatment of this subject, however, would be more proper to another treatise. It is because the blood stands in greater need of discrimination after the absorption of food that sleep occurs, and it continues until the purer part of the blood is separated upwards, and the more turbid

343

downwards ; when this has happened the animals awake, being liberated from the heaviness induced by food.

We have thus explained the cause of sleep : that it is the recession of the solid matter which is carried upwards by the inherent heat in a mass to the primary sense organ ; as for what sleep is, we have shown that it is a paralysis of the first sense organ to prevent it from functioning, and is a necessary process (for an animal cannot exist apart from the conditions which fulfil its nature), its purpose being the conservation of the animal ; for rest has a conservative effect.

Summary.

ΠΕΡΙ ΕΝΥΠΝΙΩΝ

I. Μετὰ δὲ ταῦτα περὶ ἐνυπνίου ζητητέον, καὶ
458 b πρῶτον τίνι τῶν τῆς ψυχῆς φαίνεται, καὶ πότερον
τοῦ νοητικοῦ τὸ πάθος ἐστὶ τοῦτο ἢ τοῦ αἰσθητικοῦ·
τούτοις γὰρ μόνοις τῶν ἐν ἡμῖν γνωρίζομέν τι.

Εἰ δὲ χρῆσις ὄψεως ὅρασις καὶ ἀκοῆς τὸ ἀκούειν
5 καὶ ὅλως αἰσθήσεως τὸ αἰσθάνεσθαι, κοινὰ δ' ἐστὶ
τῶν αἰσθήσεων οἷον σχῆμα καὶ μέγεθος καὶ κίνησις
καὶ τἆλλα τὰ τοιαῦτα, ἴδια δ' οἷον χρῶμα ψόφος
χυμός, ἀδυνατεῖ δὲ πάντα μύοντα καὶ καθεύδοντα
ὁρᾶν, ὁμοίως δὲ καὶ ἐπὶ καὶ λοιπῶν, ὥστε[1] δῆλον
ὅτι οὐκ αἰσθανόμεθα οὐδὲν ἐν τοῖς ὕπνοις· οὐκ ἄρα
γε τῇ αἰσθήσει τὸ ἐνύπνιον αἰσθανόμεθα.

10 Ἀλλὰ μὴν οὐδὲ τῇ δόξῃ. οὐ γὰρ μόνον τὸ προσιόν
φαμεν ἄνθρωπον ἢ ἵππον εἶναι, ἀλλὰ καὶ λευκὸν ἢ
καλόν· ὧν ἡ δόξα ἄνευ αἰσθήσεως οὐδὲν ἂν φήσειεν,
οὔτ' ἀληθῶς οὔτε ψευδῶς. ἐν δὲ τοῖς ὕπνοις
συμβαίνει τὴν ψυχὴν τοῦτο ποιεῖν· ὁμοίως γὰρ ὅτι
15 ἄνθρωπος καὶ ὅτι λευκὸς ὁ προσιὼν δοκοῦμεν ὁρᾶν.
ἔτι παρὰ τὸ ἐνύπνιον ἐννοοῦμεν ἄλλο τι, καθάπερ
ἐν τῷ ἐγρηγορέναι αἰσθανόμενοί τι· περὶ οὗ γὰρ

---

[1] ὥστε om. LSU, Bekker.

ON DREAMS

I. Our next inquiry is concerned with the dream; in the first place, to what faculty of the soul it appears, *i.e.*, whether the affection belongs to the intellectual or to the sensitive faculty; for these are the only faculties within us by which we can attain to knowledge.

Now if the employment of the visual faculty is seeing, and that of the auditory faculty is hearing, and generally that of the sensitive faculty is perceiving; and if some sensibles such as shape, size, movement, etc., are common to all the senses, while others such as colour, sound and flavours are peculiar; and if everything that has its eyes shut and is asleep is incapable of seeing, and similarly with the other senses, so that clearly we have no perception in sleep at all: then it follows that it is not by sense-perception that we see our dreams.

Nor is it by opinion. For we do not merely say that the thing approaching is a man or a horse, but also that it is white or handsome; but on these points opinion could not pronounce, either truly or falsely, without perception. Yet the soul actually does this in sleep; for we seem to see that the approaching object is white no less than that it is a man. Moreover, besides the dream we think something else, just as when we are percipient while awake; for we often cogitate about what we per-

Dreaming is not perception,

nor opinion.

458 b

αἰσθανόμεθα, πολλάκις καὶ διανοούμεθά τι. οὕτω
καὶ ἐν τοῖς ὕπνοις παρὰ τὰ φαντάσματα ἐνίοτε
ἄλλα ἐννοοῦμεν, φανείη δ' ἄν τῳ τοῦτο, εἴ τις προσ-
20 έχοι τὸν νοῦν καὶ πειρῷτο μνημονεύειν ἀναστάς.
ἤδη δέ τινες καὶ ἑωράκασιν ἐνύπνια τοιαῦτα οἷον
οἱ δοκοῦντες κατὰ τὸ μνημονικὸν παράγγελμα
τίθεσθαι τὰ προβαλλόμενα· συμβαίνει γὰρ αὐτοῖς
πολλάκις ἄλλο τι παρὰ τὸ ἐνύπνιον τίθεσθαι πρὸ
ὀμμάτων εἰς τὸν τόπον φάντασμα. ὥστε δῆλον ὅτι
25 οὔτε ἐνύπνιον πᾶν τὸ ἐν ὕπνῳ φάντασμα, καὶ ὅτι
ὃ ἐννοοῦμεν τῇ δόξῃ δοξάζομεν.

Δῆλον δὲ περὶ τούτων ἁπάντων τό γε τοσοῦτον,
ὅτι τῷ αὐτῷ ᾧ καὶ ἐγρηγορότες ἐν ταῖς νόσοις
ἀπατώμεθα, ὅτι τοῦτ' αὐτὸ καὶ ἐν τῷ ὕπνῳ ποιεῖ
τὸ πάθος. καὶ ὑγιαίνουσι δὲ καὶ εἰδόσιν ὅμως ὁ
ἥλιος ποδιαῖος εἶναι δοκεῖ. ἀλλ' εἴτε δὴ ταὐτὸν
30 εἴθ' ἕτερον τὸ φανταστικὸν τῆς ψυχῆς καὶ τὸ
αἰσθητικόν, οὐδὲν ἧττον οὐ γίνεται ἄνευ τοῦ ὁρᾶν
καὶ αἰσθάνεσθαί τι· τὸ γὰρ παρορᾶν καὶ παρ-
ακούειν ὁρῶντος ἀληθές τι καὶ ἀκούοντος, οὐ μέντοι τοῦτο ὃ οἴεται. ἐν δὲ τῷ ὕπνῳ ὑπόκειται μη-
459 a δὲν ὁρᾶν μηδ' ἀκούειν μηδ' ὅλως αἰσθάνεσθαι. ἆρ'
οὖν τὸ μὲν μηδὲν ὁρᾶν ἀληθές, τὸ δὲ μηδὲν πάσχειν
τὴν αἴσθησιν οὐκ ἀληθές, ἀλλ' ἐνδέχεται καὶ τὴν
ὄψιν πάσχειν τι καὶ τὰς ἄλλας αἰσθήσεις, ἕκαστον
5 δὲ τούτων ὥσπερ ἐγρηγορότος προσβάλλει μέν
πως τῇ αἰσθήσει, οὐχ οὕτω δὲ ὥσπερ ἐγρηγορότος·
καὶ ὁτὲ μὲν ἡ δόξα λέγει ὅτι ψεῦδος τὸ ὁρώμενον,
ὥσπερ ἐγρηγορόσιν, ὁτὲ δὲ κατέχεται καὶ ἀκολουθεῖ τῷ φαντάσματι.

ceive. And so in sleep we sometimes have other thoughts besides the mental pictures. This will become obvious to anyone if he concentrates and tries to remember his dream immediately upon rising. Indeed some men have seen such dreams, *e.g.*, those who think that they are arranging suggested subjects according to some principle of memorizing ; for they often find themselves envisaging some other imaginary concept, apart from the dream, into position. Thus it is clear that not every image seen in sleep is a dream, and that what we think at such times is the effect of our opinion.

On the whole question this much at any rate is clear, that the same faculty by which we are deceived in illness when we are awake causes this affection in sleep also. The sun appears to measure a foot across even to men who are in health, and know its real measurement. But whether the imaginative part of the soul is the same as the sensitive part or not, in any case this affection does not occur without some act of vision and perception ; for to see or hear incorrectly implies that the subject sees or hears something real, but not what he thinks it is ; but in sleep it is assumed that one neither sees nor hears nor perceives anything at all. Perhaps that one sees nothing is true, but that the sense is unaffected is not true, and it is possible that the sight and the other senses are somehow affected, each of these making some impression upon the sense-faculty, as when one is awake, but not in quite the same way. And sometimes opinion states, as it does to those who are awake, that what is seen is false ; at other times opinion is fettered, and assents to the mental picture.

Yet some form of perception is necessary.

459 a

Ὅτι μὲν οὖν οὐκ ἔστι τοῦ δοξάζοντος οὐδὲ τοῦ
διανοουμένου τὸ πάθος τοῦτο ὃ καλοῦμεν ἐνυπνιά-
10 ζειν φανερόν. ἀλλ' οὐδὲ τοῦ αἰσθανομένου ἁπλῶς·
ὁρᾶν γὰρ ἂν ἦν καὶ ἀκούειν ἁπλῶς. ἀλλὰ πῶς
δὴ καὶ τίνα τρόπον, ἐπισκεπτέον. ὑποκείσθω δ',
ὅπερ ἐστὶ καὶ φανερόν, ὅτι τοῦ αἰσθητικοῦ τὸ
πάθος, εἴπερ καὶ ὁ ὕπνος· οὐ γὰρ ἄλλῳ μέν τινι
τῶν ζῴων ὑπάρχει ὁ ὕπνος, ἄλλῳ δὲ τὸ ἐνυπνιάζειν,
15 ἀλλὰ τῷ αὐτῷ. ἐπεὶ δὲ περὶ φαντασίας ἐν τοῖς
περὶ ψυχῆς εἴρηται, καὶ ἔστι μὲν τὸ αὐτὸ τῷ
αἰσθητικῷ τὸ φανταστικόν, τὸ δ' εἶναι φανταστι-
κῷ καὶ αἰσθητικῷ ἕτερον, ἔστι δὲ φαντασία ἡ
ὑπὸ τῆς κατ' ἐνέργειαν αἰσθήσεως γινομένη κίνησις,
τὸ δ' ἐνύπνιον φάντασμά τι φαίνεται εἶναι (τὸ γὰρ
20 ἐν ὕπνῳ φάντασμα ἐνύπνιον λέγομεν, εἴθ' ἁπλῶς
εἴτε τρόπον τινὰ γινόμενον), φανερὸν ὅτι τοῦ
αἰσθητικοῦ μέν ἐστι τὸ ἐνυπνιάζειν, τούτου δ' ᾗ
τὸ φανταστικόν.

II. Τί δ' ἐστὶ τὸ ἐνύπνιον καὶ πῶς γίνεται, ἐκ
τῶν περὶ τὸν ὕπνον συμβαινόντων μάλιστ' ἂν
25 θεωρήσαιμεν. τὰ γὰρ αἰσθητὰ καθ' ἕκαστον
αἰσθητήριον ἡμῖν ἐμποιοῦσιν αἴσθησιν, καὶ τὸ
γινόμενον ὑπ' αὐτῶν πάθος οὐ μόνον ἐνυπάρχει ἐν
τοῖς αἰσθητηρίοις ἐνεργουσῶν τῶν αἰσθήσεων, ἀλλὰ
καὶ ἀπελθουσῶν.

Παραπλήσιον γὰρ τὸ πάθος ἐπί τε τούτων καὶ
ἐπὶ τῶν φερομένων ἔοικεν εἶναι. καὶ γὰρ ἐπὶ τῶν
30 φερομένων τοῦ κινήσαντος οὐκέτι θιγγάνοντος
κινεῖται· τὸ γὰρ κινῆσαν ἐκίνησεν ἀέρα τινά, καὶ
πάλιν οὗτος κινούμενος ἕτερον. καὶ τοῦτον δὴ

ᵃ De An. 429 a 1.

Now it is clear that the experience which we call It is closely
dreaming does not belong to the opinionative or to connected
the intellective faculty ; nor to the sensitive faculty, imagina-
in the normal sense ; for in that case it would be tion.
possible ⟨in a dream⟩ to see and hear normally. But
we must consider how and in what way dreaming
occurs. Let us first lay down what is quite obvious,
that it is an experience of the sensitive faculty just
as sleep is ; for we do not find that sleep is the characteristic of one animal and dreaming of another,
but both are found in the same subject. But since
we have discussed imagination in the treatise *On the
Soul*,[a] and the imaginative is the same as the sensitive
faculty, although the imaginative and sensitive are
different in essence ; and since imagination is the
process set up by a sense-faculty in a state of activity,
and a dream appears to be some sort of mental image
(for an image which appears in sleep, whether simply
or in a special sense, we call a dream) ; it is clear that
dreaming belongs to the sensitive faculty, but belongs
to it *qua* imaginative.

II. What a dream is and how it originates, we can
best study from the circumstances which attend sleep.
For the sensible objects corresponding to each sense
organ produce sensation in us, and the affection produced by them persists in the sense organs not only
while the sensations are active, but also after they
have gone.

The affection in these cases seems to be similar to
that observable in projectiles. For in the case of the A
latter movement continues even when the moving mechanical
agent is no longer in contact with them ; for the produces an
moving agent imparts motion to a portion of the air effect which
which, being moved, again moves another portion. after the
original

τὸν τρόπον, ἕως ἂν στῇ, ποιεῖται τὴν κίνησιν καὶ
459 b ἐν ἀέρι καὶ ἐν τοῖς ὑγροῖς.

Ὁμοίως δ᾽ ὑπολαβεῖν τοῦτο δεῖ καὶ ἐπ᾽ ἀλ-
λοιώσεως· τὸ γὰρ θερμανθὲν ὑπὸ τοῦ θερμοῦ τὸ
πλησίον θερμαίνει, καὶ τοῦτο διαδίδωσιν ἕως τῆς
ἀρχῆς, ὥστε καὶ ἐν ᾧ τὸ αἰσθάνεσθαι, ἐπειδή ἐστιν
5 ἀλλοίωσίς τις ἡ κατ᾽ ἐνέργειαν αἴσθησις, ἀνάγκη
τοῦτο συμβαίνειν. διὸ τὸ πάθος ἐστὶν οὐ μόνον
ἐν αἰσθανομένοις τοῖς αἰσθητηρίοις, ἀλλὰ καὶ ἐν
πεπαυμένοις, καὶ ἐν βάθει καὶ ἐπιπολῆς. φανερὸν
δ᾽ ὅταν συνεχῶς αἰσθανώμεθά τι· μεταφερόντων
γὰρ τὴν αἴσθησιν ἀκολουθεῖ τὸ πάθος, οἷον ἐκ
10 τοῦ ἡλίου εἰς τὸ σκότος· συμβαίνει γὰρ μηδὲν ὁρᾶν
διὰ τὴν ἔτι ὑποῦσαν κίνησιν ἐν τοῖς ὄμμασιν ὑπὸ
τοῦ φωτός. κἂν πρὸς ἓν χρῶμα πολὺν χρόνον
βλέψωμεν ἢ λευκὸν ἢ χλωρόν, τοιοῦτον φαίνεται
ἐφ᾽ ὅπερ ἂν τὴν ὄψιν μεταβάλωμεν. κἂν πρὸς
τὸν ἥλιον βλέψαντες ἢ ἄλλο τι λαμπρὸν μύσωμεν,
15 παρατηρήσασι φαίνεται κατ᾽ εὐθυωρίαν, ᾗ συμβαίνει τὴν ὄψιν ὁρᾶν, πρῶτον μὲν τοιοῦτον τὴν
χρόαν, εἶτα μεταβάλλει εἰς φοινικοῦν κἄπειτα
πορφυροῦν, ἕως ἂν εἰς τὴν μέλαιναν ἔλθῃ χρόαν
καὶ ἀφανισθῇ. καὶ αἱ ἀπὸ τῶν κινουμένων δὲ
μεταβάλλουσιν, οἷον ἀπὸ τῶν ποταμῶν, μάλιστα
20 δ᾽ ἀπὸ τῶν τάχιστα ῥεόντων· φαίνεται γὰρ τὰ
ἠρεμοῦντα κινούμενα. γίνονται δὲ καὶ ἀπὸ τῶν
μεγάλων ψόφων δύσκωφοι καὶ ἀπὸ τῶν ἰσχυρῶν
ὀσμῶν δύσοσμοι, καὶ ἐπὶ τῶν ὁμοίων. ταῦτά γε
δὴ φανερῶς συμβαίνει τοῦτον τὸν τρόπον.

This is the way in which objects continue moving, *impulse has ceased.* both in air and in liquids, until they come to a standstill.

One must suppose that something similar takes *So too with heat* place in change of state as well; for that which is heated by the hot in turn heats that which is near it, and this hands on the succession until the beginning is reached again. So in the case of sensation, since sensation in active operation is a kind of change of state, this must also happen. Consequently this affection persists in the sense organs, both deep down and on the surface, not only while they are perceiving, but also when they have ceased to do so. This is *and light.* clear in cases of continuous perception; for even when we change our sensation the affection goes on, for instance, when we turn from sunlight to darkness; the result is that we see nothing, because the movement produced in our eyes by the light still persists. Again if we look for a long time at one colour—say white or green—any object to which we shift our gaze appears to be that colour. And if, after looking at the sun or some other bright object, we shut our eyes, then, if we watch carefully, it appears in the same direct line as we saw it before, first of all in its own proper colour; then it changes to red, and then to purple, until it fades to black and disappears. The same persistence of vision occurs when we turn our gaze from moving objects—*e.g.*, rivers, especially when they flow very rapidly; for then objects really at rest appear to be moving. So men become deafened by loud noises and have their sense of smell damaged by strong odours, and so on. These affections obviously come about in the way described above.

Ὅτι δὲ ταχὺ τὰ αἰσθητήρια καὶ μικρᾶς διαφορᾶς
25 αἰσθάνεται, σημεῖον τὸ ἐπὶ τῶν ἐνόπτρων γινό-
μενον· περὶ οὗ κἂν καθ' αὑτὸ[1] ἐπιστήσας σκέψαιτό
τις ἂν καὶ ἀπορήσειεν. ἅμα δ' ἐξ αὐτοῦ δῆλον ὅτι
ὥσπερ καὶ ἡ ὄψις πάσχει, οὕτω καὶ ποιεῖ τι. ἐν
γὰρ τοῖς ἐνόπτροις τοῖς σφόδρα καθαροῖς, ὅταν
τῶν καταμηνίων ταῖς γυναιξὶ γινομένων ἐμ-
30 βλέψωσιν εἰς τὸ κάτοπτρον, γίνεται τὸ ἐπιπολῆς
τοῦ ἐνόπτρου οἷον νεφέλη αἱματώδης· κἂν μὲν
καινὸν ᾖ τὸ κάτοπτρον, οὐ ῥᾴδιον ἐκμάξαι τὴν
τοιαύτην κηλῖδα, ἐὰν δὲ παλαιόν, ῥᾷον. αἴτιον δ',
460 a ὥσπερ εἴπομεν, ὅτι οὐ μόνον πάσχει τι ἡ ὄψις
ὑπὸ τοῦ ἀέρος, ἀλλὰ καὶ ποιεῖ τι καὶ κινεῖ, ὥσπερ
καὶ τὰ λαμπρά· καὶ γὰρ ἡ ὄψις τῶν λαμπρῶν καὶ
ἐχόντων χρῶμα. τὰ μὲν οὖν ὄμματα εὐλόγως,
5 ὅταν ᾖ τὰ καταμήνια, διάκειται ὥσπερ καὶ ἕτερον
μέρος ὁτιοῦν· καὶ γὰρ φύσει τυγχάνουσι φλεβώ-
δεις ὄντες. διὸ γινομένων τῶν καταμηνίων διὰ
ταραχὴν καὶ φλεγμασίαν αἱματικὴν ἡμῖν μὲν ἡ ἐν
τοῖς ὄμμασι διαφορὰ ἄδηλος, ἔνεστι δέ (ἡ γὰρ
αὐτὴ φύσις σπέρματος καὶ καταμηνίων), ὁ δ' ἀὴρ
10 κινεῖται ὑπ' αὐτῶν, καὶ τὸν ἐπὶ τῶν κατόπτρων
ἀέρα συνεχῆ ὄντα ποιόν τινα ποιεῖ καὶ τοιοῦτον
οἷον αὐτὸς πάσχει· ὁ δὲ τοῦ κατόπτρου τὴν ἐπι-
φάνειαν. ὥσπερ δὲ τῶν ἱματίων τὰ μάλιστα
καθαρὰ τάχιστα κηλιδοῦται· τὸ γὰρ καθαρὸν ἀκριβῶς δηλοῖ ὅ τι ἂν δέξηται, καὶ τὸ[2] μάλιστα τὰς ἐλα-
15 χίστας κινήσεις. ὁ δὲ χαλκὸς διὰ μὲν τὸ λεῖος
εἶναι ὁποιασοῦν ἁφῆς μάλιστα αἰσθάνεται (δεῖ δὲ
νοῆσαι οἷον τρίψιν οὖσαν τὴν τοῦ ἀέρος ἁφὴν καὶ

---

[1] κἂν καθ' αὑτὸ(ν) ELSU : καὶ αὐτοῦ MY, Bekker.
[2] τὸ EML, Biehl : om. vulgo.

An example of the rapidity with which the sense organs perceive even a slight difference is found in the behaviour of mirrors ; a subject which, even considered by itself, would give scope for careful study and investigation. At the same time it is quite clear from this instance that the organ of sight not only is acted upon by its object, but acts reciprocally upon it. If a woman looks into a highly polished mirror during the menstrual period, the surface of the mirror becomes clouded with a blood-red colour (and if the mirror is a new one the stain is not easy to remove, but if it is an old one there is less difficulty). The reason for this is that, as we have said, the organ of sight not only is acted upon by the air, but also sets up an active process, just as bright objects do ; for the organ of sight is itself a bright object possessing colour. Now it is reasonable to suppose that at the menstrual periods the eyes are in the same state as any other part of the body ; and there is the additional fact that they are naturally full of blood-vessels. Thus when menstruation takes place, as the result of a feverish disorder of the blood, the difference of condition in the eyes, though invisible to us, is none the less real (for the nature of the menses and of the semen is the same) ; and the eyes set up a movement in the air. This imparts a certain quality to the layer of air extending over the mirror, and assimilates it to itself ; and this layer affects the surface of the mirror. Now the cleaner one's clothes are, the more readily they become stained, because a clean object exhibits distinctly any mark which it receives ; and the cleaner the object, the better it exhibits even the slightest effects produced upon it. In the same way the bronze surface of the mirror, being smooth, is peculiarly sensitive to any impact (and one must regard the impact of the air as a kind

Vision affects the object seen.

of friction or impression or washing [a]) ; and because the surface is clean, any impact upon it is clearly apparent. The reason why the stain does not readily come off from a new mirror is that the surface is clean and smooth ; in such cases the stain penetrates deeply all over, deeply because the surface is clean, and all over because it is smooth. The reason why it does not persist in old mirrors is that it does not penetrate so deeply, but is comparatively superficial. All this proves first that movement is produced even by minute differences, secondly that perception is very rapid, and thirdly that the sense organ which perceives colours is not only affected by colours, but also in turn affects them. This conclusion is further supported by what occurs with wines and with the preparation of perfumes. For oil which has been prepared quickly takes on the scent of what is near it, and wines are affected in the same way ; for they acquire the smell not merely of what is put into them, or mixed in small quantities with them, but even of that which is placed or grows near the vessels which contain them.

With regard to our original inquiry, one fact, which is clear from what we have said, may be laid down— that the sensation still remains perceptible even after the external object perceived has gone, and moreover that we are easily deceived about our perceptions when we are in emotional states, some in one state and others in another ; *e.g.*, the coward in his fear, the lover in his love ; so that even from a very faint resemblance the coward thinks that he sees his enemy, and the lover his loved one ; and in proportion to his excitement, his imagination is stimulated by a

Persistence of vision as an explanation of dreams.

---

[a] *i.e.*, washing *on*, as of a pigment. There is no question of cleaning. The surface is *ex hypothesi* clean.

αὐτὸν δὲ τρόπον καὶ ἐν ὀργαῖς καὶ ἐν πάσαις ἐπι-
10 θυμίαις εὐαπάτητοι γίνονται πάντες, καὶ μᾶλλον
ὅσῳ ἂν μᾶλλον ἐν τοῖς πάθεσιν ὦσιν. διὸ καὶ τοῖς
πυρέττουσιν ἐνίοτε φαίνεται ζῷα ἐν τοῖς τοίχοις
ἀπὸ μικρᾶς ὁμοιότητος τῶν γραμμῶν συντιθεμέ-
νων. καὶ ταῦτ' ἐνίοτε συνεπιτείνει τοῖς πάθεσιν
οὕτως ὥστ' ἐὰν μὲν μὴ σφόδρα κάμνωσι, μὴ
15 λανθάνειν ὅτι ψεῦδος, ἐὰν δὲ μεῖζον ᾖ τὸ πάθος,
καὶ κινεῖσθαι πρὸς αὐτά. αἴτιον δὲ τοῦ συμβαίνειν
ταῦτα τὸ μὴ κατὰ τὴν αὐτὴν δύναμιν κρίνειν τὸ¹
κύριον καὶ ᾧ τὰ φαντάσματα γίνεται. τούτου
δὲ σημεῖον ὅτι φαίνεται μὲν ὁ ἥλιος ποδιαῖος,
ἀντίφησι δὲ πολλάκις ἕτερόν τι πρὸς τὴν φαντα-
20 σίαν. καὶ τῇ ἐπαλλάξει τῶν δακτύλων τὸ ἓν δύο
φαίνεται, ἀλλ' ὅμως οὔ φαμεν δύο· κυριωτέρα γὰρ
τῆς ἁφῆς ἡ ὄψις. εἰ δ' ἦν ἡ ἁφὴ μόνη, κἂν
ἐκρίνομεν τὸ ἓν δύο. τοῦ δὲ διεψεῦσθαι αἴτιον
ὅτι οὐ μόνον τοῦ αἰσθητοῦ κινοῦντος² φαίνεται
ἀδήποτε, ἀλλὰ καὶ τῆς αἰσθήσεως κινουμένης αὐ-
25 τῆς, ἐὰν ὡσαύτως κινῆται ὥσπερ καὶ ὑπὸ τοῦ
αἰσθητοῦ· λέγω δ' οἷον ἡ γῆ δοκεῖ τοῖς πλέουσι
κινεῖσθαι κινουμένης τῆς ὄψεως ὑπ' ἄλλου.

III. Ἐκ δὴ τούτων φανερὸν ὅτι οὐ μόνον ἐγρηγορό-
των αἱ κινήσεις αἱ ἀπὸ τῶν αἰσθημάτων γινόμεναι
30 τῶν τε θύραθεν καὶ τῶν ἐκ τοῦ σώματος ἐνυπαρχου-
σῶν, ἀλλὰ καὶ ὅταν γένηται τὸ πάθος τοῦτο ὃ καλεῖται ὕπνος, καὶ μᾶλλον τότε φαίνονται. μεθ' ἡμέραν
461 a μὲν γὰρ ἐκκρούονται ἐνεργουσῶν τῶν αἰσθήσεων

---

¹ τὸ] τό τε LSU, Bekker.
² κινοῦντος Bywater : κινουμένου.

more remote resemblance. Similarly in fits of anger and in all forms of desire all are easily deceived, and the more easily, the more they are under the influence of emotion. So men in fever sometimes think that they see animals on the walls from the slight resemblance of marks in a pattern. Sometimes the illusion corresponds to the degree of emotion, so that those who are not very ill are aware that the impression is false, but if the malady is more severe, they move themselves in accordance with what they think they see. The reason why this happens is that the controlling sense does not judge these things by the same faculty as that by which sense images occur. This is proved by the fact that the sun appears to measure a foot across, but something else often contradicts this impression. So again when the fingers are crossed one object ⟨between them⟩ appears to be two, but yet we deny that there are two ; for sight has more authority than touch. But if touch were the only criterion, we should judge the one as two. The cause of this deception is that appearances of any kind may come to us, not only when the object of sense supplies the stimulus, but also when the sense is stimulated by itself, provided that it is stimulated in the same way as by an object of sense ; I mean for instance that to those who are sailing past it the land seems to move, though really the eye is being moved by something else.

III. It is evident from the foregoing that stimuli arising from sense-impressions, both those which are derived from without and those which have their origin within the body, occur not only when we are awake, but also when the affection we call sleep supervenes, and even more at that time. In the daytime, when the senses and the mind are active, they are

Explanations by analogy.

461 a

καὶ τῆς διανοίας, καὶ ἀφανίζονται ὥσπερ παρὰ
πολὺ πῦρ ἔλαττον καὶ λῦπαι καὶ ἡδοναὶ μικραὶ
παρὰ μεγάλας, παυσαμένων δ' ἐπιπολάζει καὶ τὰ
μικρά· νύκτωρ δὲ δι' ἀργίαν τῶν κατὰ μόριον
5 αἰσθήσεων καὶ ἀδυναμίαν τοῦ ἐνεργεῖν, διὰ τὸ ἐκ
τῶν ἔξω εἰς τὸ ἐντὸς γίνεσθαι τὴν τοῦ θερμοῦ
παλίρροιαν, ἐπὶ τὴν ἀρχὴν τῆς αἰσθήσεως κατα-
φέρονται καὶ γίνονται φανεραὶ καθισταμένης τῆς
ταραχῆς. δεῖ δ' ὑπολαβεῖν ὥσπερ τὰς μικρὰς
10 δίνας τὰς ἐν τοῖς ποταμοῖς γινομένας, οὕτω τὴν
κίνησιν ἑκάστην γίνεσθαι συνεχῶς, πολλάκις μὲν
ὁμοίως, πολλάκις δὲ διαλυομένας εἰς ἄλλα σχήματα
διὰ τὴν ἀντίκρουσιν. διὸ καὶ μετὰ τὴν τροφὴν
καὶ πάμπαν νέοις οὖσιν, οἷον τοῖς παιδίοις, οὐ
γίνεται ἐνύπνια· πολλὴ γὰρ ἡ κίνησις διὰ τὴν ἀπὸ
τῆς τροφῆς θερμότητα. ὥστε καθάπερ ἐν ὑγρῷ,
15 ἐὰν σφόδρα κινῇ τις, ὁτὲ μὲν οὐδὲν φαίνεται
εἴδωλον ὁτὲ δὲ φαίνεται μὲν διεστραμμένον δὲ
πάμπαν, ὥστε φαίνεσθαι ἀλλοῖον ἢ οἷόν ἐστιν,
ἠρεμήσαντος δὲ καθαρὰ καὶ φανερά, οὕτω καὶ ἐν
τῷ καθεύδειν τὰ φαντάσματα καὶ αἱ ὑπόλοιποι
κινήσεις αἱ συμβαίνουσαι ἀπὸ τῶν αἰσθημάτων
20 ὁτὲ μὲν ὑπὸ μείζονος οὔσης τῆς εἰρημένης κινήσεως
ἀφανίζονται πάμπαν, ὁτὲ δὲ τεταραγμέναι φαίνονται αἱ ὄψεις καὶ τερατώδεις καὶ οὐκ ἐρρωμένα τὰ
ἐνύπνια, οἷον τοῖς μελαγχολικοῖς καὶ πυρέττουσι
καὶ οἰνωμένοις· πάντα γὰρ τὰ τοιαῦτα πάθη πνευ-
25 ματώδη ὄντα πολλὴν ποιεῖ κίνησιν καὶ ταραχήν.
καθισταμένου δὲ καὶ διακρινομένου τοῦ αἵματος
ἐν τοῖς ἐναίμοις, σωζομένη τῶν αἰσθημάτων ἡ
κίνησις ἀφ' ἑκάστου τῶν αἰσθητηρίων ἐρρωμένα
τε ποιεῖ τὰ ἐνύπνια, καὶ φαίνεσθαί τι καὶ δοκεῖν

thrust aside or obscured just in the same way as a smaller fire is obscured by a greater, and small pains and pleasures by great, although when the latter have ceased even the small ones come to the surface; but at night, because the particular senses are at rest and cannot function, owing to the heat's reversing its flow and passing from the outside to the inside, these stimuli reach the starting-point of sensation and become noticeable, as the bustle subsides. One may suppose that, like the eddies often seen in rivers, each movement takes place continuously, often with unchanging pattern, but often again dividing into other shapes owing to some obstruction. For this reason no dreams occur after food or to the very young such as infants; for the movement is considerable owing to the heat arising from the food. Hence, just as in a liquid, if one disturbs it violently, sometimes no image appears, and sometimes it appears but is entirely distorted, so that it seems quite different from what it really is, although when the movement has ceased, the reflections are clear and plain; so also in sleep, the images or residuary movements that arise from the sense-impressions are altogether obscured owing to the aforesaid movement when it is too great, and sometimes the visions appear confused and monstrous, and the dreams are morbid, as occurs with the melancholic, the feverish and the intoxicated; for all these affections, being spirituous, produce much movement and confusion. In animals that have blood, as the blood becomes quiet and its purer elements separate, the persistence of the sensory stimulus derived from each of the sense organs makes the dreams healthy, and causes an image to appear, and the dreamer to think, because of the

461 a

διὰ μὲν τὰ ἀπὸ τῆς ὄψεως καταφερόμενα ὁρᾶν,
διὰ δὲ τὰ ἀπὸ τῆς ἀκοῆς ἀκούειν. ὁμοιοτρόπως
30 δὲ καὶ ἀπὸ τῶν ἄλλων αἰσθητηρίων· τῷ μὲν γὰρ
ἐκεῖθεν ἀφικνεῖσθαι τὴν κίνησιν πρὸς τὴν ἀρχὴν
461 b καὶ ἐγρηγορὼς δοκεῖ ὁρᾶν καὶ ἀκούειν καὶ αἰσθάνεσθαι, καὶ διὰ τὸ τὴν ὄψιν ἐνίοτε κινεῖσθαι δοκεῖν
οὐ κινουμένην ὁρᾶν φαμέν, καὶ τῷ τὴν ἀφὴν δύο
κινήσεις εἰσαγγέλλειν τὸ ἓν δύο δοκεῖν. ὅλως
5 γὰρ τὸ ἀφ' ἑκάστης αἰσθήσεώς φησιν ἡ ἀρχή, ἐὰν
μὴ ἑτέρα κυριωτέρα ἀντιφῇ. φαίνεται μὲν οὖν
πάντως, δοκεῖ δὲ οὐ πάντως τὸ φαινόμενον, ἀλλ'
ἐὰν τὸ ἐπικρῖνον κατέχηται ἢ μὴ κινῆται τὴν
οἰκείαν κίνησιν. ὥσπερ δ' εἴπομεν ὅτι ἄλλοι δι'
ἄλλο πάθος εὐαπάτητοι, οὕτως ὁ καθεύδων διὰ
10 τὸν ὕπνον καὶ τὸ κινεῖσθαι τὰ αἰσθητήρια καὶ
τἆλλα τὰ συμβαίνοντα περὶ τὴν αἴσθησιν, ὥστε
τὸ μικρὰν ἔχον ὁμοιότητα φαίνεται ἐκεῖνο. ὅταν
γὰρ καθεύδῃ, κατιόντος τοῦ πλείστου αἵματος ἐπὶ
τὴν ἀρχὴν συγκατέρχονται αἱ ἐνοῦσαι κινήσεις, αἱ
μὲν δυνάμει αἱ δὲ ἐνεργείᾳ. οὕτω δ' ἔχουσιν ὥστε
15 ἐν τῇ κινήσει τῃδὶ ἥδε ἐπιπολάσει ἐξ αὐτοῦ ἡ
κίνησις, ἂν δ' αὕτη φθαρῇ, ἥδε. καὶ πρὸς ἀλλήλας
δ' ἔχουσιν ὥσπερ οἱ πεπλασμένοι βάτραχοι οἱ
ἀνιόντες ἐν τῷ ὕδατι τηκομένου τοῦ ἁλός. οὕτως
ἔνεισι δυνάμει, ἀνειμένου δὲ τοῦ κωλύοντος ἐνερ-
γοῦσιν· καὶ λυόμεναι ἐν ὀλίγῳ τῷ λοιπῷ αἵματι
τῷ ἐν τοῖς αἰσθητηρίοις κινοῦνται, ἔχουσαι ὁμοιό-

---

[a] Sc., of the blood.

364

data supplied from the organs of sight and hearing, that he really sees and hears. The same is true of the other sense organs. Even in the waking state one's belief that he sees, hears or perceives is due to the fact that the stimulus comes to the starting-point from the sense organs ; and it is because the vision seems at times to be stimulated, though really it is not, that we say that we see, and because the sense of touch reports two stimuli that one object seems to be two. For, speaking generally, the controlling power affirms the report given by each sense, unless another power still more authoritative contradicts it. In every case the sense-impression is there, but the sense-impression does not in every case seem real, but only when the judging faculty is restrained, or is not moving with its proper movement. And just as we have said that some are liable to be deceived by one affection and others by another, so the sleeper owing to his sleep, and the stimulation of the sense organs, and the other factors which are connected with sensation, is liable to be deceived so that an image which has only a remote resemblance seems to be the object itself. For when a man is asleep, as most of the blood sinks to its source, the internal movements, some potential and some actual, travel with it. They are so conditioned that at a given disturbance [a] one will detach itself and rise to the surface, and if this is destroyed another will do so. Their relation to each other is similar to that of artificial frogs, which rise to the surface of the water as the salt is dissolved. In a similar way these movements reside in us potentially, but are actualized when the preventing cause is removed ; and as they are freed, in the little blood which remains in the sense organs they begin to move, bearing a resemblance, as cloud-shapes do, which in their rapid changes we liken now to men and now to centaurs. Each of them, as has been said, is a residue of the actual impression ; though the real impression has gone, this still inheres, and it is true to say that it is (for instance) like Coriscus, though it is not Coriscus. At the time of perception the controlling and discriminating sense did not declare it to be Coriscus, but on account of it stated that the genuine Coriscus was that man yonder. Thus the impulse which it, when actually perceiving, so describes (unless entirely repressed by the blood), it now, as though still perceiving, receives from the movements residing in the sense organs ; and the likeness seems to be the true impression. And the power of sleep is such that it makes one unaware of this. If a man is unaware that a finger is being pressed below his eye, not only will one thing seem to be two, but he will think that it is two, whereas, if he is not unaware, it will still appear to be two, but he will not think that it is two. Just in the same way in sleep, if a man is conscious that he is asleep, *i.e.*, of the sleeping state in which the perception occurs, the appearance is there, but something within him tells him that although it appears to be Coriscus, it is not really Coriscus (for often when a man is asleep something in his soul tells him that what appears to him is a dream) ; but if he is unaware that he is asleep there is nothing to contradict the imagination.

It becomes quite clear that our account is true, and that there are imaginative stimuli in the sense organs, if one tries attentively to remember how we are affected when dropping off to sleep or waking up ; Residual images in the sense organs.

462 a

εἴδωλα καθεύδοντι φωράσει ἐγειρόμενος κινήσεις
οὔσας ἐν τοῖς αἰσθητηρίοις· ἐνίοις γὰρ τῶν νεωτέ-
ρων καὶ πάμπαν διαβλέπουσιν, ἐὰν ᾖ σκότος, φαίνεται εἴδωλα πολλὰ κινούμενα, ὥστ' ἐγκαλύπτεσθαι
15 πολλάκις φοβουμένους.

Ἐκ δὴ τούτων ἁπάντων δεῖ συλλογίσασθαι ὅτι
ἐστὶ τὸ ἐνύπνιον φάντασμα μέν τι καὶ ἐν ὕπνῳ·
τὰ γὰρ ἄρτι λεχθέντα εἴδωλα οὐκ ἔστιν ἐνύπνια,
οὐδ' εἴ τι ἄλλο λελυμένων τῶν αἰσθήσεων φαίνεται·
οὐδὲ τὸ ἐν ὕπνῳ φάντασμα πᾶν. πρῶτον μὲν γὰρ
20 ἐνίοις συμβαίνει καὶ αἰσθάνεσθαί πῃ καὶ ψόφων καὶ
φωτὸς καὶ χυμοῦ καὶ ἁφῆς, ἀσθενικῶς μέντοι
καὶ οἷον πόρρωθεν· ἤδη γὰρ ἐν τῷ καθεύδειν
ὑποβλέποντες, ὃ ἠρέμα ἑώρων φῶς τοῦ λύχνου
καθεύδοντες, ὡς ᾤοντο, ἔπειτ' ἐγερθέντες[1] εὐθὺς
ἐγνώρισαν τὸ τοῦ λύχνου ὄν, καὶ ἀλεκτρυόνων καὶ
25 κυνῶν φωνὴν ἠρέμα ἀκούοντες ἐγερθέντες σαφῶς
ἐγνώρισαν. ἔνιοι δὲ καὶ ἀποκρίνονται ἐρωτώμενοι·
ἐνδέχεται γὰρ τοῦ ἐγρηγορέναι καὶ καθεύδειν ἁπλῶς
θατέρου ὑπάρχοντος θάτερόν πῃ ὑπάρχειν. ὧν
οὐδὲν ἐνύπνιον φατέον. οὐδ' ὅσαι δὴ ἐν τῷ ὕπνῳ
γίνονται ἀληθεῖς ἔννοιαι παρὰ τὰ φαντάσματα,
30 ἀλλὰ τὸ φάντασμα τὸ ἀπὸ τῆς κινήσεως τῶν
αἰσθημάτων, ὅταν ἐν τῷ καθεύδειν ᾖ, ᾗ καθεύδει,
τοῦτ' ἐστὶν ἐνύπνιον.

Ἤδη δέ τισι συμβέβηκεν ὥστε μηδὲν ἐνύπνιον
ἑωρακέναι κατὰ τὸν βίον. σπάνιον μὲν οὖν τὸ
462 b τοιοῦτόν ἐστι, συμβαίνει δ' ὅμως. καὶ τοῖς μὲν
ὅλως διετέλεσεν, ἐνίοις δὲ καὶ προελθοῦσι πόρρω

[1] ἔπειτ' ἐγερθέντες ΕΜΥ : ἐπεγερθέντες.

for sometimes images which appear to one when asleep one will in the act of waking detect to be movements in the sense organs ; for to some young people even when their eyes are wide open, if it is dark, many moving images appear, so that they often cover their heads in fright.

One must conclude from all this that a dream is one form of mental image, which occurs in sleep ; for the images just referred to are not dreams, nor is any other image which appears when the senses are free ; nor is every mental picture that occurs in sleep a dream. For in the first place some people actually have a certain perception of sounds, light, flavour, and touch, but faintly and as it were remotely : for it has happened that people while asleep but with their eyes partly open saw dimly in their sleep the light of the lamp, as they supposed, which later on awakening they instantly recognized to be really the light of the lamp, and others hearing faintly the crowing of cocks, or barking of dogs, they recognized them definitely when they awoke. Some ⟨sleepers⟩ again even answer when they are asked a question ; for it is quite possible, in the case of waking and sleeping, that when one is fully present the other should also in some sense be present. But none of these phenomena can be called a dream. Nor can those true thoughts which occur in sleep besides the mental pictures. It is the mental picture which arises from the movement of sense-impressions when one is asleep, in so far as this condition exists, that is a dream. *Dreams and true perception in sleep.*

But it has been the lot of some that they have never seen a dream throughout their life. Such a condition is uncommon, but nevertheless it occurs. With some the condition lasts all through their lives, but with others dreams come to them when they are far ad- *Dreamers.*

462 b

τῆς ἡλικίας ἐγένετο, πρότερον οὐδὲν ἐνύπνιον ἑωρα-
κόσιν. τὸ δ' αἴτιον τοῦ μὴ γίνεσθαι παραπλήσιόν
τι δεῖ νομίζειν, ὅτι οὐδὲ μετὰ τὴν τροφὴν καθ-
5 υπνώσασιν οὐδὲ τοῖς παιδίοις γίνεται ἐνύπνιον·
ὅσοις γὰρ τοῦτον τὸν τρόπον συνέστηκεν ἡ φύσις
ὥστε πολλὴν προσπίπτειν ἀναθυμίασιν πρὸς τὸν
ἄνω τόπον, ἢ πάλιν καταφερομένη ποιεῖ πλῆθος
κινήσεως, εὐλόγως τούτοις οὐδὲν φαίνεται φάντασμα. προϊούσης δὲ τῆς ἡλικίας οὐδὲν ἄτοπον
10 φανῆναι ἐνύπνιον· μεταβολῆς γάρ τινος γενομένης
ἢ καθ' ἡλικίαν ἢ κατὰ πάθος ἀναγκαῖον συμβῆναι
τὴν ἐναντίωσιν ταύτην.

vanced in life, never having seen one before. One must suppose that the reason of their not having dreams is much the same as that which accounts for dreams not appearing to those who sleep after food, or to children. For those whose nature is so constituted that much upward evaporation takes place, which when it descends again produces a considerable movement, naturally see no mental picture. But it is not surprising that a dream should appear to them as their age advances ; for as change takes place in them conformably with advancing age or some other experience, this reversal of a disposition to its contrary must necessarily occur.

ΠΕΡΙ ΤΗΣ ΚΑΘ᾽ ΥΠΝΟΝ
ΜΑΝΤΙΚΗΣ

I. Περὶ δὲ τῆς μαντικῆς τῆς ἐν τοῖς ὕπνοις γινομένης καὶ λεγομένης συμβαίνειν ἀπὸ τῶν ἐνυπνίων,
οὔτε καταφρονῆσαι ῥᾴδιον οὔτε πεισθῆναι. τὸ μὲν
15 γὰρ πάντας ἢ πολλοὺς ὑπολαμβάνειν ἔχειν τι
σημειῶδες τὰ ἐνύπνια παρέχεται πίστιν ὡς ἐξ
ἐμπειρίας λεγόμενον, καὶ τὸ περὶ ἐνίων εἶναι τὴν
μαντικὴν ἐν τοῖς ἐνυπνίοις οὐκ ἄπιστον· ἔχει γάρ
τινα λόγον, διὸ καὶ περὶ τῶν ἄλλων ἐνυπνίων
ὁμοίως ἄν τις οἰηθείη. τὸ δὲ μηδεμίαν αἰτίαν
20 εὔλογον ὁρᾶν, καθ᾽ ἣν ἂν γίνοιτο, τοῦτο διαπιστεῖν
ποιεῖ· τό τε γὰρ θεὸν εἶναι τὸν πέμποντα, πρὸς
τῇ ἄλλῃ ἀλογίᾳ, καὶ τὸ μὴ τοῖς βελτίστοις καὶ
φρονιμωτάτοις ἀλλὰ τοῖς τυχοῦσι πέμπειν ἄτοπον.
ἀφαιρεθείσης δὲ τῆς ἀπὸ τοῦ θεοῦ αἰτίας οὐδεμία
25 τῶν ἄλλων εὔλογος εἶναι φαίνεται αἰτία· τὸ γὰρ
περὶ τῶν ἐφ᾽ Ἡρακλείαις στήλαις ἢ τῶν ἐν
Βορυσθένει προορᾶν τινάς, ὑπὲρ τὴν ἡμετέραν εἶναι
δόξειεν ἂν σύνεσιν εὑρεῖν τούτων τὴν ἀρχήν.

Ἀνάγκη δ᾽ οὖν τὰ ἐνύπνια ἢ αἴτια εἶναι ἢ σημεῖα
τῶν γιγνομένων ἢ συμπτώματα, ἢ πάντα ἢ ἔνια
τούτων ἢ ἓν μόνον. λέγω δ᾽ αἴτιον μὲν οἷον τὴν

374

ON PROPHECY IN SLEEP

I. As for prophecy which takes place in sleep and which is said to proceed from dreams, it is not an easy matter either to despise it or to believe in it. The fact that all, or at least many, suppose that dreams have a significance inclines one to believe the theory, as based on experience, nor is it incredible that on some subjects there should be divination in dreams ; for it has some show of reason, and therefore one might suppose the same of other dreams as well. But the fact that one can see no reasonable cause why it should be so, makes one distrust it ; for apart from its improbability on other grounds, it is absurd to hold that it is God who sends such dreams, and yet that He sends them not to the best and wisest, but to any chance persons. But, if we dismiss the theory of causation by God, none of the other causes seems probable ; for it seems beyond our understanding to find any reason why anyone should foresee things occurring at the Pillars of Heracles or on the Borysthenes.

Now dreams must be either causes or signs of events which occur, or else coincidences ; either all or some of these, or one only. I use the word " cause "

462 b

30 σελήνην τοῦ ἐκλείπειν τὸν ἥλιον καὶ τὸν κόπον τοῦ
πυρετοῦ, σημεῖον δὲ τῆς ἐκλείψεως τὸ τὸν ἀστέρα
εἰσελθεῖν, τὴν δὲ τραχύτητα τῆς γλώττης τοῦ
πυρέττειν, σύμπτωμα δὲ τὸ βαδίζοντος ἐκλείπειν
463 a τὸν ἥλιον· οὔτε γὰρ σημεῖον τοῦ ἐκλείπειν τοῦτ᾽
ἐστὶν οὔτ᾽ αἴτιον, οὔθ᾽ ἡ ἔκλειψις τοῦ βαδίζειν.
διὸ τῶν συμπτωμάτων οὐδὲν οὔτ᾽ ἀεὶ γίνεται οὔθ᾽
ὡς ἐπὶ τὸ πολύ. ἆρ᾽ οὖν ἐστι τῶν ἐνυπνίων τὰ
μὲν αἴτια, τὰ δὲ σημεῖα, οἷον τῶν περὶ τὸ σῶμα
5 συμβαινόντων; λέγουσι γοῦν καὶ τῶν ἰατρῶν οἱ
χαρίεντες ὅτι δεῖ σφόδρα προσέχειν τοῖς ἐνυπνίοις·
εὔλογον δ᾽ οὕτως ὑπολαβεῖν καὶ τοῖς μὴ τεχνίταις
μέν, σκοπουμένοις δέ τι καὶ φιλοσοφοῦσιν.

Αἱ γὰρ μεθ᾽ ἡμέραν γινόμεναι κινήσεις, ἂν μὴ
σφόδρα μεγάλαι ὦσι καὶ ἰσχυραί, λανθάνουσι παρὰ
10 μείζους τὰς ἐγρηγορικὰς κινήσεις. ἐν δὲ τῷ
καθεύδειν τοὐναντίον· καὶ γὰρ αἱ μικραὶ μεγάλαι
δοκοῦσιν εἶναι. δῆλον δ᾽ ἐπὶ τῶν συμβαινόντων
κατὰ τοὺς ὕπνους πολλάκις· οἴονται γὰρ κεραυνοῦ-
σθαι καὶ βροντᾶσθαι μικρῶν ἤχων ἐν τοῖς ὠσὶ
γινομένων, καὶ μέλιτος καὶ γλυκέων χυμῶν ἀπο-
15 λαύειν ἀκαριαίου φλέγματος καταρρέοντος, καὶ
βαδίζειν διὰ πυρὸς καὶ θερμαίνεσθαι σφόδρα μικρᾶς
θερμασίας περί τινα μέρη γιγνομένης. ἐπεγειρο-
μένοις δὲ ταῦτα φανερὰ τοῦτον ἔχοντα τὸν τρόπον.
ὥστ᾽ ἐπεὶ μικραὶ πάντων αἱ ἀρχαί, δῆλον ὅτι καὶ
τῶν νόσων καὶ τῶν ἄλλων παθημάτων τῶν ἐν τοῖς
20 σώμασι μελλόντων γίνεσθαι. φανερὸν οὖν ὅτι
ταῦτα ἀναγκαῖον ἐν τοῖς ὕπνοις εἶναι καταφανῆ
μᾶλλον ἢ ἐν τῷ ἐγρηγορέναι.

in the sense in which the moon is the cause of an
eclipse of the sun, or fatigue is the cause of fever ;
the fact that a star comes into view I call a " sign "
of the eclipse, and the roughness of the tongue a
" sign " of fever ; but the fact that someone is walking when the sun is eclipsed is a coincidence. For
this is neither a sign nor a cause of the eclipse, any
more than the eclipse is a cause or a sign of a man's
walking. So no coincidence occurs invariably or even
commonly. Is it true then that some dreams are
causes and others signs—of what happens in the body
for instance ? At any rate even accomplished
physicians say that close attention should be paid to
dreams ; and it is natural for those to suppose so,
who are not skilled, but who are inquirers and lovers
of truth.

Stimuli occurring in the daytime, if they are not Exaggera-
very great and powerful, pass unnoticed because of dreams.
greater waking impulses. But in the time of sleep
the opposite takes place ; for then small stimuli seem
to be great. This is clear from what often happens
in sleep ; men think that it is lightening and thunder-
ing, when there are only faint echoes in their ears,
and that they are enjoying honey and sweet flavours,
when only a drop of phlegm is slipping down ⟨their
throats⟩, and that they are walking through fire and
are tremendously hot, when there is only a slight
heating about certain parts ; but the true state of
affairs becomes obvious when they wake up. Since
the beginnings of all things are small, obviously the
beginnings of diseases and other distempers, which
are about to visit the body, must also be small.
Clearly then these must be more evident in sleep
than in the waking state.

463 a

Ἀλλὰ μὴν καὶ ἔνιά γε τῶν καθ' ὕπνον φαντασμάτων αἴτια εἶναι τῶν οἰκείων ἑκάστῳ πράξεων οὐκ
ἄλογον· ὥσπερ γὰρ μέλλοντες πράττειν καὶ ἐν ταῖς
25 πράξεσιν ὄντες ἢ πεπραχότες πολλάκις εὐθυονειρίᾳ
τούτοις σύνεσμεν καὶ πράττομεν (αἴτιον δ' ὅτι προ-
ωδοποιημένη τυγχάνει ἡ κίνησις ἀπὸ τῶν μεθ'
ἡμέραν ἀρχῶν), οὕτω πάλιν ἀναγκαῖον καὶ τὰς καθ'
ὕπνον κινήσεις πολλάκις ἀρχὴν εἶναι τῶν μεθ'
ἡμέραν πράξεων διὰ τὸ προωδοποιῆσθαι πάλιν καὶ
30 τούτων τὴν διάνοιαν ἐν τοῖς φαντάσμασι τοῖς
νυκτερινοῖς. οὕτω μὲν οὖν ἐνδέχεται τῶν ἐνυπνίων
ἔνια καὶ σημεῖα καὶ αἴτια εἶναι.

463 b

Τὰ δὲ πολλὰ συμπτώμασιν ἔοικε, μάλιστα δὲ τά
τε ὑπερβατὰ πάντα καὶ ὧν μὴ ἐν αὐτοῖς ἡ ἀρχὴ
οἷον[1] περὶ ναυμαχίας καὶ τῶν πόρρω συμβαινόντων
ἐστίν· περὶ γὰρ τούτων τὸν αὐτὸν τρόπον ἔχειν
εἰκὸς οἷον ὅταν μεμνημένῳ τινὶ περί τινος τύχῃ
5 τοῦτο γινόμενον· τί γὰρ κωλύει καὶ ἐν τοῖς ὕπνοις
οὕτως; μᾶλλον δ' εἰκὸς πολλὰ τοιαῦτα συμβαίνειν.
ὥσπερ οὖν οὐδὲ τὸ μνησθῆναι περὶ τοῦδε σημεῖον
οὐδ' αἴτιον τοῦ παραγενέσθαι αὐτόν, οὕτως οὐδ'
ἐκεῖ τοῦ ἀποβῆναι τὸ ἐνύπνιον τῷ ἰδόντι οὔτε
σημεῖον οὔτ' αἴτιον, ἀλλὰ σύμπτωμα. διὸ καὶ
10 πολλὰ τῶν ἐνυπνίων οὐκ ἀποβαίνει· τὰ γὰρ συμπτώματα οὔτ' ἀεὶ οὔθ' ὡς ἐπὶ τὸ πολὺ γίνεται.

II. Ὅλως δ' ἐπεὶ καὶ τῶν ἄλλων ζῴων ὀνειρώτ-
τει τινά, θεόπεμπτα μὲν οὐκ ἂν εἴη τὰ ἐνύπνια,
οὐδὲ γέγονε τούτου χάριν, δαιμόνια μέντοι· ἡ γὰρ
15 φύσις δαιμονία, ἀλλ' οὐ θεία. σημεῖον δέ· πάνυ

---

[1] ἀλλὰ B.

---

[a] i.e., divination.

378

Again it is not unreasonable to suppose that some They may cause action. of the mental pictures which appear in sleep are the causes of the actions associated with them ; for just as, when we are contemplating or engaged in some action, or have already completed it, we are often associated with this act and perform it in a vivid dream (the reason being that the stimulus arising from the first causes in the daytime has paved the way for it), so conversely stimuli in sleep must often be the first cause of actions in the daytime, because the way has been paved for the intention to do these actions in dreams at night. In this way then it is possible for some dreams to be both signs and causes.

But most of them resemble coincidences, especially Some dreams are certainly coincidences. those which are extravagant, and those in which the initiation of fulfilment does not lie with the dreamers, *e.g.*, in the case of naval battles and far-off events ; in these cases it seems likely that what happens is much the same as when something which has just been mentioned comes to pass. Is there anything to prevent this occurring in sleep ? On the contrary, probably many things of this kind occur. Just as the act of mentioning was neither a sign nor a cause of the man's approach, so in our case the dream was neither a sign nor a cause of its fulfilment to the man who saw it, but only a coincidence. Consequently many dreams have no fulfilment ; for coincidences do not occur invariably or even usually.

II. Generally speaking, since some of the lower Normally dreams are not of divine origin. animals also dream, dreams cannot be sent by God, nor is this *a* the purpose of their appearance, but they have a divine origin ; for nature is divinely ordained, though not itself divine. There is proof of this ; for

379

γὰρ εὐτελεῖς ἄνθρωποι προορατικοί εἰσι καὶ εὐθυ-
όνειροι, ὡς οὐ θεοῦ πέμποντος, ἀλλ᾽ ὅσων ὥσπερ ἂν
εἰ λάλος ἡ φύσις ἐστὶ καὶ μελαγχολική, παντοδαπὰς
ὄψεις ὁρῶσιν· διὰ γὰρ τὸ πολλὰ καὶ παντοδαπὰ
κινεῖσθαι ἐπιτυγχάνουσιν ὁμοίοις θεωρήμασιν, ἐπι-
20 τυχεῖς ὄντες ἐν τούτοις ὥσπερ ἔνιοι ἀρτιάζοντες[1]·
ὥσπερ γὰρ καὶ λέγεται " ἂν πολλὰ βάλλῃς, ἄλλοτ᾽
ἀλλοῖον βαλεῖς " καὶ ἐπὶ τούτων τοῦτο συμβαίνει.

Ὅτι δ᾽ οὐκ ἀποβαίνει πολλὰ τῶν ἐνυπνίων, οὐδὲν
ἄτοπον· οὐδὲ γὰρ τῶν ἐν τοῖς σώμασι σημείων καὶ
τῶν οὐρανίων, οἷον τὰ τῶν ὑδάτων καὶ τὰ τῶν
25 πνευμάτων· ἂν γὰρ ἄλλη κυριωτέρα ταύτης συμβῇ
κίνησις, ἀφ᾽ ἧς μελλούσης ἐγένετο τὸ σημεῖον, οὐ
γίνεται. καὶ πολλὰ βουλευθέντα καλῶς τῶν πραχ-
θῆναι δεόντων διελύθη δι᾽ ἄλλας κυριωτέρας ἀρχάς.
ὅλως γὰρ οὐ πᾶν γίνεται τὸ μελλῆσαν, οὐδὲ ταὐτὸ
30 τὸ ἐσόμενον καὶ τὸ μέλλον· ἀλλ᾽ ὅμως ἀρχάς τε
λεκτέον εἶναι, ἀφ᾽ ὧν οὐκ ἐπετελέσθη, καὶ σημεῖα
πέφυκε ταῦτα τινῶν οὐ γινομένων.[2]

Περὶ δὲ τῶν μὴ τοιαύτας ἐχόντων ἀρχὰς ἐνυπνίων
οἵας εἴπομεν, ἀλλ᾽ ὑπερορίας ἢ τοῖς χρόνοις ἢ τοῖς
τόποις ἢ τοῖς μεγέθεσιν, ἢ τούτων μὲν μηδέν, μὴ
μέντοι γε ἐν αὑτοῖς ἐχόντων τὰς ἀρχὰς τῶν ἰδόντων
τὸ ἐνύπνιον, εἰ μὴ γίνεται τὸ προορᾶν ἀπὸ συμπτώ-
5 ματος, τοιόνδ᾽ ἂν εἴη μᾶλλον ἢ ὥσπερ λέγει Δημόκριτος εἴδωλα καὶ ἀπορροίας αἰτιώμενος. ὥσπερ
γὰρ ὅταν κινήσῃ τι τὸ ὕδωρ ἢ τὸν ἀέρα, τοῦθ᾽
ἕτερον ἐκίνησε, καὶ παυσαμένου ἐκείνου συμβαίνει

---

[1] ἀρτιάζοντες ci. Bekker : ἄρτια μερίζοντες.
[2] γινομένων EMYS : γενομένων.

quite common men have prescience and vivid dreams, which shows that these are not sent by God ; but that men whose nature is as it were garrulous or melancholic see all kinds of sights ; for since they respond often to any kind of stimulus, they chance upon visions similar to events, doing so by sheer luck, like men playing odd and even ; for just as the saying goes " If you throw often enough your luck will change," so it happens in the case we are discussing.

It is in no way surprising that many dreams have no fulfilment ; for the same is true of bodily symptoms and weather-signs, as we see in the case of rain and wind ; for if another impulse supervenes more powerful than that from which, as yet unfulfilled, the sign arose, the event does not follow. Many things that require doing, though excellently planned, fail owing to more powerful causes. Speaking quite generally not all probabilities occur, nor is that which shall be the same as that which is now likely to be ; but all the same we must call these beginnings, although no result ensues from them, and they are natural signs of certain things which, however, do not happen. *Sometimes no result occurs.*

As for dreams which entail beginnings [a] not such as we have described, but such as are extravagant in time or place or size, or which are extravagant in none of these respects, but yet are outside the control of those who see the dream ; unless the prediction is purely a coincidence, such dreams seem to have the following explanation rather than that given by Democritus, who attributes them to images and emanations. When anything has stirred water or air, the part moved moves another, and when the first impulse has ceased, a similar movement still *Why dreams come to men of inferior intelligence.*

[a] *Sc.*, of events.

381

464 a

τὴν τοιαύτην κίνησιν προϊέναι μέχρι τινός, τοῦ
κινήσαντος οὐ παρόντος, οὕτως οὐδὲν κωλύει
10 κίνησίν τινα καὶ αἴσθησιν ἀφικνεῖσθαι πρὸς τὰς
ψυχὰς τὰς ἐνυπνιαζούσας, ἀφ' ὧν ἐκεῖνος τὰ εἴδωλα
ποιεῖ καὶ τὰς ἀπορροίας, καὶ ὅπῃ δὴ ἔτυχεν ἀφικνου-
μένας μᾶλλον αἰσθητὰς εἶναι νύκτωρ διὰ τὸ μεθ'
ἡμέραν φερομένας διαλύεσθαι μᾶλλον (ἀταραχω-
15 δέστερος γὰρ ὁ ἀὴρ τῆς νυκτὸς διὰ τὸ νηνεμωτέρας
εἶναι τὰς νύκτας), καὶ ἐν τῷ σώματι ποιεῖν αἴσθησιν
διὰ τὸν ὕπνον, διὰ τὸ καὶ τῶν μικρῶν κινήσεων τῶν
ἐντὸς αἰσθάνεσθαι καθεύδοντας μᾶλλον ἢ ἐγρηγο-
ρότας. αὗται δ' αἱ κινήσεις φαντάσματα ποιοῦσιν,
ἐξ ὧν προορῶσι τὰ μέλλοντα περὶ τῶν τοιούτων.
20 καὶ διὰ ταῦτα συμβαίνει τὸ πάθος τοῦτο τοῖς
τυχοῦσι καὶ οὐ τοῖς φρονιμωτάτοις. μεθ' ἡμέραν
τε γὰρ ἐγίνετ' ἂν καὶ τοῖς σοφοῖς, εἰ θεὸς ἦν ὁ
πέμπων· οὕτω δ' εἰκὸς τοὺς τυχόντας προορᾶν· ἡ
γὰρ διάνοια τῶν τοιούτων οὐ φροντιστικὴ ἀλλ'
ὥσπερ ἔρημος καὶ κενὴ πάντων, καὶ κινηθεῖσα κατὰ
τὸ κινοῦν ἄγεται.
25 Καὶ τοῦ ἐνίους τῶν ἐκστατικῶν προορᾶν αἴτιον
ὅτι αἱ οἰκεῖαι κινήσεις οὐκ ἐνοχλοῦσιν ἀλλ' ἀπορ-
ραπίζονται· τῶν ξενικῶν οὖν μάλιστα αἰσθάνονται.
τὸ δέ τινας εὐθυονείρους εἶναι καὶ τὸ τοὺς γνωρίμους περὶ τῶν γνωρίμων μάλιστα προορᾶν συμβαίνει διὰ τὸ μάλιστα τοὺς γνωρίμους ὑπὲρ ἀλλήλων
30 φροντίζειν· ὥσπερ γὰρ πόρρω ὄντων μάλιστα[1] γνω-
ρίζουσι καὶ αἰσθάνονται, οὕτω καὶ τῶν κινήσεων·
αἱ γὰρ τῶν γνωρίμων γνωριμώτεραι κινήσεις. οἱ
δὲ μελαγχολικοὶ διὰ τὸ σφόδρα, ὥσπερ βάλλοντες

---

[1] τάχιστα LSU, Bekker.

continues up to a point, where the moving agent is not present ; in the same way it is quite possible that some movement and perception should come to souls that are asleep from the objects from which Democritus derives his images and emanations ; and however they come, that they should be more easily perceived by night, because in the daytime they are apt to be borne away and dissipated (for the air is less disturbed at night because the nights are calmer) ; and that they should cause sensation in the body owing to sleep, because men are more conscious of even slight internal stimuli when asleep than when awake. And these stimuli produce mental pictures, from which men predict what will happen about such events. This is why this affection occurs more readily to ordinary men, and not to those who are specially intelligent. If it were God who sent them they would appear by day also, and to the wise ; but, as it is, it is natural that ordinary men should foresee ; for the minds of such men are not given to deep thought, but are empty and vacant of all thoughts, and when once stimulated are carried away by the impulse.

The reason why some whose minds are unstable foresee, is that their own mental impulses do not occlude others, but are driven off by them. So they are very sensitive to outside impulses. For some cases of vivid dreams there are particular explanations ; *e.g.*, the fact that men have special foresight about their friends is because those who are great friends care deeply for each other : for just as they are especially apt to perceive and recognize each other at a distance, so too in the case of impulses ; for the impulses of familiar friends are themselves more familiar. Choleric people, because of their impetuosity, are (to use a metaphor) good marksmen

Some dreams have special causes.

383

when shooting from a distance ; and because of their liability to change, the next image in the series comes rapidly before them ; for just as the insane recite and con over the poems of Philaegides, such as the *Aphrodite*, in which the ideas are all associated ; so the choleric pursue the series of impulses. Also owing to their impetuosity one impulse is not easily banished from their consciousness by another.

The most skilful judge of dreams is the man who possesses the ability to detect likenesses ; for anyone can judge the vivid dream. By likenesses I mean that the mental pictures are like reflections in water, as we have said before. In the latter case, if there is much movement, the reflection is not like the original, nor the images like the real object. Thus he would indeed be a clever interpreter of reflections who could quickly discriminate, and envisage these scattered and distorted fragments of the images as representing a man, say, or a horse or any other object. Now in the other case too the dream has a somewhat similar result ; for the movement destroys the clarity of the dream. We have now explained what sleep and dreams are, why each of them occurs, and also about prophecy from dreams.

Special powers necessary to explain dreams.

INTRODUCTION

THE remaining sections return to the sphere of physiology ; they discuss how life is maintained in the organism, and what the causes are which lead to deterioration and dissolution. Although the sections have received separate titles, the discussion is really continuous, and may be conveniently analysed as a whole.

Maintenance of Life. The chapters " On Length and Shortness of Life " form a general introduction to the subject, and state the principles upon which Aristotle's physiology is based. All matter as it occurs in nature exhibits contrary qualities which act upon and tend to destroy one another. (This doctrine runs all through Greek physical speculation. The pre-Socratics had meant by " contraries " not qualities but matter endowed with those qualities. Aristotle understood the distinction well enough, but he seems often—at any rate when discussing traditional or popular views—to have used the term in its primitive sense.) In a healthy organism a constant relation is maintained between these contraries ; but this does not mean that they must be equally balanced. In any given pair it is probable that one will dominate the other, and a proportion of the weaker contrary will be constantly extruded as " residue " or waste matter. If the loss is not made good in some way, the weaker contrary will steadily lose ground until the physiological balance is so disturbed that the organism ceases to function and dies.

Two of the chief factors which maintain the proper proportion or " temperament " of the contraries are food and environment. There is also a third which will be mentioned presently.

The contraries which are most important for physiology are also the most ultimate and elemental, *viz.*, Hot and Cold, Wet and Dry (corresponding to the elements Fire, Air, Water and Earth). Since warmth and moisture are clearly essential to life, Aristotle infers that Hot and Wet must be dominant in the living organism. It follows that a warm humid climate is conducive to longevity. So is large size, because a large animal normally contains more moisture and does not easily become dry. Fat is also helpful ; it contains warmth and moisture. (Aristotle is on more doubtful ground in asserting that it contains air, which, being the purest element after fire, helps it to resist decay.) Finally, excessive excretion or secretion is detrimental to life ; partly because it entails loss of moisture, partly because the " residue " contains a recalcitrant quality (otherwise it would not have been expelled) which affects the health of the organism.

Plants generally tend to live longer than animals because their moisture (sap) is oily and viscous, and so persistent. Their great advantage, however, lies in their capacity for self-renewal, their vital principle being so uniformly distributed that every part potentially contains both root and stem.

Vital Heat. Broadly speaking, all organisms have three distinct parts : mouth, body and fundament.

(The divisibility of plants and some of the lower animals, which somewhat obscures the application of this law, is due to their being agglomerations rather than biological units.) The heart or its analogue is found in the centre of the body ; being the seat of the soul, it is the dominant organ, and supplies the heat which is necessary to maintain life.

There are two ways in which this heat may fail : by extinction, when it has too little fuel (*i.e.*, food), and by exhaustion, when it has too much ; or rather when the fuel is consumed too quickly. To prevent over-rapid combustion some form of refrigeration is necessary.

Refrigeration and Respiration. Among the lower forms of life combustion is retarded in various ways. Plants normally obtain enough refrigeration from the air or water which surrounds them, and the same is true of most very small and bloodless " animals." Buzzing insects have a peculiar method : they expand and contract the air contained in their bodies, thus forcing it through their narrow waists, where it is cooled by the surrounding air ; in its passage to and fro it sets up friction against a membrane (the nature and position of which is not clear) and so produces a musical note. Fishes regulate their vital heat by taking in water through their gills. Animals generally, however, obtain refrigeration through respiration, although some which have bloodless and spongy lungs (*e.g.*, many ovipara) do not respire much ; the lungs when once charged with air remain cool for a long time.

Aristotle's account of the mechanics of respiration is close enough to the truth and calls for little comment, apart from the all-important fact that the process is conceived and described as purely refrigerative. One might have supposed that blood is actually cooled in the lungs, but it seems clear from *Hist. An.* 496 a 32 and the description of the analogous process in the case of fishes (478 b 10 ff.) that the inhaled air is supposed to pass from the lungs to the heart (through the pulmonary arteries ?) and so to temper the vital heat at its very centre. This leads us on to the explanation of another phenomenon.

Pulsation. Knowing nothing of the circulation of the blood or of the distinction between veins and arteries, Aristotle accounts for pulsation by an ingenious theory. We have already seen in the Introduction to *On the Soul* (p. 7) the function of the σύμφυτον πνεῦμα or " connatural breath " in transmitting psychical stimuli to the physical parts of an organism. Unhappily Aristotle never explains clearly what he understands by this πνεῦμα. Obviously it is not the same thing as ordinary breath. The epithet σύμφυτον is usually attached to make the distinction clear ; but it is often omitted. One might have supposed that πνεῦμα is peculiar to living organisms ; but in *G.A.* 762 a 20 we are told that " water is present in earth, and πνεῦμα in water, and soul-heat in all πνεῦμα, so that in a sense all things are full of soul." It is especially connected with semen : *G.A.* 736 a 1 " semen is a compound of πνεῦμα and water—πνεῦμα is hot air " ; and again in 736 b 30 ff. the fertility of semen is ascribed to the presence in it of the hot substance which is not fire but " the πνεῦμα which is enclosed within the semen or foam-like stuff, and the natural substance which is in the πνεῦμα ; and this substance is analogous to the element which belongs to the stars " (Peck). This element is the *aether* of the *De Caelo*, where we are told that it is ungenerated, indestructible and divine, and endowed with circular motion (269 a 31, 270 a 12, b 10, 289 a 15). It would seem, then, that πνεῦμα is found in all liquids ; it is characterized by heat and (perhaps) foam or froth ; it is associated with boiling (479 b 32) ; it forms a link between the immaterial and the material ; and it is a kind of motive force. All this suggests something very like steam, or perhaps the capacity of liquid to be evaporated and expanded by heat.[a]

At any rate the pulsation of the heart and blood-vessels is attributed by Aristotle to the aeration or pneumatization of the liquid food as it reaches the heart and is converted into blood, and is expanded by the intense soul-heat which is concentrated in that region. Probably the expansion is conceived as spasmodic, being continually checked as fresh supplies of cold liquid food arrive ; but the theory may have been suggested by the vibration of a cooking-vessel when liquid is boiled in it.

In spite of its many shortcomings and obvious errors, it is difficult not to be impressed by the ingenuity of Aristotle's physiological theory. The wonder is, not that he made mistakes, but that so

[a] Of course this applies only to the physical conception of πνεῦμα. No doubt the term had many primitive associations of a religious and semi-mystical character. The doctrine of Pneumatism has been discussed at length by Wellman : "Die pneumatische Schule bis auf Archigenes" (*Philologische Untersuchungen* xiv. (1895), Jaeger : "Das Pneuma im Lykeion" (*Hermes* xlviii. (1913), pp. 29-74 and Allbutt : *Greek Medicine in Rome* (1921), pp. 224 ff. All these accounts tend to go beyond the evidence and should be accepted with caution.

many of his conjectures, **often** made with very imperfect opportunities for observation, came so near to the truth. We are still unable to judge with any certainty how much of his teaching on this subject was original and how much was merely a synthesis of existing doctrines; but at least he must have the credit of having set out a fairly coherent system. If later and lesser thinkers, by attributing to his doctrines a finality which they did not deserve, hindered rather than helped the advance of knowledge, it is surely unfair to hold Aristotle himself responsible.

ΠΕΡΙ ΜΑΚΡΟΒΙΟΤΗΤΟΣ
ΚΑΙ ΒΡΑΧΥΒΙΟΤΗΤΟΣ

I. Περὶ δὲ τοῦ τὰ μὲν εἶναι μακρόβια τῶν ζῴων
20 τὰ δὲ βραχύβια, καὶ περὶ ζωῆς ὅλως μήκους καὶ
βραχύτητος ἐπισκεπτέον τὰς αἰτίας. ἀρχὴ δὲ τῆς
σκέψεως ἀναγκαία πρῶτον ἐκ τοῦ διαπορῆσαι περὶ
αὐτῶν. οὐ γάρ ἐστι δῆλον πότερον ἕτερον ἢ τὸ
αὐτὸ αἴτιον πᾶσι τοῖς ζῴοις καὶ φυτοῖς τοῦ τὰ μὲν
25 εἶναι μακρόβια τὰ δὲ βραχύβια· καὶ γὰρ τῶν φυτῶν
τὰ μὲν ἐπέτειον τὰ δὲ πολυχρόνιον ἔχει τὴν ζωήν.
ἔτι δὲ πότερον ταὐτὰ μακρόβια καὶ τὴν φύσιν
ὑγιεινὰ τῶν φύσει συνεστώτων, ἢ κεχώρισται καὶ
τὸ βραχύβιον καὶ τὸ νοσῶδες, ἢ κατ' ἐνίας μὲν
νόσους ἐπαλλάττει τὰ νοσώδη τὴν φύσιν σώματα
30 τοῖς βραχυβίοις, κατ' ἐνίας δ' οὐδὲν κωλύει νοσώ-
δεις εἶναι μακροβίους ὄντας.

Περὶ μὲν οὖν ὕπνου καὶ ἐγρηγόρσεως εἴρηται
πρότερον, περὶ δὲ ζωῆς καὶ θανάτου λεκτέον
ὕστερον, ὁμοίως δὲ καὶ περὶ νόσου καὶ ὑγιείας,
465 a ὅσον ἐπιβάλλει τῇ φυσικῇ φιλοσοφίᾳ· νῦν δὲ περὶ
τῆς αἰτίας τοῦ τὰ μὲν εἶναι μακρόβια τὰ δὲ βραχύβια, καθάπερ εἴρηται πρότερον, θεωρητέον. ἔστι
δ' ἔχοντα τὴν διαφορὰν ταύτην ὅλα τε πρὸς ὅλα

ON LENGTH AND SHORTNESS
OF LIFE

Our task is now to consider the reasons why Connexion between some living creatures are long-lived and others short-lived, and generally to inquire into the causes of length and shortness of life. The necessary starting-point of our inquiry is a statement of the difficulties that arise on the subject. For it is not clear whether the reason why some animals and plants are long-lived, and others short-lived, is the same in all cases or different. For some plants last only for one year, while others live for a long time. Secondly, are longevity and good health the same thing in the case of all natural structures, or are short life and disease unconnected, or again in some diseases do disease and short life go together, while in others there is nothing to prevent the diseased from being also long-lived?

We have previously discussed sleep and the waking state, and later on we must speak of life and death, and similarly of disease and health, as far as they appertain to physical philosophy. But our present inquiry, as I have said before, is concerned with the reasons why some living things live long, and some are short-lived. This difference is found not only between genera as

465 a

γένη, καὶ τῶν ὑφ' ἓν εἶδος ἕτερα πρὸς ἕτερα. λέγω
5 δὲ κατὰ γένος μὲν διαφέρειν οἷον ἄνθρωπον πρὸς
ἵππον (μακροβιώτερον γὰρ τὸ τῶν ἀνθρώπων γένος
ἢ τὸ τῶν ἵππων), κατ' εἶδος δ' ἄνθρωπον πρὸς
ἄνθρωπον· εἰσὶ γὰρ καὶ ἄνθρωποι οἱ μὲν μακρόβιοι
οἱ δὲ βραχυβίοι ἕτεροι καθ' ἑτέρους τόπους διεστῶ-
τες· τὰ μὲν γὰρ ἐν τοῖς θερμοῖς τῶν ἐθνῶν μακρο-
10 βιώτερα, τὰ δ' ἐν τοῖς ψυχροῖς βραχυβιώτερα.
καὶ τῶν τὸν αὐτὸν δὲ τόπον οἰκούντων διαφέρουσιν
ὁμοίως τινὲς ταύτην πρὸς ἀλλήλους τὴν διαφοράν.

II. Δεῖ δὴ λαβεῖν τί τὸ εὔφθαρτον ἐν τοῖς φύσει
συνεστῶσι καὶ τί τὸ οὐκ εὔφθαρτον. πῦρ γὰρ καὶ
15 ὕδωρ καὶ τὰ τούτοις συγγενῆ, οὐκ ἔχοντα τὴν
αὐτὴν δύναμιν, τυγχάνει γενέσεως καὶ φθορᾶς
αἴτια ἀλλήλοις, ὥστε καὶ τῶν ἄλλων ἕκαστον ἐκ
τούτων ὄντα καὶ συνεστῶτα **μετέχειν** τῆς τούτων
φύσεως εὔλογον, ὅσα μὴ συνθέσει ἐκ πολλῶν ἐστίν,
οἷον οἰκία. περὶ μὲν οὖν τῶν ἄλλων ἕτερος λόγος·
20 εἰσὶ γὰρ ἴδιαι φθοραὶ πολλοῖς τῶν ὄντων, οἷον
ἐπιστήμη καὶ ὑγιείᾳ καὶ νόσῳ· ταῦτα γὰρ φθείρεται
καὶ μὴ φθειρομένων τῶν δεκτικῶν ἀλλὰ σωζομένων,
οἷον ἀγνοίας μὲν φθορὰ ἀνάμνησις καὶ μάθησις,
ἐπιστήμης δὲ λήθη καὶ ἀπάτη. κατὰ συμβεβηκὸς
25 δ' ἀκολουθοῦσι τοῖς φυσικοῖς αἱ τῶν ἄλλων φθοραί·
φθειρομένων γὰρ τῶν ζῴων φθείρεται καὶ ἡ ἐπιστήμη καὶ ἡ ὑγίεια ἡ ἐν τοῖς ζῴοις.

Διὸ καὶ περὶ ψυχῆς συλλογίσαιτ' ἄν τις ἐκ τού-
των· εἰ γάρ ἐστι μὴ φύσει ἀλλ' ὥσπερ ἐπιστήμη ἐν
ψυχῇ, οὕτω καὶ ψυχὴ ἐν σώματι, εἴη ἄν τις αὐτῆς

wholes, but also between groups of individuals included under one species. By differences in genus I mean *e.g.* that between man and horse (for the genus man is longer-lived than the genus horse), and by difference within a species that between man and man ; for of men too some are long-lived, and some short-lived, differing according to their different localities ; for races living in hot countries live longer than those in cold countries. Some even of those who live in the same locality exhibit similar differences from one another.

II. We must of course grasp what it is among natural objects that makes them easily destroyed or the reverse. For fire, water, and the kindred elements, not all having the same power, are reciprocal causes of each other's generation and destruction, so that it is only natural that anything proceeding from or consisting of these elements should share in their nature, except for things formed by a combination of numerous parts, such as a house. With regard to all other things it is a different story ; for many things, such as knowledge, health, and disease, have their own peculiar forms of destruction. They can be destroyed even when what contains them is not destroyed, but continues to exist : for instance, learning and recollection destroy ignorance, and forgetfulness and error destroy knowledge. But in an accidental sense the destruction of these other things follows the destruction of natural objects, for when living creatures are destroyed, the knowledge or health that is in them is destroyed also.

From these facts one might come to some conclusion about the soul, for if the soul is not by its own nature contained in a body, but only in the way in which knowledge is contained in the soul, it might be

The sources of destruction.

Soul and body.

ARISTOTLE

30 καὶ ἄλλη φθορὰ παρὰ τὴν φθορὰν ἣν φθείρεται
φθειρομένου τοῦ σώματος. ὥστ' ἐπεὶ οὐ φαίνεται
τοιαύτη οὖσα, ἄλλως ἂν ἔχοι πρὸς τὴν τοῦ σώματος
κοινωνίαν.

III. Ἴσως δ' ἄν τις ἀπορήσειεν εὐλόγως, ἆρ'
ἔστιν οὗ ἄφθαρτον ἔσται τὸ φθαρτόν, οἷον τὸ πῦρ
ἄνω, οὗ μή ἐστι τὸ ἐναντίον. φθείρεται γὰρ τὰ
μὲν ὑπάρχοντα τοῖς ἐναντίοις κατὰ συμβεβηκός,
5 τῷ ἐκεῖνα φθείρεσθαι· ἀναιρεῖται γὰρ τἀναντία ὑπ'
ἀλλήλων· κατὰ συμβεβηκὸς δ' οὐθὲν τῶν ἐν ταῖς
οὐσίαις ἐναντίων φθείρεται, διὰ τὸ μηθενὸς ὑποκειμένου κατηγορεῖσθαι τὴν οὐσίαν. ὥσθ' ᾧ μή
ἐστιν ἐναντίον καὶ ὅπου μή ἐστιν, ἀδύνατον ἂν εἴη
φθαρῆναι· τί γὰρ ἔσται τὸ φθεροῦν, εἴπερ ὑπ' ἐν-
10 αντίων μὲν φθείρεσθαι συμβαίνει μόνων, τοῦτο δὲ
μὴ ὑπάρχει, ἢ ὅλως ἢ ἐνταῦθα; ἢ τοῦτο τῇ μὲν
ἀληθές ἐστι τῇ δ' οὔ· ἀδύνατον γὰρ τῷ ὕλην ἔχοντι
μὴ ὑπάρχειν πως τὸ ἐναντίον. πάντῃ μὲν γὰρ
ἐνεῖναι τὸ θερμὸν ἢ τὸ εὐθὺ ἐνδέχεται, πᾶν δ' εἶναι
ἀδύνατον ἢ θερμὸν ἢ εὐθὺ ἢ λευκόν· ἔσται γὰρ τὰ
15 πάθη κεχωρισμένα. εἰ οὖν, ὅταν ἅμα ᾖ τὸ ποιητικὸν καὶ τὸ παθητικόν, ἀεὶ τὸ μὲν ποιεῖ τὸ δὲ
πάσχει, ἀδύνατον μὴ μεταβάλλειν. ἔτι καὶ εἰ
ἀνάγκη περίττωμα ποιεῖν, τὸ δὲ περίττωμα ἐν-
αντίον· ἐξ ἐναντίου γὰρ ἀεὶ ἡ μεταβολή, τὸ δὲ
περίττωμα ὑπόλειμμα τοῦ προτέρου. εἰ δὲ πᾶν
20 ἐξελαύνει τὸ ἐνεργείᾳ ἐναντίον, κἂν ἐνταῦθ' ἄ-

liable to some other form of destruction beyond that which overtakes it when the body is damaged. Since this is evidently not the case, the association of the soul with the body must be on a different principle.

III. It might reasonably be asked whether there Is inde-structibility is any place in which the destructible will be inde-impossible? structible, as fire is in the upper regions, where it has no contrary. For attributes belonging to contraries are destroyed accidentally by the destruction of the contraries ; for contraries eliminate one another ; but none of the contraries which exist in substances is destroyed accidentally, because substance cannot be predicated of any subject. It would therefore be impossible for anything to be destroyed which has no contrary, or where its contrary is not present. For what would there be to destroy it, if things can only be destroyed by their contraries, and if such a contrary does not exist, either at all, or at that particular place ? Perhaps this statement is partly true and partly untrue ; for everything which possesses matter must have a contrary in some sense. Qualities such as heat or straightness may be present anywhere, but nothing can consist solely of heat, straightness, or whiteness ; for in that case affections could exist in isolation. If then, whenever the active and passive are found together, the one always acts and the other is acted upon, it is impossible that there should be no change. This is so again if a waste product must be formed, and a waste product is a contrary ; for all change proceeds from a contrary, and the waste product is what remains of the former state. But if an object expels all that is actually contrary to it, in that

ARISTOTLE

φθαρτον ἂν εἴη. ἢ οὔ, ἀλλ' ὑπὸ τοῦ περιέχοντος φθείρεται.

Εἰ μὲν οὖν,[1] ἱκανὸν ἐκ τῶν εἰρημένων· εἰ δὲ μή, ὑποθέσθαι δεῖ ὅτι ἔνεστί τι ἐνεργείᾳ ἐναντίον, καὶ περίττωμα γίνεται. διὸ ἡ ἐλάττων φλὸξ κατακάε- ται ὑπὸ τῆς πολλῆς κατὰ συμβεβηκός, ὅτι ἦν τρο- 25 φὴν[2] ἐκείνη ἐν πολλῷ χρόνῳ ἀναλίσκει τὸν καπνόν, ταύτην ἡ πολλὴ φλὸξ ταχύ. διὸ πάντα ἀεὶ ἐν κινήσει ἐστί, καὶ γίνεται ἢ φθείρεται. τὸ δὲ περι- έχον ἢ συμπράττει ἢ ἀντιπράττει. καὶ διὰ τοῦτο μετατιθέμενα πολυχρονιώτερα μὲν γίνεται καὶ ὀλιγοχρονιώτερα τῆς φύσεως, ἀΐδια δ' οὐδαμοῦ 30 ὅσοις ἐναντία ἐστίν· εὐθὺς γὰρ ἡ ὕλη τὸ ἐναντίον ἔχει. ὥστ' εἰ μὲν τοῦ ποῦ, κατὰ τόπον μεταβάλλει, εἰ δὲ τοῦ ποσοῦ, κατ' αὔξησιν καὶ φθίσιν· εἰ δὲ πάθους, ἀλλοιοῦται.

466 a IV. Ἔστι δ' οὔτε τὰ μέγιστα ἀφθαρτότερα (ἵππος γὰρ ἀνθρώπου βραχυβιώτερον) οὔτε τὰ μικρά (ἐπέτεια γὰρ τὰ πολλὰ τῶν ἐντόμων) οὔτε 5 τὰ φυτὰ ὅλως τῶν ζῴων (ἐπέτεια γὰρ ἔνια τῶν φυτῶν) οὔτε τὰ ἔναιμα (μέλιττα γὰρ πολυχρονιώ- τερον ἐνίων ἐναίμων) οὔτε τὰ ἄναιμα (τὰ γὰρ μαλάκια ἐπέτεια μέν, ἄναιμα δέ) οὔτε τὰ ἐν τῇ γῇ (καὶ γὰρ φυτὰ ἐπέτειά ἐστι καὶ ζῷα πεζά) οὔτε τὰ ἐν τῇ θαλάττῃ (καὶ γὰρ ἐκεῖ βραχύβια καὶ τὰ ὀστρακηρὰ καὶ τὰ μαλάκια). ὅλως δὲ τὰ μακρο- 10 βιώτατα ἐν τοῖς φυτοῖς ἐστιν, οἷον ὁ φοῖνιξ· εἶτ'

[1] virgulam quae vulgo post ἱκανὸν retinetur ante ἱκανὸν posuit Biehl. [2] ἦν τροφὴν Bywater: ἡ τροφὴ ἦν.

400

case too it would be indestructible. Surely not ; it is destroyed by its environment.

If this is so, it has been sufficiently established by what we have said ; if it is not so, one must assume that some actual contrary is present,[a] and that refuse is produced. So the lesser flame is consumed by the greater accidentally, because the nourishment—namely the smoke—which the lesser flame exhausts after a long time, is exhausted rapidly by the greater. So everything is in a state of movement, and is being either generated or destroyed. The environment either works with it or against it. For this reason things which change their locality may become longer or shorter lived than their own nature allows, but in no case can they be everlasting, if they have contraries ; for their matter directly entails contrariety. So that if the contrariety is of place, the change is one of locality ; if of quantity, the change takes place by growth and decay ; if of affection, a change of state results. *Nothing can last for ever.*

IV. Very large creatures are not less liable to destruction (for a horse does not live as long as a man), nor are small ones (for most of the insects only live a year), nor are plants generally less liable than animals (for some plants are annuals), nor are sanguineous animals less liable (for the bee lives longer than some sanguineous animals), nor the bloodless (for the molluscs, which are bloodless, only live for a year), nor terrestrial creatures (for there are terrestrial plants and animals which only live for a year), nor marine creatures (for in the sea the testacea and molluscs are both short-lived). Generally speaking, the longest lived are to be found among plants, *e.g.*, the date-palm ; secondly, longevity is commoner *How far is longevity dependent on size ?*

---

[a] *Sc.*, in the changing object.

ἐν τοῖς ἐναίμοις ζῴοις μᾶλλον ἢ ἐν τοῖς ἀναίμοις,
καὶ ἐν τοῖς πεζοῖς ἢ ἐν τοῖς ἐνύδροις. ὥστε καὶ
συνδυασθέντων ἐν τοῖς ἐναίμοις καὶ πεζοῖς τὰ
μακροβιώτατα τῶν ζῴων ἐστίν, οἷον ἄνθρωπος καὶ
ἐλέφας. καὶ δὴ καὶ τὰ μείζω ὡς ἐπὶ τὸ πολὺ
15 εἰπεῖν τῶν ἐλαττόνων μακροβιώτερα· καὶ γὰρ καὶ
τοῖς ἄλλοις συμβέβηκε τοῖς μακροβιωτάτοις μέγεθος, ὥσπερ καὶ τοῖς εἰρημένοις.

V. Τὴν δ' αἰτίαν περὶ τούτων ἁπάντων ἐντεῦθεν
ἄν τις θεωρήσειεν. δεῖ γὰρ λαβεῖν ὅτι τὸ ζῷον
ἐστι φύσει ὑγρὸν καὶ θερμόν, καὶ τὸ ζῆν τοιοῦτον,
20 τὸ δὲ γῆρας ψυχρὸν καὶ ξηρόν, καὶ τὸ τεθνηκός·
φαίνεται γὰρ οὕτως. ὕλη δὲ τῶν σωμάτων τοῖς
οὖσι ταῦτα, τὸ θερμὸν καὶ τὸ ψυχρόν, καὶ τὸ ξηρὸν
καὶ τὸ ὑγρόν. ἀνάγκη τοίνυν γηράσκοντα ξηραί-
νεσθαι. διὸ δεῖ μὴ εὐξήραντον εἶναι τὸ ὑγρόν. καὶ
διὰ τοῦτο τὰ λιπαρὰ ἄσηπτα. αἴτιον δ' ὅτι ἀέρος·
25 ὁ δ' ἀὴρ πρὸς τἆλλα πῦρ· πῦρ δ' οὐ γίνεται σαπρόν.
οὐδ' αὖ ὀλίγον δεῖ εἶναι τὸ ὑγρόν· εὐξήραντον γὰρ
καὶ τὸ ὀλίγον. διὸ καὶ τὰ μεγάλα καὶ ζῷα καὶ
φυτὰ ὡς ὅλως εἰπεῖν μακροβιώτερα, καθάπερ
ἐλέχθη πρότερον· εὔλογον γὰρ τὰ μείζω πλέον
ἔχειν ὑγρόν. οὐ μόνον δὲ διὰ τοῦτο μακροβιώτερα·
30 δύο γὰρ τὰ αἴτια, τό τε ποσὸν καὶ τὸ ποιόν, ὥστε
δεῖ μὴ μόνον πλῆθος εἶναι ὑγροῦ, ἀλλὰ τοῦτο καὶ
θερμόν, ἵνα μήτε εὔπηκτον μήτε εὐξήραντον ᾖ.
καὶ διὰ τοῦτο ἄνθρωπος μακρόβιον μᾶλλον ἐνίων
μειζόνων· μακροβιώτερα γὰρ τὰ λειπόμενα τῷ
466 b πλήθει τοῦ ὑγροῦ, ἐὰν πλείονι λόγῳ ὑπερέχῃ κατὰ

among the animals with blood than among those
without, and among land animals rather than water
animals. Thus combining these two characteristics
we find the longest lived of all among sanguineous
terrestrial creatures, *e.g.*, man and the elephant.
Speaking generally, the larger animals are longer
lived than the smaller. For size is an attribute of the
other very long-lived creatures as it is of the two
we have mentioned.

V. One might view the reason for all these facts as
follows. We must grasp that the living creature is
naturally moist and warm, and that life too is of this
nature, whereas old age is cold and dry, and so is a
dead body. This is an observed fact. Now the matter
of which all things are composed consists of hot and
cold, dry and moist. Thus as they grow old they
must dry up, and so the moist must not be easily dried.
For this reason fatty things are not liable to decay.
The reason for this is that they contain air ; and air
bears the same relation to the other elements as fire
does ; and fire does not become corrupt. Nor must
the moisture exist in small quantity ; for a small
quantity is easily dried. This is why the large creatures—both animals and plants—are, generally speaking, longer lived, as has been said above ; for the
larger animal naturally contains more moisture. But
this is not the only reason for their greater longevity ;
there are two causes, quantity and quality, so that
the moisture must not only be present in quantity,
but this must also be hot, in order that it may not
easily be either frozen or dried. This is why man
is longer lived than some larger creatures ; for those
creatures with a smaller quantity of moisture will be
longer lived if their excess in quality is greater than

403

τὸ ποιὸν ἢ λείπεται κατὰ τὸ ποσόν. ἔστι δ' ἐνίοις
μὲν τὸ θερμὸν τὸ λιπαρόν, ὃ ἅμα ποιεῖ τό τε μὴ
εὐξήραντον καὶ τὸ μὴ εὔψυκτον· ἐνίοις δ' ἄλλον
ἔχει χυμόν.

5 Ἔτι δεῖ τὸ μέλλον εἶναι μὴ εὔφθαρτον μὴ περιτ-
τωματικὸν εἶναι. ἀναιρεῖ γὰρ τὸ τοιοῦτον ἢ νόσῳ
ἢ φύσει· ἐναντία δ' ἡ τοῦ περιττώματος δύναμις
καὶ φθαρτικὴ ἡ μὲν τῆς φύσεως ἡ δὲ μορίου. διὸ
τὰ ὀχευτικὰ καὶ πολύσπερμα γηράσκει ταχύ· τὸ
γὰρ σπέρμα περίττωμα, καὶ ἔτι ξηραίνει ἀπιόν.
10 καὶ διὰ τοῦτο ἡμίονος μακροβιώτερος ἵππου καὶ
ὄνου, ἐξ ὧν ἐγένετο, καὶ τὰ θήλεα τῶν ἀρρένων,
ἐὰν ὀχευτικὰ ᾖ τὰ ἄρρενα· διὸ οἱ στρουθοὶ οἱ ἄρρενες
βραχυβιώτεροι τῶν θηλειῶν. ἔτι δὲ καὶ ὅσα
πονητικὰ τῶν ἀρρένων, καὶ διὰ τὸν πόνον γηράσκει
μᾶλλον· ξηραίνει γὰρ ὁ πόνος, τὸ δὲ γῆρας ξηρόν
15 ἐστιν. φύσει δὲ καὶ ὡς ἐπὶ τὸ πᾶν εἰπεῖν τὰ ἄρρενα
τῶν θηλειῶν μακροβιώτερα· αἴτιον δ' ὅτι θερμότερον ζῷον τὸ ἄρρεν ἐστὶ τοῦ θήλεος. τὰ δ' αὐτὰ
ἐν τοῖς ἀλεεινοῖς μακροβιώτερά ἐστιν ἢ ἐν τοῖς
ψυχροῖς τόποις, διὰ τὴν αὐτὴν αἰτίαν δι' ἥνπερ καὶ
μείζω.[1] καὶ μάλιστ' ἐπίδηλον τὸ μέγεθος τῶν τὴν
20 φύσιν ψυχρῶν ζῴων· διὸ οἵ τ' ὄφεις καὶ αἱ σαῦραι
καὶ τὰ φολιδωτὰ μεγάλα ἐν τοῖς θερμοῖς τόποις,
καὶ ἐν τῇ θαλάττῃ τῇ ἐρυθρᾷ τὰ ὀστρακόδερμα· τῆς
τε γὰρ αὐξήσεως ἡ θερμὴ ὑγρότης αἰτία καὶ τῆς
ζωῆς. ἐν δὲ τοῖς ψυχροῖς τόποις ὑδατωδέστερον τὸ
ὑγρὸν τὸ ἐν τοῖς ζῴοις ἐστίν· διὸ εὔπηκτον, ὥστε
25 τὰ μὲν οὐ γίνεται ὅλως τῶν ζῴων τῶν ὀλιγαίμων ἢ
ἀναίμων ἐν τοῖς πρὸς τὴν ἄρκτον τόποις (οὔτε τὸ
πεζὰ ἐν τῇ γῇ οὔτε τὰ ἔνυδρα ἐν τῇ θαλάττῃ), τὰ

---

[1] καὶ μείζω P : τὰ μείζω.

their defect in quantity. In some again the hot matter is of a fatty nature, which makes it neither easily dried nor easily chilled; but in others it has some other flavour.

Moreover, that which is to resist destruction must not be productive of waste matter. Such matter causes death either by disease or naturally; for it has a contrary force, destructive sometimes of the vital nature, at other times of some part of it. Hence animals that are salacious and have much seed age quickly; for the seed is a waste product, and also by its emission causes dryness. This is why the mule is longer lived than the horse and the ass from which it was born, and females are longer lived than males, if the males are salacious; so also the male sparrow lives a shorter time than the female. Further this is true of all males that work hard, and grow old more rapidly because of their toil; for toil produces dryness, and old age is dry. But by nature, as a general rule, the male lives longer than the female; the reason being that the male is a warmer creature than the female. The same kinds of animals are longer lived in hot countries than in cold, for the same reason as that which makes them larger. The size of animals which have a cold nature makes this obvious; so snakes, lizards, and reptiles with horny scales are large in hot countries, and so are testacea in the Red Sea; for the warm moisture causes growth as well as life. But in cold countries the moisture in animals is more watery; consequently it congeals easily, so that in the northern climes animals with little or no blood either do not occur at all (neither terrestrial animals on land, nor water creatures in the sea), or

Waste products.

if they do occur they are smaller and shorter lived ; for the frost robs them of growth.

Both plants and animals die if they do not receive nourishment, for they consume themselves ; for just as a great flame burns up and destroys the small flame by exhausting what feeds it, so the natural warmth, the prime cause of digestion, exhausts the matter in which it resides. Water animals are less long-lived than land animals, not merely because they are moist, but because they are watery ; this kind of moisture is easily destroyed, because it is cold and easily congealed. For the same reason a bloodless animal is easily destroyed, unless it is protected by its size ; for it has neither fat nor sweetness. For in animals the fat is sweet ; for this reason bees are longer lived than other larger animals.

VI. It is among plants that the greatest longevity is found, rather than among animals : first because they are less watery, so that the moisture is not readily congealed ; secondly, they have oiliness and viscosity, and though dry and earthy, yet have a moisture which is not easily dried. But we must discover the reason why trees have an enduring nature ; for in relation to animals, except the insects, they have a nature peculiar to themselves. For plants are always being reborn ; that is why they last so long. For some shoots are always new, while others grow old. The same is true of their roots, but not simultaneously ; at times the trunk and branches alone are destroyed, and others grow up beside them ; when this is so other roots grow from the surviving part, and so the tree continues, part dying and part being born ; hence their long life. As has been said above, plants resemble insects ; for they live even

Longevity in plants.

20 δύο καὶ πολλὰ γίνεται ἐξ ἑνός. τὰ δ' ἔντομα μέχρι
μὲν τοῦ ζῆν ἦλθεν, πολὺν δ' οὐ δύναται χρόνον· οὐ
γὰρ ἔχει ὄργανα, οὐδὲ δύναται ποιεῖν αὐτὰ ἡ ἀρχὴ
ἡ ἐν ἑκάστῳ. ἡ δ' ἐν τῷ φυτῷ δύναται· πανταχῇ
γὰρ ἔχει καὶ ῥίζαν καὶ καυλὸν δυνάμει. διὸ ἀπὸ
ταύτης ἀεὶ προέρχεται τὸ μὲν νέον τὸ δὲ γηράσκον,
25 μικρόν τι διαφέροντα τοῦ[1] εἶναι μακρόβια, οὕτως
ὥσπερ τὰ ἀποφυτευόμενα. καὶ γὰρ ἐν τῇ ἀποφυτείᾳ τρόπον τινὰ φαίη ἄν τις ταὐτὰ συμβαίνειν·
μόριον γάρ τι τὸ ἀποφυτευθέν. ἐν μὲν οὖν τῇ
ἀποφυτείᾳ χωριζομένων συμβαίνει τοῦτο, ἐκεῖ δὲ
διὰ τοῦ συνεχοῦς. αἴτιον δ' ὅτι ἐνυπάρχει πάντῃ ἡ
30 ἀρχὴ δυνάμει ἐνοῦσα.

Συμβαίνει δὲ ταὐτὸ ἐπί τε τῶν ζῴων καὶ φυτῶν.
ἔν τε γὰρ τοῖς ζῴοις τὰ ἄρρενα μακροβιώτερα ὡς
ἐπὶ τὸ πολύ· τούτων δὲ τὰ ἄνω μείζω ἢ τὰ κάτω
(νανωδέστερον γὰρ τοῦ θήλεος τὸ ἄρρεν), ἐν δὲ τῷ
ἄνω τὸ θερμόν, καὶ τὸ ψυχρὸν ἐν τῷ κάτω· καὶ τῶν
467 b φυτῶν τὰ κεφαλοβαρῆ μακροβιώτερα. τοιαῦτα δὲ
τὰ μὴ ἐπέτεια ἀλλὰ δενδρώδη· τὸ γὰρ ἄνω τοῦ
φυτοῦ καὶ κεφαλὴ ἡ ῥίζα ἐστί, τὰ δ' ἐπέτεια ἐπὶ
τὸ κάτω καὶ τὸν καρπὸν λαμβάνει τὴν αὔξησιν.
5 ἀλλὰ περὶ μὲν τούτων[2] καὶ καθ' αὑτὰ ἐν τοῖς περὶ
φυτῶν διορισθήσεται· νῦν δὲ περὶ τῶν ἄλλων ζῴων
εἴρηται τὸ αἴτιον περί τε μεγέθους ζωῆς καὶ βραχυ-
βιότητος. λοιπὸν δ' ἡμῖν θεωρῆσαι περί τε νεότητος καὶ γήρως καὶ ζωῆς καὶ θανάτου· τούτων γὰρ
διορισθέντων τέλος ἂν ἡ περὶ τῶν ζῴων ἔχοι
μέθοδος.

---

[1] τοῦ ci. Ross : τὸ S : τῷ vulgo. [2] τούτων MZ : τούτου.

when divided, and two or more are from one. Insects,
however, though they contrive to live, cannot do so
for long ; for they have not the necessary organs,
nor can the vital principle in each section supply
them. But in the plant it can ; for the plant possesses
potential root and stalk in every part of it. Consequently there is always proceeding from it a new part
besides that which is growing old (a case of virtual
longevity), just as when slips are taken. For one
might say that the same happens, in a sense, in taking
slips ; for the slip is part of the plant. Thus in taking
slips the phenomenon occurs when they are separated,
but in the other case by continuity. The reason is
that the vital principle exists potentially in every
part of the plant.

The same thing occurs both in animals and in
plants. For in animals the males are longer lived as
a general rule ; in their case the upper parts are
larger than the lower (for the male is more dwarflike
than the female), and the warmth resides in the upper
parts, and the cold in the lower ; and among plants
those with heavy heads are longer lived. Of such a
nature are those plants which are not annual, but
are like trees ; for the root is really the upper part
and head of the plant, and annuals grow downwards
and towards the fruit. We shall discuss this matter
further, as a separate subject, in our treatise on
plants [a] ; but now as far as animals are concerned
we have discussed the reason for length or shortness
of life. It remains for us to discuss youth and age, life
and death ; for when our examination of these subjects is complete, our whole inquiry about animals will
be complete.

ΠΕΡΙ ΝΕΟΤΗΤΟΣ ΚΑΙ ΓΗΡΩΣ
ΠΕΡΙ ΖΩΗΣ ΚΑΙ ΘΑΝΑΤΟΥ

I. Περὶ δὲ νεότητος καὶ γήρως καὶ περὶ ζωῆς
καὶ θανάτου λεκτέον νῦν· ἅμα δὲ καὶ περὶ ἀναπνοῆς
ἀναγκαῖον ἴσως τὰς αἰτίας εἰπεῖν· ἐνίοις γὰρ τῶν
ζῴων διὰ τοῦτο συμβαίνει τὸ ζῆν καὶ τὸ μὴ ζῆν.
ἐπεὶ δὲ περὶ ψυχῆς ἐν ἑτέροις διώρισται, καὶ δῆλον
ὅτι οὐχ οἷόν τ' εἶναι σῶμα τὴν οὐσίαν αὐτῆς, ἀλλ'
ὅμως ὅτι γ' ἔν τινι τοῦ σώματος ὑπάρχει μορίῳ,
φανερόν, καὶ ἐν τούτῳ τινὶ τῶν ἐχόντων δύναμιν ἐν
τοῖς μορίοις. τὰ μὲν οὖν ἄλλα τῆς ψυχῆς ἢ μόρια
ἢ δυνάμεις, ὁποτέρως ποτὲ δεῖ καλεῖν, ἀφείσθω
τὰ νῦν· ὅσα δὲ ζῷα λέγεται καὶ ζῆν, ἐν μὲν τοῖς
ἀμφοτέρων τούτων τετυχηκόσι (λέγω δ' ἀμφοτέρων
τοῦ τε ζῷον εἶναι καὶ τοῦ ζῆν) ἀνάγκη ταὐτὸν εἶναι
καὶ ἓν μόριον καθ' ὅ τε ζῇ καὶ καθ' ὃ προσ-
αγορεύομεν αὐτὸ ζῷον. τὸ μὲν γὰρ ζῷον ᾗ ζῷον,
ἀδύνατον μὴ ζῆν· ᾗ δὲ ζῇ, ταύτῃ ζῷον ὑπάρχειν
οὐκ ἀναγκαῖον· τὰ γὰρ φυτὰ ζῇ μέν, οὐκ ἔχει δ'
αἴσθησιν, τῷ δ' αἰσθάνεσθαι τὸ ζῷον πρὸς τὸ μὴ
ζῷον διορίζομεν.

Ἀριθμῷ μὲν οὖν ἀναγκαῖον ἓν εἶναι καὶ τὸ αὐτὸ
τοῦτο τὸ μόριον, τῷ δ' εἶναι πλείω καὶ ἕτερα· οὐ

412

ON YOUTH AND OLD AGE
ON LIFE AND DEATH

I. We must now discuss youth and old age, life The essential to life is sensation. and death. At the same time we shall presumably have also to describe the causes of respiration ; for in some animals this constitutes the difference between living and not living. In another treatise we have given a detailed account of the soul, and while it is clear that its essence cannot be corporeal, yet it is equally clear that it resides in some part of the body which is among those which have control over the rest. We may now neglect the other parts or functions of the soul, whichever we should call them ; but as for what we call animals and living, in anything to which both terms (*viz.*, animal and living) are applicable, there must be one identical part in virtue of which the creature lives and we call it an animal. For the animal, in so far as it is such, cannot fail to live, but it need not be an animal because it lives ; for plants live, but have no sensation, and it is by sensation that we distinguish the animal from that which is not animal.

Numerically, then, this part must be one and the same, but in essence more than one and differentiated ;

413

467 b

γὰρ ταὐτὸ τὸ ζῴῳ τε εἶναι καὶ τὸ ζῆν. ἐπεὶ οὖν
τῶν ἰδίων αἰσθητηρίων ἕν τι κοινόν ἐστιν αἰσθητήριον, εἰς ὃ τὰς κατ᾽ ἐνέργειαν αἰσθήσεις ἀναγκαῖον
30 ἀπαντᾶν, τοῦτο δ᾽ ἂν εἴη μέσον τοῦ πρόσθεν καλου-
μένου καὶ ὄπισθεν (ἔμπροσθεν μὲν γὰρ λέγεται ἐφ᾽
ὃ ἐστιν ἡμῖν ἡ αἴσθησις, ὄπισθεν δὲ τὸ ἀντι-
κείμενον), ἔτι δὲ διῃρημένου τοῦ σώματος τῶν μὲν
ζώντων πάντων τῷ τ᾽ ἄνω καὶ κάτω (πάντα γὰρ
ἔχει τὸ ἄνω καὶ τὸ κάτω, ὥστε καὶ τὰ φυτά),
468 a δῆλον ὅτι τὴν θρεπτικὴν ἀρχὴν ἔχοι ἂν ἐν μέσῳ
τούτων· καθ᾽ ὃ μὲν γὰρ εἰσέρχεται μόριον ἡ τροφή,
ἄνω καλοῦμεν, πρὸς αὐτὸ βλέποντες ἀλλ᾽ οὐ πρὸς
τὸ περιέχον ὅλον, κάτω δὲ καθ᾽ ὃ τὸ περίττωμα
ἀφίησι τὸ πρῶτον. ἔχει δ᾽ ἐναντίως τοῖς φυτοῖς
5 τοῦτο καὶ τοῖς ζῴοις· τῷ μὲν γὰρ ἀνθρώπῳ διὰ τὴν
ὀρθότητα μάλιστα ὑπάρχει τοῦτο τῶν ζῴων, τὸ
ἔχειν τὸ ἄνω μόριον πρὸς τὸ τοῦ παντὸς ἄνω, τοῖς
δ᾽ ἄλλοις μεταξύ· τοῖς δὲ φυτοῖς ἀκινήτοις οὖσι καὶ
λαμβάνουσιν ἐκ τῆς γῆς τὴν τροφὴν ἀναγκαῖον ἀεὶ
10 κάτω τοῦτ᾽ ἔχειν τὸ μόριον. ἀνάλογον γάρ εἰσιν
αἱ ῥίζαι τοῖς φυτοῖς καὶ τὸ καλούμενον στόμα τοῖς
ζῴοις, δι᾽ οὗ τὴν τροφὴν τὰ μὲν ἐκ τῆς γῆς λαμβάνει, τὰ δὲ δι᾽ αὐτῶν.

II. Τριῶν δὲ μερῶν ὄντων εἰς ἃ διαιρεῖται πάντα
τὰ τέλεια τῶν ζῴων, ἑνὸς μὲν ᾗ δέχεται τὴν τροφήν,
15 ἑνὸς δ᾽ ᾗ τὸ περίττωμα προΐεται, τρίτου δὲ τοῦ
μέσου τούτων, τοῦτο ἐν μὲν τοῖς μεγίστοις τῶν
ζῴων καλεῖται στῆθος, ἐν δὲ τοῖς ἄλλοις τὸ ἀνάλογον. διήρθρωται δὲ μᾶλλον ἑτέροις ἑτέρων. ὅσα
δ᾽ αὐτῶν ἐστι πορευτικά, πρόσκειται καὶ μόρια τὰ

ᵃ Presumably A. means parasitic plants such as mistletoe.

for the fact of being an animal is not the same thing as living. Since then the individual sense organs have one common sense organ in which the senses when actualized must meet, and this must lie in between the part called " before " and that called " behind " (" before " means in the direction of the sensation, and " behind " the opposite), and the bodies of all living creatures being divided into " upper " and " lower " (for all have upper and lower parts ; hence so have plants), clearly the nutritive principle lies between these. The part by which food enters we call the upper part, considering it by itself, and not in relation to the surrounding universe ; the lower part is that by which the primary waste product is voided. Plants are contrary to animals in this respect. Because of his erect carriage man of all living creatures has this characteristic most conspicuously, that his upper part is also upper in relation to the whole universe, while in other animals it is midway ; but plants, as they are immovable and take their food from the soil, must always have this part below. For the roots in plants correspond to what is called the mouth in animals ; the part, that is, by which some absorb their food from the earth, and others from each other.[a] *The nutritive part of the soul.*

II. There are three parts into which all the perfect animals are divided, one by which they absorb food, one by which the waste products are evacuated, and the third between the other two. This latter part is what is called the chest in the largest animals, in the others it is the analogue of this. It is better distinguished in some than in others. Those animals which have the power of locomotion have in addition *The divisions of the living body.*

415

468 a

πρὸς ταύτην τὴν ὑπηρεσίαν, οἷς τὸ πᾶν οἴσουσι
20 κύτος, σκέλη τε καὶ πόδες καὶ τὰ τούτοις ἔχοντα
τὴν αὐτὴν δύναμιν. ἀλλ' ἥ γε τῆς θρεπτικῆς ἀρχὴ
ψυχῆς ἐν τῷ μέσῳ τῶν τριῶν μορίων καὶ κατὰ τὴν
αἴσθησιν οὖσα φαίνεται καὶ κατὰ τὸν λόγον· πολλὰ
γὰρ τῶν ζῴων ἀφαιρουμένου ἑκατέρου τῶν μορίων,
τῆς τε καλουμένης κεφαλῆς καὶ τοῦ δεκτικοῦ τῆς
25 τροφῆς, ζῇ μεθ' οὗπερ ἂν ᾖ τὸ μέσον. δῆλον δ'
ἐπὶ τῶν ἐντόμων, οἷον σφηκῶν τε καὶ μελιττῶν,
τοῦτο συμβαῖνον· καὶ τῶν μὴ ἐντόμων δὲ πολλὰ
διαιρούμενα δύναται ζῆν διὰ τὸ θρεπτικόν. τὸ δὲ
τοιοῦτον μόριον ἐνεργείᾳ μὲν ἔχει ἕν, δυνάμει δὲ
πλείω· τὸν αὐτὸν γὰρ συνέστηκε τοῖς φυτοῖς τρό-
30 πον· καὶ γὰρ τὰ φυτὰ διαιρούμενα ζῇ χωρίς, καὶ
γίνεται πολλὰ ἀπὸ μιᾶς ἀρχῆς δένδρα. δι' ἣν δ'
αἰτίαν τὰ μὲν οὐ δύναται διαιρούμενα ζῆν, τὰ δ'
468 b ἀποφυτεύεται τῶν φυτῶν, ἕτερος ἔσται λόγος.
ἀλλ' ὁμοίως ἔχει κατά γε τοῦτο τά τε φυτὰ καὶ τὸ
τῶν ἐντόμων γένος. ἀνάγκη δὴ[1] καὶ τὴν θρεπτικὴν
ψυχὴν ἐνεργείᾳ μὲν ἐν τοῖς ἔχουσιν εἶναι μίαν,
δυνάμει δὲ πλείους. ὁμοίως δὲ καὶ τὴν αἰσθητικὴν
5 ἀρχήν· φαίνεται γὰρ ἔχοντα αἴσθησιν τὰ διαιρούμενα αὐτῶν. ἀλλὰ πρὸς τὸ σώζεσθαι τὴν φύσιν,
τὰ μὲν φυτὰ δύναται, ταῦτα δ' οὐ δύναται διὰ τὸ
μὴ ἔχειν ὄργανα πρὸς σωτηρίαν, ἐνδεᾶ τ' εἶναι τὰ
μὲν τοῦ ληψομένου τὰ δὲ τοῦ δεξομένου τὴν τροφήν,
τὰ δ' ἄλλων τε καὶ τούτων ἀμφοτέρων. ἐοίκασι
10 γὰρ τὰ τοιαῦτα τῶν ζῴων πολλοῖς ζῴοις συμ-
πεφυκόσιν· τὰ δ' ἄριστα συνεστηκότα τοῦτ' οὐ
πάσχει τῶν ζῴων διὰ τὸ εἶναι τὴν φύσιν αὐτῶν ὡς
ἐνδέχεται μάλιστα μίαν. διὸ καὶ μικρὰν αἴσθησιν
ἔνια ποιεῖ διαιρούμενα τῶν μορίων, ὅτι ἔχει τι

___

[1] δὴ Susemihl : δέ.

parts designed for that service, whereby they can
carry the whole trunk ; these are legs and feet, and
parts which perform the same functions as these.
But the seat of the nutritive faculty of the soul resides
in the middle of the three parts, as is evident from
sensation, and is in itself reasonable.   For many
animals, when one of the other two parts is lost, that
is to say what are called respectively the head and
the food receptacle, continue to live with the part attached to the middle.   This clearly happens with such
insects as wasps and bees ; and many animals which
are not insects can live when divided by means of their
nutritive part.   In them this part is actually one, but
potentially more, for their constitution is similar to
that of plants ; for plant-sections live separately
and many trees arise from one origin.   The reason
why some plants cannot live when divided, and others
can be grown from slips, will be the subject of another
discussion.   In the above respect, however, plants
and insects are alike.   Thus the nutritive soul must
be actually one in things which possess it, but potentially more than one ; and the same is true of the
sensitive first principle, for the divided parts evidently
have sensation.   But whereas plants are able to conserve their nature, insects are unable to do so, because
they have not the organs necessary to conserve it,
and are lacking, some in an organ to seize and some
in an organ to receive food, and others again in both.
Animals of this kind are like a concretion of several
animals ; but the best constituted animals do not
show this defect, because their nature is one in so far
as it can be.   Hence some parts when divided show
a slight sensitive power because they retain some

468 b

ψυχικὸν πάθος· χωριζομένων γὰρ τῶν σπλάγχνων
15 ποιεῖται κίνησιν, οἷον καὶ αἱ χελῶναι τῆς καρδίας
ἀφῃρημένης.

III. Ἔτι δὲ ἐπί τε τῶν φυτῶν δῆλον καὶ ἐπὶ τῶν
ζῴων, τῶν μὲν φυτῶν τήν τ᾿ ἐκ τῶν σπερμάτων
γένεσιν ἐπισκοποῦσι καὶ τὰς ἐμφυτείας τε καὶ τὰς
ἀποφυτείας. ἡ μὲν[1] γὰρ ἐκ[2] τῶν σπερμάτων γένε-
20 σις συμβαίνει πᾶσιν ἐκ τοῦ μέσου· διθύρων γὰρ
ὄντων πάντων, ᾗ συμπέφυκεν ἔχεται, καὶ τὸ μέσον
ἐστὶν ἑκατέρου τῶν μορίων· ἐντεῦθεν γὰρ ὅ τε
καυλὸς ἐκφύεται καὶ ἡ ῥίζα τῶν φυομένων, ἡ δ᾿
ἀρχὴ τὸ μέσον αὐτῶν ἐστιν. ἔν τε ταῖς ἐμφυτείαις
καὶ ταῖς ἀποφυτείαις μάλιστα τοῦτο συμβαίνει περὶ
25 τοὺς ὄζους· ἔστι γὰρ ἀρχή τις ὁ ὄζος τοῦ κλάδου,
ἅμα δὲ καὶ μέσον, ὥστε ἢ τοῦτο ἀφαιροῦσιν ἢ εἰς
τοῦτο ἐμβάλλουσιν, ἵνα ἢ ὁ κλάδος ἢ αἱ ῥίζαι ἐκ
τούτων γίνωνται, ὡς οὔσης τῆς ἀρχῆς ἐκ τοῦ μέσου
καυλοῦ καὶ ῥίζης.

Καὶ τῶν ζῴων τῶν ἐναίμων ἡ καρδία γίνεται
πρῶτον. τοῦτο δὲ δῆλον ἐξ ὧν ἐν τοῖς ἐνδεχομένοις
30 ἔτι γιγνομένοις ἰδεῖν τεθεωρήκαμεν. ὥστε καὶ ἐν
τοῖς ἀναίμοις ἀναγκαῖον τὸ ἀνάλογον τῇ καρδίᾳ
γίνεσθαι πρῶτον. ἡ δὲ καρδία ὅτι ἐστὶν ἀρχὴ τῶν
φλεβῶν, ἐν τοῖς περὶ τὰ μέρη τῶν ζῴων εἴρηται
469 a πρότερον· καὶ ὅτι τὸ αἷμα τοῖς ἐναίμοις ἐστὶ τελευ-
ταία τροφή, ἐξ οὗ γίνεται τὰ μόρια. φανερὸν
τοίνυν ὅτι μίαν μέν τινα ἐργασίαν ἡ τοῦ στόματος
λειτουργεῖ δύναμις, ἑτέραν δ᾿ ἡ τῆς κοιλίας περὶ
5 τὴν τροφήν· ἡ δὲ καρδία κυριωτάτη, καὶ τὸ τέλος
ἐπιτίθησιν. ὥστ᾿ ἀνάγκη καὶ τῆς αἰσθητικῆς καὶ
τῆς θρεπτικῆς ψυχῆς ἐν τῇ καρδίᾳ τὴν ἀρχὴν εἶναι
τοῖς ἐναίμοις· τὰ γὰρ τῶν ἄλλων μορίων ἔργα περὶ

---

[1] μὲν] τε LP. [2] ἐκ om. SP.

of the soul's influence; for the animals continue moving when deprived of their internal organs, as tortoises do after the heart is removed.

III. This fact is clear in the case of both plants Grafting and animals: in plants if we study their generation plants. from seed or by grafting and slip-taking. Generation from seed always begins from the middle: for all seeds have two cells, whose junction is at the point of attachment, and the part in between belongs to both the cells; from this part spring both the stem and the root of the plant, and the starting-point is in the middle of them. In grafts and cuttings this is well exemplified with respect to the buds; for the bud is the starting-point of the branch, and is also in the middle, so that they either cut this off, or graft at this point, in order that the branch or the root may spring from it; which shows that the starting-point lies between stem and root.

So too in sanguineous animals the heart is formed Animals. first. This is evident from observed specimens in cases where formation can be watched. Hence in bloodless animals also what corresponds to the heart must be formed first. We have stated already in our treatise *On the Parts of Animals* [a] that the heart is the source of the veins; and that, in sanguineous animals, the blood is the ultimate nutriment from which the parts are developed. Now it is clear that one function in respect of food is served by the mouth and another by the stomach; but the heart is the supreme power, and contributes the final step. So in sanguineous animals the source of both sensitive and nutritive soul must lie in the heart; for the functions of the other parts in respect of food are for the sake of the

[a] *De Part. An.* iii. 665 b 15.

469 a

τὴν τροφὴν τοῦ ταύτης ἔργου χάριν ἐστί· δεῖ γὰρ τὸ
κύριον πρὸς τὸ οὗ ἕνεκα διατελεῖν, ἀλλ' οὐκ ἐν τοῖς
10 τούτου ἕνεκα, οἷον ἰατρὸς πρὸς τὴν ὑγίειαν. ἀλλὰ
μὴν τό γε κύριον τῶν αἰσθήσεων ἐν ταύτῃ τοῖς
ἐναίμοις πᾶσιν· ἐν τούτῳ γὰρ ἀναγκαῖον εἶναι τὸ
πάντων τῶν αἰσθητηρίων κοινὸν αἰσθητήριον. δύο
δὲ φανερῶς ἐνταῦθα συντεινούσας ὁρῶμεν, τήν τε
γεῦσιν καὶ τὴν ἁφήν, ὥστε καὶ τὰς ἄλλας ἀναγκαῖον·
15 ἐν τούτῳ μὲν γὰρ τοῖς ἄλλοις αἰσθητηρίοις ἐνδέχεται ποιεῖσθαι τὴν κίνησιν, ταῦτα δ' οὐδὲν συν-
τείνει πρὸς τὸν ἄνω τόπον. χωρὶς δὲ τούτων, εἰ
τὸ ζῆν ἐν τούτῳ τῷ μορίῳ πᾶσίν ἐστι, δῆλον ὅτι καὶ
τὴν αἰσθητικὴν ἀρχὴν ἀναγκαῖον· ᾗ μὲν γὰρ ζῷον,
ταύτῃ καὶ ζῆν φαμέν, ᾗ δ' αἰσθητικόν, ταύτῃ τὸ
20 σῶμα ζῷον εἶναι λέγομεν. διὰ τί δ' αἱ μὲν τῶν
αἰσθήσεων φανερῶς συντείνουσι πρὸς τὴν καρδίαν,
αἱ δ' εἰσὶν ἐν τῇ κεφαλῇ (διὸ καὶ δοκεῖ τισὶν αἰσθάνεσθαι τὰ ζῷα διὰ τὸν ἐγκέφαλον), τὸ αἴτιον
τούτων ἐν ἑτέροις εἴρηται χωρίς.

IV. Κατὰ μὲν οὖν τὰ φαινόμενα δῆλον ἐκ τῶν
25 εἰρημένων ὅτι ἐν τούτῳ τε καὶ ἐν τῷ μέσῳ τοῦ
σώματος τῶν τριῶν μορίων ἥ τε τῆς αἰσθητικῆς
ἀρχὴ ψυχῆς ἐστι καὶ ἡ τῆς αὐξητικῆς καὶ τῆς
θρεπτικῆς. κατὰ δὲ τὸν λόγον, ὅτι τὴν φύσιν
ὁρῶμεν ἐν πᾶσιν ἐκ τῶν δυνατῶν ποιοῦσαν τὸ
30 κάλλιστον. ἐν τῷ μέσῳ δὲ τῆς οὐσίας τῆς ἀρχῆς
οὔσης ἑκατέρας μάλιστα μὲν ἀποτελεῖ τῶν μορίων
ἑκάτερον τὸ αὑτοῦ ἔργον, τό τε κατεργαζόμενον
τὴν ἐσχάτην τροφὴν καὶ τὸ δεκτικόν· πρὸς ἑκατέρῳ
γὰρ αὐτῶν οὕτως ἔσται, καὶ ἔστιν ἡ τοῦ τοιούτου
469 b μέση χώρα ἄρχοντος χώρα. ἔτι τὸ χρώμενον καὶ

---

^a *De Part. An.* 656 b 5.

420

heart's function ; for the dominant force must be directed towards the final aim, as is the physician's relation to health, and not reside in subordinate processes. Moreover, in all sanguineous animals the supreme organ of the sense-faculties lies in the heart ; for in this part must lie the common sensorium of all the sense-organs. We can clearly see that two, taste and touch, centre in the heart, so that all the others must do so too ; for in this part it is possible for the other sense organs to effect an impulse, but taste and touch do not extend to the upper region. Apart from this, if in all creatures life resides in this part, clearly so too must the origin of sensation ; for we say that a creature is alive in so far as it is an animal, and an animal in so far as it is sensitive. As to why some of the senses evidently extend to the heart, and others reside in the head (in consequence of which some suppose that sensation in animals is due to the brain), the reason for this has been given separately elsewhere.[a]

IV. Thus in the light of observed facts it is clear from what we have said that the source of the sensitive and of the growth-producing and nutritive parts of the soul lies here ; that is, in the middle of the three parts of the body. This is evident by deduction also, because we see that in every case nature produces the best of the possible results ; and if both these principles reside in the middle of the substance, each part, both that which receives the food and that which reduces it to its final form, will best perform its own function ; for the soul will thus be close to both, and the central position of such a faculty will be a position of control. Again, that which employs an instrument

ᾧ χρῆται, δεῖ διαφέρειν—ὥσπερ δὲ τὴν δύναμιν, οὕτως, ἂν ἐνδέχηται,[1] καὶ τὸν τόπον,[2] ὥσπερ οἵ τ' αὐλοὶ καὶ τὸ κινοῦν τοὺς αὐλούς, ἡ χείρ. εἴπερ οὖν 5 τὸ ζῷον ὥρισται τῷ τὴν αἰσθητικὴν ἔχειν ψυχήν, τοῖς μὲν ἐναίμοις ἀναγκαῖον ἐν τῇ καρδίᾳ ταύτην ἔχειν τὴν ἀρχήν, τοῖς δ' ἀναίμοις ἐν τῷ ἀνάλογον μορίῳ.

Πάντα δὲ τὰ μόρια καὶ πᾶν τὸ σῶμα τῶν ζῴων ἔχει τινὰ σύμφυτον θερμότητα φυσικήν· διὸ ζῶντα μὲν φαίνεται θερμά, τελευτῶντα δὲ καὶ στερισκό-10 μενα τοῦ ζῆν τοὐναντίον. ἀναγκαῖον δὴ ταύτης τὴν ἀρχὴν τῆς θερμότητος ἐν τῇ καρδίᾳ τοῖς ἐναίμοις εἶναι, τοῖς δ' ἀναίμοις ἐν τῷ ἀνάλογον· ἐργάζεται γὰρ καὶ πέττει τῷ φυσικῷ θερμῷ τὴν τροφὴν πάντα, μάλιστα δὲ τὸ κυριώτατον. διὸ τῶν μὲν ἄλλων μορίων ψυχομένων ὑπομένει τὸ ζῆν, τοῦ δ' ἐν ταύτῃ φθείρεται πάμπαν, διὰ τὸ 15 τὴν ἀρχὴν ἐντεῦθεν τῆς θερμότητος ἠρτῆσθαι πᾶσι, καὶ τῆς ψυχῆς ὥσπερ ἐμπεπυρευμένης ἐν τοῖς μορίοις τούτοις, τῶν μὲν ἀναίμων ἐν τῷ ἀνάλογον, ἐν δὲ τῇ καρδίᾳ τῶν ἐναίμων. ἀνάγκη τοίνυν ἅμα τό τε ζῆν ὑπάρχειν καὶ τὴν τοῦ θερμοῦ τούτου σωτηρίαν, καὶ τὸν καλούμενον θάνατον εἶναι 20 τὴν τούτου φθοράν.

V. Ἀλλὰ μὴν πυρός γε δύο ὁρῶμεν φθοράς, μάρανσίν τε καὶ σβέσιν. καλοῦμεν δὲ τὴν μὲν ὑφ' αὑτοῦ μάρανσιν, τὴν δ' ὑπὸ τῶν ἐναντίων σβέσιν, τὴν μὲν γήρᾳ, τὴν δὲ βίαιον. συμβαίνει δ' ἀμφοτέρας διὰ ταὐτὸ γίνεσθαι τὰς φθοράς· ὑπο-25 λειπούσης γὰρ τῆς τροφῆς, οὐ δυναμένου λαμβάνειν τοῦ θερμοῦ τὴν τροφήν, φθορὰ γίνεται τοῦ

---

[1] ἐνδέχηται MP : ἐνδέχεται SZ : ἐνδέχοιτο vulgo.
[2] τόπον LZ : τρόπον.

must be distinct from that which it employs (distinct spatially, if possible, no less than in function), as the flute is distinct from the hand which plays it. Since then the animal is defined by the possession of a sensitive soul, in animals with blood this principle must be in the heart, and in bloodless animals in some part corresponding to it.

Now all parts and indeed the whole body of living creatures contain within them some connate heat ; so when alive they are perceptibly warm, but when dead and deprived of life the opposite of this. Now the source of this warmth must lie in the heart in animals with blood, and in the bloodless in some corresponding part ; for while every part reduces and digests the food by means of its natural heat, this is most true of the part with most control. So when other parts get chilled life remains, but when the region of the heart gets cold, the whole body is destroyed, because the principle of heat in all the other parts depends on the heat therein, and the soul is so to speak fired in this organ, which in sanguineous animals is the heart, and in the bloodless that which corresponds to the heart. Thus life must coincide with the conservation of this heat, and what we know as death must be the destruction of this heat.

Function of the heart.

Destruction of heat.

V. Now we can see two ways in which fire is destroyed—by dying out and by extinction. By dying out we mean a decay arising from itself, and by extinction a decay caused by opposites : one is death by old age and the other by violence. But in fact both forms of destruction arise from the same cause ; for in both cases fire dies from the failure of what feeds it, that is when it cannot get food for the heat.

423

πυρός. τὸ μὲν γὰρ ἐναντίον παῦον τὴν πέψιν
κωλύει τρέφεσθαι· ὁτὲ δὲ μαραίνεσθαι συμβαίνει,
πλείονος ἀθροιζομένου θερμοῦ διὰ τὸ μὴ ἀναπνεῖν
μηδὲ καταψύχεσθαι· ταχὺ γὰρ καὶ οὕτω κατανα-
30 λίσκει τὴν τροφὴν πολὺ συναθροιζόμενον τὸ θερμόν,
καὶ φθάνει καταναλίσκον πρὶν ἐπιστῆναι τὴν ἀναθυμίασιν. διόπερ οὐ μόνον μαραίνεται τὸ ἔλαττον
παρὰ τὸ πλεῖον πῦρ, ἀλλὰ καὶ αὐτὴ καθ' αὑτὴν ἡ
470 a τοῦ λύχνου φλὸξ ἐντιθεμένη πλείονι φλογὶ κατα-
κάεται, καθάπερ ὁτιοῦν ἄλλο τῶν καυστῶν. αἴτιον
δ' ὅτι τὴν μὲν οὖσαν ἐν τῇ φλογὶ τροφὴν φθάνει
λαμβάνουσα ἡ μείζων φλὸξ πρὶν ἐπελθεῖν ἑτέραν,
τὸ δὲ πῦρ ἀεὶ διατελεῖ γινόμενον καὶ ῥέον ὥσπερ
5 ποταμός, ἀλλὰ λανθάνει διὰ τὸ τάχος.

Δῆλον τοίνυν ὡς εἴπερ δεῖ σώζεσθαι τὸ θερμόν
(τοῦτο δ' ἀναγκαῖον, εἴπερ μέλλει ζῆν), δεῖ γίνεσθαί
τινα τοῦ θερμοῦ τοῦ ἐν τῇ ἀρχῇ κατάψυξιν. παρά-
δειγμα δὲ τούτου λαβεῖν ἔστι τὸ συμβαῖνον ἐπὶ
τῶν καταπνιγομένων ἀνθράκων· ἂν μὲν γὰρ ὦσι
10 περιπεπωμασμένοι τῷ καλουμένῳ πνιγεῖ συνεχῶς,
ἀποσβέννυνται ταχέως· ἂν δὲ παρ' ἄλληλά τις
ποιῇ πυκνὰ τὴν ἀφαίρεσιν καὶ τὴν ἐπίθεσιν, μένουσι
πεπυρωμένοι πολὺν χρόνον. ἡ δ' ἔγκρυψις σώζει
τὸ πῦρ· οὔτε γὰρ ἀποπνεῖν κωλύεται διὰ μανότητα
τῆς τέφρας, ἀντιφράττει τε τῷ πέριξ ἀέρι πρὸς τὸ
15 μὴ σβεννύναι τῷ πλήθει τῆς ἐνυπαρχούσης αὐτῷ
θερμότητος. ἀλλὰ περὶ μὲν τῆς αἰτίας ταύτης,
ὅτι τὸ ἐναντίον συμβαίνει τῷ ἐγκρυπτομένῳ καὶ
καταπνιγομένῳ πυρί (τὸ μὲν γὰρ μαραίνεται, τὸ
δὲ διαμένει πλείω χρόνον), εἴρηται ἐν τοῖς προ-
βλήμασιν.

⟨In extinction⟩ the opposite prevents the fire from being fed by checking its assimilation ; but sometimes the fire dies out, when there is too much concentration of heat due to the absence of respiration and cooling ; for in this case too the great concentration of heat quickly uses up the food, and finishes doing so before the evaporation comes to a stand.[a] Hence not only does a smaller fire fade out in the presence of a greater one, but the flame of a lamp, when placed in a larger one, is burned up entirely by itself, just like any other combustible. The reason is that the greater flame seizes the food available in the flame before other food can take its place ; and fire is always coming into being and flowing like a river, but its speed is so great that it is not noticed.

It is obvious therefore that if the heat is to be conserved (and it must be if the creature is to continue to live) there must be some cooling down of the heat in its very source. We can find an illustration of this in what occurs when coals are damped down. If they are covered continuously with a lid, which we call a choke, they are very soon quenched ; but if one puts on and takes off the lid in quick alternation, they remain alight for a long time. And banking up a fire preserves it ; for the air is not prevented from getting to it owing to the porous nature of the ashes, and the banking protects the fire from the surrounding air, so that it is not extinguished by the quantity of heat in it. But we have discussed in the *Problems*[b] the reason why contrary effects occur in a banked and in a choked fire (the latter being quenched, while the former has its life prolonged).

[a] *Cf.* 456 b 20.
[b] The passage in question is not extant.

VI. Everything that lives has a soul, and this, as we have said, cannot exist without natural heat. The assistance which plants get through food and the surrounding air is sufficient for the preservation of their natural heat. For the entry of food causes a cooling, just as it does in man when he first admits it ; but fasting causes heat and thirst ; for the air when it is motionless always grows hot, but, moving as it does when food enters it, it becomes cool, until digestion is complete. If the surrounding air is excessively cold owing to the season of the year, when hard frosts occur, the plant withers ; if on the other hand the heat is great in summer, and the moisture drawn out of the ground cannot effect cooling, the heat wastes away and perishes. Trees so affected at these seasons are said to be blighted and star-struck. For this reason men put under the roots certain kinds of stones and water in pots, that the roots of the plants may be cooled. As for animals, since some live in water and others spend their lives in air, they achieve the cooling from and by these elements— the former using water and the latter air. In what manner and how this occurs we must describe after a closer study of the subject.

427

ΠΕΡΙ ΑΝΑΠΝΟΗΣ

I. Περὶ γὰρ ἀναπνοῆς ὀλίγοι μέν τινες τῶν πρότερον φυσικῶν εἰρήκασιν· τίνος μέντοι χάριν ὑπάρχει τοῖς ζῴοις, οἱ μὲν οὐδὲν ἀπεφήναντο, οἱ δὲ εἰρήκασι μέν, οὐ καλῶς δ᾽ εἰρήκασιν ἀλλ᾽ ἀπειρο-
10 τέρως τῶν συμβαινόντων. ἔτι δὲ πάντα τὰ ζῷα φασιν ἀναπνεῖν· τοῦτο δ᾽ οὐκ ἔστιν ἀληθές. ὥστ᾽ ἀναγκαῖον περὶ τούτων πρῶτον ἐπελθεῖν, ὅπως μὴ δοκῶμεν ἀπόντων κενὴν κατηγορεῖν.

Ὅτι μὲν οὖν ὅσα πλεύμονα ἔχει τῶν ζῴων ἀναπνεῖ πάντα, φανερόν. ἀλλὰ καὶ τούτων αὐτῶν ὅσα μὲν ἄναιμον ἔχει τὸν πλεύμονα καὶ σομφόν,
15 ἧττον δέονται τῆς ἀναπνοῆς· διὸ πολὺν χρόνον ἐν τοῖς ὕδασι δύνανται διαμένειν παρὰ τὴν τοῦ σώματος ἰσχύν. τὸν δὲ πλεύμονα σομφὸν ἔχει πάντα τὰ ᾠοτοκοῦντα, οἷον τὸ τῶν βατράχων γένος. ἔτι δὲ αἱ ἐμύδες τε καὶ χελῶναι πολὺν χρόνον μένουσιν
20 ἐν τοῖς ὑγροῖς· ὁ γὰρ πλεύμων ὀλίγην ἔχει θερ-μότητα· ὀλίγαιμον γὰρ ἔχουσιν αὐτόν· ἐμφυσώ-μενος οὖν αὐτὸς τῇ κινήσει καταψύχει καὶ ποιεῖ διαμένειν πολὺν χρόνον. ἐὰν μέντοι βιάζηταί τις λίαν κατέχων πολὺν χρόνον, ἀποπνίγονται πάντα· οὐδὲν γὰρ τῶν τοιούτων δέχεται τὸ ὕδωρ ὥσπερ οἱ ἰχθύες.[1] τὰ δ᾽ ἔναιμον ἔχοντα τὸν πλεύμονα

---

[1] ἰχθῦς LS, Bekker.

ON RESPIRATION

I. A FEW of the earlier natural philosophers have dealt with respiration ; some of them have offered no explanation why this phenomenon occurs in living creatures ; others have discussed it without much insight, and with insufficient experience of the facts. Again they say that all living creatures breathe ; but this is not true. We must then first essay these questions, so that we may not be thought to be passing a baseless censure on men who are not there to defend themselves.

It is of course quite obvious that all animals with lungs breathe ; but of these same animals such as have a bloodless or spongy lung need breathing less than the others ; this is why they can remain in water a long time for their bodily strength. All oviparous animals, such as the genus frog, have a spongy lung. Again both the fresh-water and sea tortoises live under water for a long time ; for the lung, containing little blood, has little heat ; so when inflated it itself causes by its movement a cooling effect, and enables the tortoise to remain under water for a long time. But if one holds it down too long, an animal of this kind is drowned ; for none of these animals can take in water like the fish. All animals whose lung contains

431

25 πάντα μᾶλλον δεῖται τῆς ἀναπνοῆς διὰ τὸ πλῆθος
τῆς θερμότητος· τῶν δ' ἄλλων ὅσα μὴ ἔχει πλεύμονα, οὐδὲν ἀναπνεῖ.

II. Δημόκριτος μὲν οὖν ὁ Ἀβδηρίτης καί τινες
ἄλλοι τῶν περὶ ἀναπνοῆς εἰρηκότων οὐδὲν περὶ
30 τῶν ἄλλων διωρίκασι ζῴων, ἐοίκασι μέντοι λέγειν
ὡς πάντων ἀναπνεόντων· Ἀναξαγόρας δὲ καὶ
Διογένης, πάντα φάσκοντες ἀναπνεῖν, περὶ τῶν
ἰχθύων καὶ τῶν ὀστρείων λέγουσι τίνα τρόπον
ἀναπνέουσιν. καὶ φησὶν Ἀναξαγόρας μέν, ὅταν
471 a ἀφῶσι τὸ ὕδωρ διὰ τῶν βραγχίων, τὸν ἐν τῷ
στόματι γινόμενον ἀέρα ἕλκοντας ἀναπνεῖν τοὺς
ἰχθῦς· οὐ γὰρ εἶναι κενὸν οὐδέν· Διογένης δ' ὅταν
ἀφῶσι τὸ ὕδωρ διὰ τῶν βραγχίων, ἐκ τοῦ περὶ
τὸ στόμα περιεστῶτος ὕδατος ἕλκειν τῷ κενῷ τῷ
5 ἐν τῷ στόματι τὸν ἀέρα, ὡς ἐνόντος ἐν τῷ ὕδατι
ἀέρος. ταῦτα δ' ἐστὶν ἀδύνατα. πρῶτον μὲν γὰρ
τὸ ἥμισυ τοῦ πράγματος ἀφαιροῦσι, διὰ τὸ τὸ
κοινὸν ἐπὶ θατέρου λέγεσθαι μόνον. ἀναπνοὴ γὰρ
καλεῖται, ταύτης δὲ τὸ μὲν ἐκπνοή ἐστι τὸ δ'
εἰσπνοή· περὶ ἧς οὐθὲν λέγουσι, τίνα τρόπον
10 ἐκπνέουσι τὰ τοιαῦτα τῶν ζῴων. οὐδ' ἐνδέχεται
αὐτοῖς εἰπεῖν· ὅταν γὰρ ἀναπνεύσωσι, ταύτῃ ᾗ
ἀνέπνευσαν πάλιν δεῖ ἐκπνεῖν, καὶ τοῦτο ποιεῖν
ἀεὶ παραλλάξ, ὥστε συμβαίνει ἅμα δέχεσθαι τὸ
ὕδωρ κατὰ τὸ στόμα καὶ ἐκπνεῖν. ἀνάγκη δ'
ἀπαντῶντα ἐμποδίζειν θάτερον θατέρῳ. εἶτα ὅταν
15 ἀφῶσι τὸ ὕδωρ, τότε ἐκπνέουσι κατὰ τὸ στόμα ἢ
κατὰ τὰ βράγχια, ὥστε συμβαίνει ἅμα ἐκπνεῖν
καὶ ἀναπνεῖν· τότε γάρ φασιν αὐτὸ ἀναπνεῖν. ἅμα
δ' ἀναπνεῖν καὶ ἐκπνεῖν ἀδύνατον. ὥστ' εἰ ἀνάγκη

much blood depend more upon breathing because of the amount of their heat ; but none of the other creatures which have no lung breathe at all.

II. Democritus of Abdera and certain others who have discussed breathing have not treated these other animals in detail, but appear to speak as if they all breathed ; Anaxagoras and Diogenes, asserting that all creatures breathe, describe the way in which fishes and oysters breathe. Anaxagoras says that fishes breathe by drawing in the air which enters their mouth when they discharge water through their gills, for there can be no vacuum ; Diogenes says that when they let out water through the gills by means of the vacuum in the mouth they draw in the air from the water surrounding the mouth ; which implies that there is air in the water. But all this is impossible. First of all they leave out half the facts, by confining to one process a term which covers two. For what is called respiration includes exhalation as well as inhalation ; but of the former they make no mention ; how, that is, such animals exhale. Nor can they explain it ; for when animals have respired, they must exhale again by the same way by which they respired, and so on alternately, so that it follows that they take in water by the mouth and exhale by it at the same time. But the one operation must clearly meet and obstruct the other. Again, when they discharge the water, they are at the same time exhaling by the mouth or by the gills, so that they must exhale and respire at the same time ; for they say that this is the time at which they respire. But it is impossible to respire and exhale at the same time. The con-

471 a

τὰ ἀναπνέοντα ἐκπνεῖν καὶ εἰσπνεῖν, ἐκπνεῖν δὲ μὴ
ἐνδέχεται μηδὲν αὐτῶν, φανερὸν ὡς οὐδ' ἀναπνεῖ
αὐτῶν οὐδέν.

20 III. Ἔτι δὲ τὸ φάναι τὸν ἀέρα ἕλκειν ἐκ τοῦ
στόματος ἢ ἐκ τοῦ ὕδατος διὰ τοῦ στόματος
ἀδύνατον· οὐ γὰρ ἔχουσιν ἀρτηρίαν διὰ τὸ πλεύμονα μὴ ἔχειν, ἀλλ' εὐθὺς ἡ κοιλία πρὸς τῷ
στόματί ἐστιν, ὥστ' ἀναγκαῖον τῇ κοιλίᾳ ἕλκειν.
τοῦτο δὲ κἂν τἆλλα ἐποίει ζῷα· νῦν δὲ οὐ ποιοῦσιν.
25 κἂν ἐκεῖνα δ' ἔξω τοῦ ὑγροῦ ὄντα ἐπιδήλως ἂν
αὐτὸ ἐποίει· φαίνεται δ' οὐ ποιοῦντ' αὐτό. ἔτι
πάντων τῶν ἀναπνεόντων καὶ ἑλκόντων τὸ πνεῦμα
ὁρῶμεν γινομένην κίνησίν τινα τοῦ μορίου τοῦ
ἕλκοντος, ἐπὶ δὲ τῶν ἰχθύων οὐ συμβαῖνον· οὐδὲν
γὰρ φαίνονται κινοῦντες τῶν περὶ τὴν κοιλίαν,
30 ἀλλ' ἢ τὰ βράγχια μόνον, καὶ ἐν τῷ ὑγρῷ καὶ εἰς
τὸ ξηρὸν ἐκπεσόντες, ὅταν σπαίρωσιν. ἔτι ὅταν
471 b ἀποθνῄσκῃ πνιγόμενα ἐν τοῖς ὑγροῖς πάντα τὰ ἀναπνέοντα, γίνονται πομφόλυγες τοῦ πνεύματος ἐξιόντος βιαίως, οἷον ἐάν τις βιάζηται χελώνας ἢ
βατράχους ἤ τι ἄλλο τῶν τοιούτων γενῶν· ἐπὶ
δὲ τῶν ἰχθύων οὐ συμβαίνει πειρωμένοις πάντα
5 τρόπον, ὡς οὐκ ἐχόντων πνεῦμα θύραθεν οὐθέν. ὅν
τε τρόπον λέγουσι γίνεσθαι τὴν ἀναπνοὴν αὐτοῖς,
ἐνδέχεται καὶ τοῖς ἀνθρώποις οὖσιν ἐν τῷ ὑγρῷ
συμβαίνειν· εἰ γὰρ καὶ οἱ ἰχθύες ἕλκουσιν ἐκ τοῦ
πέριξ ὕδατος τῷ στόματι, διὰ τί τοῦτο οὐκ ἂν
ποιοῖμεν καὶ οἱ ἄνθρωποι καὶ τἆλλα ζῷα; καὶ
10 τὸν ἐκ τοῦ στόματος δ' ἂν ἕλκοιμεν ὁμοίως τοῖς
ἰχθύσιν. ὥστ' εἴπερ κἀκεῖνα ἦν δυνατά, καὶ ταῦτ'
ἂν ἦν· ἐπεὶ δ' οὐκ ἔστι, δῆλον ὡς οὐδ' ἐπ' ἐκείνων
ἐστίν. πρὸς δὲ τούτοις διὰ τίν' αἰτίαν ἐν τῷ ἀέρι

clusion would then follow that, if respiring creatures must both exhale and inhale, and none of them can exhale, none of them can respire at all.

III. Again, to say that they draw in air from the mouth, or from the water by way of the mouth, is impossible ; for since they have no lung, they have no windpipe : the stomach is close up to the mouth, so that they must draw in the air by the stomach. But then all other living creatures would do the same thing ; which in fact they do not. Also fish would be seen to do so when they are out of the water ; but obviously they do not. Again, in the case of all creatures which respire and draw breath, we see that there is some movement of the part which draws, but this does not happen with fishes ; for we see them moving none of the parts about the stomach, but only the gills, both in water, and when they have been thrown out on to dry land, and gasp. Again, when any respiring animals die of drowning in water, bubbles of air violently expelled rise ; *e.g.*, if one holds down by force tortoises, or frogs, or anything else of such a kind ; but it does not happen in the case of fishes, try as we will ; which shows that they draw no breath from outside. But the mode of respiration which they attribute to fishes might also apply to men under water ; for if fishes draw in air from the surrounding water by the mouth, why should not we men and all other animals do the same ? We could also draw in the air from the mouth, like the fishes. If the one were possible, so also would the other be ; but since it is not so in the one case, it is clear that it is not so in the other either. Besides, why is it that

435

471 b

ἀποθνήσκουσι καὶ φαίνονται ἀσπαρίζοντα ὥσπερ
τὰ πνιγόμενα, εἴπερ ἀναπνέουσιν; οὐ γὰρ δὴ
15 τροφῆς γε ἐνδείᾳ τοῦτο πάσχουσιν. ἣν γὰρ λέγει
Διογένης αἰτίαν, εὐήθης· φησὶ γὰρ ὅτι τὸν ἀέρα
πολὺν ἕλκουσι λίαν ἐν τῷ ἀέρι, ἐν δὲ τῷ ὕδατι
μέτριον, καὶ διὰ τοῦτ' ἀποθνήσκειν. καὶ γὰρ ἐπὶ
τῶν πεζῶν ἔδει δυνατὸν εἶναι τοῦτο συμβαίνειν·
νῦν δ' οὐδὲν τῷ σφόδρα ἀναπνεῦσαι ἀποπνίγεται
20 πεζὸν ζῷον. ἔτι δ' εἰ πάντα ἀναπνεῖ, δῆλον ὅτι
καὶ τὰ ἔντομα τῶν ζῴων ἀναπνεῖ· φαίνεται δ'
αὐτῶν πολλὰ διατεμνόμενα ζῆν, οὐ μόνον εἰς δύο
μέρη ἀλλὰ καὶ εἰς πλείω, οἷον αἱ καλούμεναι σκολό-
πενδραι· ἃ πῶς ἢ τίνι ἐνδέχεται ἀναπνεῖν; αἴτιον
δὲ μάλιστα τοῦ μὴ λέγεσθαι περὶ αὐτῶν καλῶς
25 τό τε τῶν μορίων ἀπείρους εἶναι τῶν ἐντός, καὶ
τὸ[1] μὴ λαμβάνειν ἕνεκά τινος τὴν φύσιν πάντα
ποιεῖν· ζητοῦντες γὰρ τίνος ἕνεκα ἡ ἀναπνοὴ τοῖς
ζῴοις ὑπάρχει, καὶ ἐπὶ τῶν μορίων τοῦτ' ἐπισκο-
ποῦντες, οἷον ἐπὶ βραγχίων καὶ πλεύμονος, εὗρον
ἂν θᾶττον τὴν αἰτίαν.

30 IV. Δημόκριτος δ' ὅτι μὲν ἐκ τῆς ἀναπνοῆς
συμβαίνει τι τοῖς ἀναπνέουσι λέγει, φάσκων κωλύειν
472 a ἐκθλίβεσθαι τὴν ψυχήν· οὐ μέντοι ὡς τούτου γ'
ἕνεκα ποιήσασαν τοῦτο τὴν φύσιν οὐθὲν εἴρηκεν·
ὅλως γὰρ ὥσπερ καὶ οἱ ἄλλοι φυσικοί, καὶ οὗτος
οὐθὲν ἅπτεται τῆς τοιαύτης αἰτίας. λέγει δ' ὡς
ἡ ψυχὴ καὶ τὸ θερμὸν ταὐτὸν τὰ πρῶτα σχήματα
5 τῶν σφαιροειδῶν. συγκρινομένων οὖν αὐτῶν ὑπὸ
τοῦ περιέχοντος ἐκθλίβοντος, βοήθειαν γίνεσθαι
τὴν ἀναπνοήν φησιν. ἐν γὰρ τῷ ἀέρι πολὺν

[1] τὸ om. LSP.

fishes die in the air, and are seen to gasp convulsively as if they were choking, if they respire ? For their symptoms are not due to lack of food. The explanation which Diogenes gives is childish ; he says that in the air they draw in too much air, but in the water only a moderate quantity, and that this is why they die. But in that case it ought to be possible for this to happen to land animals, but in point of fact no land animal is ever choked by excessive respiration. Again, if every living creature respires, it is obvious that insects as well as other animals respire ; but many of them clearly continue living even when severed, not merely into two parts but into more, for instance the so-called centipedes ; how and with what organ can they respire ? The real reason why men have given a false account of them is that they are ignorant of their internal anatomy, and do not realize that Nature does everything with some end in view ; for if they had inquired why respiration is characteristic of animals, and had considered the question in respect of their organs, for instance the gills and lungs, they would have discovered the reason more easily.

IV. Democritus states that respiration serves a certain purpose in animals that respire ; he alleges that it prevents the soul from being crushed out ; but he never says that this is why nature evolved respiration ; for, generally speaking, he, like other natural philosophers, never touches upon any reason of this kind. But he does identify the soul with the hot, as primary shapes of his spherical particles. So he contends that when these particles are being forced together by the pressure of the surrounding air, breathing intervenes to help them. For in the

Why do any living creatures breathe ?

Refutation of Democritus.

472 a

ἀριθμὸν εἶναι τῶν τοιούτων ἃ καλεῖ ἐκεῖνος νοῦν
καὶ ψυχήν· ἀναπνέοντος οὖν καὶ εἰσιόντος τοῦ
ἀέρος συνεισιόντα ταῦτα, καὶ ἀνείργοντα τὴν
10 θλίψιν, κωλύειν τὴν ἐνοῦσαν ἐν τοῖς ζῴοις διιέναι
ψυχήν. καὶ διὰ τοῦτο ἐν τῷ ἀναπνεῖν καὶ ἐκπνεῖν
εἶναι τὸ ζῆν καὶ τὸ ἀποθνῄσκειν· ὅταν γὰρ κρατῇ
τὸ περιέχον συνθλῖβον, καὶ μηκέτι θύραθεν εἰσιὸν
δύνηται ἀνείργειν, μὴ δυναμένου ἀναπνεῖν, τότε
συμβαίνειν τὸν θάνατον τοῖς ζῴοις· εἶναι γὰρ τὸν
15 θάνατον τὴν τῶν τοιούτων σχημάτων ἐκ τοῦ
σώματος ἔξοδον ἐκ τῆς τοῦ περιέχοντος ἐκθλίψεως.
τὴν δ' αἰτίαν διὰ τί ποτε πᾶσι μὲν ἀναγκαῖον
ἀποθανεῖν, οὐ μέντοι ὅτε ἔτυχεν ἀλλὰ κατὰ φύσιν
μὲν γήρᾳ, βίᾳ δὲ παρὰ φύσιν, οὐθὲν δεδήλωκεν.
20 μενον, ὁτὲ δ' οὐ φαίνεται, πότερον τὸ αἴτιον ἔξωθέν
ἐστιν ἢ ἐντός. οὐ λέγει δὲ οὐδὲ περὶ τῆς ἀρχῆς
τοῦ ἀναπνεῖν τί τὸ αἴτιον, πότερον ἔσωθεν ἢ
ἔξωθεν· οὐ γὰρ δὴ ὁ θύραθεν νοῦς τηρεῖ τὴν βοήθειαν, ἀλλ' ἔσωθεν ἡ ἀρχὴ τῆς ἀναπνοῆς γίνεται
καὶ τῆς κινήσεως, οὐχ ὡς βιαζομένου τοῦ περι-
25 έχοντος. ἄτοπον δὲ καὶ τὸ ἅμα τὸ περιέχον συν-
θλίβειν καὶ εἰσιὸν διαστέλλειν. ἃ μὲν οὖν εἴρηκε
καὶ ὥς, σχεδὸν ταῦτ' ἐστίν.

Εἰ δὲ δεῖ νομίζειν ἀληθῆ εἶναι τὰ πρότερον λεχθέντα καὶ μὴ πάντα τὰ ζῷα ἀναπνεῖν, οὐ περὶ
παντὸς θανάτου τὴν αἰτίαν ὑποληπτέον εἰρῆσθαι
ταύτην, ἀλλὰ μόνον ἐπὶ τῶν ἀναπνεόντων. οὐ
30 μὴν οὐδ' ἐπὶ τούτων καλῶς. δῆλον δ' ἐκ τῶν

air are a large number of these particles, which he calls mind and soul; so that when an animal respires and the air enters, these enter too and, relieving the pressure, prevent the soul which is in the animal from passing out. Hence upon breathing in and breathing out depend life and death; for when the pressure of the surrounding air prevails, and none can any longer enter from outside and check it, since breathing is impossible, death then supervenes; for he considers that death is the passing out of such " shapes " from the body owing to the pressure of the surrounding air. But the reason why all living creatures must die at some time, though not at any haphazard time, but either naturally by old age or unnaturally by violence, he has nowhere explained. Since, however, it is clear that death sometimes occurs and sometimes does not, he should have explained whether the cause is external or internal. He does not even say what the cause of the beginning of respiration is, nor whether it is external or internal. As a matter of fact the external " mind " does not watch for the time to help; the origin of respiration and of its movement comes from inside, and not by pressure from the surrounding air. It is absurd, too, that the surrounding air should at the same time cause compression and by its entry expansion. Such is a rough account of his theory, and his way of explaining it.

But if we are to believe what has been said before —that not all animals respire—we must suppose that the above explanation does not apply to death in every case, but only in the case of respiring animals. It is not quite a satisfactory explanation even in their case. This is clear from the facts and from experiences

472 a

συμβαινόντων καὶ τῶν τοιούτων ὧν ἔχομεν πάντες
πεῖραν. ἐν γὰρ ταῖς ἀλέαις θερμαινόμενοι μᾶλλον
καὶ τῆς ἀναπνοῆς μᾶλλον δεόμεθα καὶ πυκνότερον
ἀναπνέομεν πάντες· ὅταν δὲ τὸ πέριξ ᾖ ψυχρὸν
καὶ συνάγῃ καὶ συμπηγνύῃ τὸ σῶμα, κατέχειν
35 συμβαίνει τὸ πνεῦμα. καίτοι τότ' ἐχρῆν τὸ
472 b θύραθεν εἰσιόν[1] κωλύειν τὴν σύνθλιψιν. νῦν δὲ
γίνεται τοὐναντίον· ὅταν γὰρ πολὺ λίαν ἀθροισθῇ
τὸ θερμὸν μὴ ἐκπνεόντων, τότε δέονται τῆς
ἀναπνοῆς· ἀναγκαῖον δ' εἰσπνεύσαντας ἀναπνεῖν.
ἀλεάζοντες δὲ πολλάκις ἀναπνέουσιν, ὡς ἀνα-
5 ψύξεως χάριν ἀναπνέοντες, ὅτε τὸ λεγόμενον ποιεῖ
πῦρ ἐπὶ πῦρ.

V. Ἡ δ' ἐν τῷ Τιμαίῳ γεγραμμένη περίωσις
περί τε τῶν ἄλλων ζώων οὐδὲν διώρικε τίνα
τρόπον αὐτοῖς ἡ τοῦ θερμοῦ γίνεται σωτηρία, πότερον τὸν αὐτὸν ἢ δι' ἄλλην τινὰ αἰτίαν· εἰ μὲν γὰρ
10 μόνοις τὸ τῆς ἀναπνοῆς ὑπάρχει τοῖς πεζοῖς, λεκτέον τὴν αἰτίαν τοῦ μόνοις· εἰ δὲ καὶ τοῖς ἄλλοις,
ὁ δὲ τρόπος ἄλλος, καὶ περὶ τούτου διοριστέον,
εἴπερ δυνατὸν ἀναπνεῖν πᾶσιν.

Ἔτι δὲ καὶ πλασματώδης ὁ τρόπος τῆς αἰτίας.
ἐξιόντος γὰρ ἔξω τοῦ θερμοῦ διὰ τοῦ στόματος,
τὸν περιέχοντα ὠθούμενον ἀέρα φερόμενον ἐμ-
15 πίπτειν εἰς τὸν αὐτὸν τόπον φησὶ διὰ μανῶν οὐσῶν
τῶν σαρκῶν, ὅθεν τὸ ἐντὸς ἐξήει θερμόν, διὰ τὸ
μηδὲν εἶναι κενὸν ἀντιπεριισταμένων ἀλλήλοις·
θερμανθέντα δὲ πάλιν ἐξιέναι κατὰ τὸν αὐτὸν
τόπον, καὶ περιωθεῖν εἴσω διὰ τοῦ στόματος τὸν
ἀέρα τὸν ἐκπίπτοντα θερμόν· καὶ τοῦτο δὴ δια-

---

[1] τὸ θύραθεν εἰσιὸν MZ : τὸν ... εἰσιόντα vulgo.

of a kind with which we are all familiar. For in hot weather as we get warmer we all feel a greater need of respiration and respire more frequently ; but when the surrounding atmosphere is cold, which contracts and freezes the body, the effect is to retard the breathing. And yet this is just the time when the air entering from the outside ought to check the compression. But in point of fact the exact opposite occurs ; for it is when too much heat is accumulated by not exhaling that we need to respire ; and we can only respire by inhaling. When people are hot they respire frequently, which implies that they do so in order to get cool, when according to Democritus's theory they would be adding fire to fire.

V. The account given in the *Timaeus* [a] of the "pushing round" of the breath does not at all explain how heat is conserved in the case of animals other than land animals—whether in the same way or for some other cause ; for if respiration occurs only in land animals, the reason for this should be stated ; while if it occurs in the rest as well, but in a different way, this too should be explained, assuming that it is possible for all animals to respire. *Plato's account of breathing. Refutation.*

Moreover, the method of explanation is fictitious. For, upon this theory, when the heat passes out through the mouth it pushes the surrounding air, which is carried round and passes through the porous parts of the flesh into the same place from which the internal heat passed out, one substance replacing the other because a vacuum is impossible ; when the air has grown hot it passes out again by the same way as before, and pushes round the expelled hot air in again through the mouth ; and this we continue to

[a] *Tim.* 79 c.

472 b

20 τελεῖν ἀεὶ ποιοῦντας, ἀναπνέοντάς τε καὶ ἐκπνέον-
τας. συμβαίνει δὲ τοῖς οὕτως οἰομένοις πρότερον
τὴν ἐκπνοὴν γίνεσθαι τῆς εἰσπνοῆς. ἔστι δὲ
τοὐναντίον. σημεῖον δέ· γίνεται μὲν γὰρ ἀλλήλοις
ταῦτα παρ᾽ ἄλληλα, τελευτῶντες δὲ ἐκπνέουσιν,
ὥστ᾽ ἀναγκαῖον εἶναι τὴν ἀρχὴν εἰσπνοήν.

Ἔτι δὲ τὸ τίνος ἕνεκα ταῦθ᾽ ὑπάρχει τοῖς ζῴοις
25 (λέγω δὲ τὸ ἀναπνεῖν καὶ τὸ ἐκπνεῖν) οὐθὲν
εἰρήκασιν οἱ τοῦτον τὸν τρόπον λέγοντες, ἀλλ᾽ ὡς
περὶ συμπτώματός τινος ἀποφαίνονται μόνον.
καίτοι γε κύρια ταῦθ᾽ ὁρῶμεν τοῦ ζῆν καὶ τελευτᾶν·
ὅταν γὰρ ἀναπνεῖν μὴ δύνωνται, τότε συμβαίνει
γίνεσθαι τὴν φθορὰν τοῖς ἀναπνέουσιν. ἔτι δὲ
30 ἄτοπον τὸ τὴν μὲν τοῦ θερμοῦ διὰ τοῦ στόματος
ἔξοδον καὶ πάλιν εἴσοδον μὴ λανθάνειν ἡμᾶς, τὴν
δ᾽ εἰς τὸν θώρακα τοῦ πνεύματος εἴσοδον καὶ
πάλιν θερμανθέντος ἔξοδον λανθάνειν. ἄτοπον δὲ
καὶ τοῦ θερμοῦ τὴν ἀναπνοὴν εἴσοδον εἶναι. φαίνεται γὰρ τοὐναντίον· τὸ μὲν γὰρ ἐκπνεόμενον εἶναι
35 θερμόν, τὸ δ᾽ εἰσπνεόμενον ψυχρόν. ὅταν δὲ θερμὸν
473 a ᾖ, ἀσθμαίνοντες ἀναπνέουσιν· διὰ γὰρ τὸ μὴ καταψύχειν ἱκανῶς τὸ εἰσιὸν πολλάκις τὸ πνεῦμα συμβαίνει σπᾶν.

VI. Ἀλλὰ μὴν οὐδὲ τροφῆς γε χάριν ὑποληπτέον
γίνεσθαι τὴν ἀναπνοήν, ὡς τρεφομένου τῷ πνεύ-
5 ματι τοῦ ἐντὸς πυρός, καὶ ἀναπνέοντος μὲν ὥσπερ
ἐπὶ πῦρ ὑπέκκαυμα ὑποβάλλεσθαι, τραφέντος δὲ
τοῦ πυρὸς γίνεσθαι τὴν ἐκπνοήν. ταὐτὰ γὰρ
ἐροῦμεν πάλιν καὶ πρὸς τοῦτον τὸν λόγον ἅπερ
καὶ¹ πρὸς τοὺς ἔμπροσθεν· καὶ γὰρ ἐπὶ τῶν ἄλλων
ζῴων ἐχρῆν τοῦτο συμβαίνειν ἢ τὸ ἀνάλογον
10 τούτῳ· πάντα γὰρ ἔχει θερμότητα ζωτικήν. ἔπειτα

442

do unceasingly, inhaling and exhaling. Now on this theory it follows that exhalation comes before inhalation. But the opposite is the truth. And here is the proof ; these actions take place alternately, but the last thing men do is to exhale, so that inhalation must be the first action.

Again, those who hold this theory have said nothing of the purpose for which animals are equipped with these functions (I mean inhalation and exhalation), but merely describe them as something coincidental. And yet we see that they control life and death ; for when respiring creatures can no longer respire, then destruction comes to them. Again, it is absurd that we should be conscious of the successive exit and entrance of heat through the mouth, but unconscious of the entrance of the breath into the chest, and its subsequent exit when it is hot. It is also absurd that inhalation should be the entrance of the hot. For the opposite appears to be true ; what is exhaled is hot and what is inhaled cool. When it is hot men breathe hard ; for, because the entering breath does not cool sufficiently, they consequently draw it in frequently.

VI. Nor can we suppose that respiration is for the sake of nourishment, on the supposition that the internal fire is nourished by breath, respiration, as it were, supplying fuel to the fire, and exhalation taking place when the fire is sufficiently fed. For we shall make the same reply to this argument that we made to the previous ones ; for this or something analogous to it should occur in the case of the other animals ; for all of them have life-giving heat. *Is breathing for the sake of food ?*

---

¹ καί om. LPS.

ARISTOTLE

καὶ τὸ γίνεσθαι θερμὸν¹ ἐκ τοῦ πνεύματος τίνα
χρὴ τρόπον λέγειν, πλασματῶδες ὄν; μᾶλλον γὰρ
ἐκ τῆς τροφῆς τοῦτο γινόμενον ὁρῶμεν. συμβαίνει τε κατὰ ταὐτὸ δέχεσθαι τὴν τροφὴν καὶ τὸ
περίττωμα ἀφιέναι· τοῦτο δ' ἐπὶ τῶν ἄλλων οὐχ
ὁρῶμεν γινόμενον.

15 VII. Λέγει δὲ περὶ ἀναπνοῆς καὶ Ἐμπεδοκλῆς,
οὐ μέντοι τίνος γ' ἕνεκα, οὐδὲ περὶ πάντων τῶν
ζῴων οὐδὲν ποιεῖ δῆλον, εἴτε ἀναπνέουσιν εἴτε
μή. καὶ περὶ τῆς διὰ τῶν μυκτήρων ἀναπνοῆς
λέγων οἴεται καὶ περὶ τῆς κυρίας λέγειν ἀναπνοῆς. ἔστι γὰρ καὶ διὰ τῆς ἀρτηρίας ἐκ τῶν
20 στηθῶν ἡ ἀναπνοή, καὶ ἡ διὰ τῶν μυκτήρων·
αὐτοῖς δὲ χωρὶς ἐκείνης οὐκ ἔστιν ἀναπνεῦσαι τοῖς
μυκτῆρσιν. καὶ τῆς μὲν διὰ τῶν μυκτήρων γινομένης ἀναπνοῆς στερισκόμενα τὰ ζῷα οὐδὲν πά-
σχουσι, τῆς δὲ κατὰ τὴν ἀρτηρίαν ἀποθνήσκουσι.
καταχρῆται γὰρ ἡ φύσις ἐν παρέργῳ τῇ διὰ τῶν
25 μυκτήρων ἀναπνοῇ πρὸς τὴν ὄσφρησιν ἐν ἐνίοις
τῶν ζῴων· διόπερ ὀσφρήσεως μὲν σχεδὸν μετέχει
πάντα τὰ ζῷα, ἔστι δ' οὐ πᾶσι τὸ αὐτὸ αἰσθητή-
473 b ριον. εἴρηται δὲ περὶ αὐτῶν ἐν ἑτέροις σαφέστερον.

Γίνεσθαι δέ φησι τὴν ἀναπνοὴν καὶ ἐκπνοὴν διὰ
τὸ φλέβας εἶναί τινας, ἐν αἷς ἔνεστι μὲν αἷμα, οὐ
μέντοι πλήρεις εἰσὶν αἵματος, ἔχουσι δὲ πόρους
εἰς τὸν ἔξω ἀέρα, τῶν μὲν τοῦ σώματος μορίων
5 ἐλάττους, τῶν δὲ τοῦ ἀέρος μείζους· διὸ τοῦ
αἵματος πεφυκότος κινεῖσθαι ἄνω καὶ κάτω, κάτω
μὲν φερομένου εἰσρεῖν τὸν ἀέρα καὶ γίνεσθαι
ἀναπνοήν, ἄνω δ' ἰόντος ἐκπίπτειν θύραζε καὶ
γίνεσθαι τὴν ἐκπνοήν, παρεικάζων τὸ συμβαῖνον
ταῖς κλεψύδραις.

¹ τὸ θερμὸν SP.

Again, what are we to say of this imaginary generation of heat from the breath ? For we can see that it is rather due to food. Besides, it would follow that the animal receives nourishment and discharges waste product by the same channel ; and this does not seem to be true in other cases.

VII. Empedocles also discusses respiration, but not its purpose, nor does he clear up the question whether all animals respire or not. Also he thinks that when he speaks of respiration through the nostrils he is speaking of respiration in its primary sense. But even respiration through the nostrils proceeds from the chest through the windpipe ; respiration through the nostrils by themselves without these is impossible. Again, animals deprived of breathing through the nostrils do not suffer at all, but when they lose their breathing through the windpipe they die. Nature employs this breathing through the nostrils for a subsidiary purpose in some animals—for smelling ; this is why nearly all animals share the sense of smell, but they have not all the same sense organ. A more detailed account of this is given in other works.[a]

Empedocles on breathing.

But Empedocles says that inhalation and exhalation occur because there are certain veins, which contain some blood but are not full of blood, but have openings to the air outside, too small for solid particles, but large enough for air ; hence since it is the nature of blood to move up and down, when it is carried down the air flows in and inhalation occurs, but, when it rises, the air is driven out, and exhalation takes place. He likens this process to what happens in water-clocks.

[a] *Cf. De An.* iii. 421 a 10, *De Sens.* 443 a 4, 444 b 7-15.

473 b

ὧδε δ' ἀναπνεῖ πάντα καὶ ἐκπνεῖ· πᾶσι λίφαιμοι
10 σαρκῶν σύριγγες πύματον κατὰ σῶμα τέτανται,
καί σφιν ἐπὶ στομίοις πυκναῖς τέτρηνται ἄλοξιν
ῥινῶν ἔσχατα τέρθρα διαμπερές, ὥστε φόνον μὲν
κεύθειν, αἰθέρι δ' εὐπορίην διόδοισι τετμῆσθαι.
ἔνθεν ἔπειθ' ὁπόταν μὲν ἀπαΐξῃ τέρεν αἷμα,
15 αἰθὴρ παφλάζων καταβήσεται οἴδματι μάργῳ,
εὖτε δ' ἀναθρώσκῃ, πάλιν ἐκπνέει, ὥσπερ ὅταν
παῖς
κλεψύδρῃ παίζουσα διειπετέος χαλκοῖο,
εὖτε μὲν αὐλοῦ πορθμὸν ἐπ' εὐειδεῖ χερὶ θεῖσα
εἰς ὕδατος βάπτῃσι τέρεν δέμας ἀργυφέοιο,
20 οὐ τότ' ἐς ἄγγος δ' ὄμβρος ἐσέρχεται, ἀλλά μιν
εἴργει
ἀέρος ὄγκος ἔσωθε πεσὼν ἐπὶ τρήματα πυκνά,
εἰσόκ' ἀποστεγάσῃ πυκινὸν ῥόον· αὐτὰρ ἔπειτα
πνεύματος ἐλλείποντος ἐσέρχεται αἴσιμον ὕδωρ.
ὡς δ' αὔτως ὅθ' ὕδωρ μὲν ἔχῃ κατὰ βένθεα
χαλκοῦ
25 πορθμοῦ χωσθέντος βροτέῳ χροῒ ἠδὲ πόροιο,
αἰθὴρ δ' ἐκτὸς ἔσω λελιημένος ὄμβρον ἐρύκει
ἀμφὶ πύλας ἰσθμοῖο δυσηχέος, ἄκρα κρατύνων,
474 a εἰσόκε χειρὶ μεθῇ· τότε δ' αὖ πάλιν, ἔμπαλιν ἢ
πρίν,
πνεύματος ἐμπίπτοντος ὑπεκθέει αἴσιμον ὕδωρ.
ὡς δ' αὔτως τέρεν αἷμα κλαδασσόμενον διὰ γυίων
ὁππότε μὲν παλίνορσον ἀπαΐξειε μυχόνδε,
5 αἰθέρος εὐθὺς ῥεῦμα κατέρχεται οἴδματι θῦον,
εὖτε δ' ἀναθρώσκῃ, πάλιν ἐκπνέει ἶσον ὀπίσσω.

Λέγει μὲν οὖν ταῦτα περὶ τοῦ ἀναπνεῖν· ἀναπνεῖ
δ', ὥσπερ εἴπομεν, τὰ φανερῶς ἀναπνέοντα διὰ τῆς
446

" Thus all things breathe in and out : all have in their flesh bloodless pipes reaching to the verge of the body, and these are pierced at their mouths with many passages right through the surface of the skin,[a] so that they keep in the blood, but an easy passage is cleft for the air. Thence whenever the gentle blood retreats, the rushing air will descend with raging tide, but when the blood leaps up again, the air again blows out, just as when a maid plays with a water-clock of gleaming bronze. When placing on her shapely hand the channel of the tube she dips it into the delicate body of water silver white, not then does the shower flow into the vessel, but the mass of the air pressing from within on the crowded holes checks it, until she sets free the dense stream. Then the air gives way and the water duly enters. So in the same way, when the water lies in the depths of the bronze vessel, the passage and channel being blocked by the human hand, the air outside craving entrance keeps the water back about the gates of the resounding channel, holding fast its surface, until the maid lets go with the hand ; then back again in the reverse way, as the air rushes in, the water duly flows away. In just this way whenever the gentle blood coursing through the limbs retreats back again to its recesses, at once a stream of air flows in with rushing tide, but when the blood leaps up, a like amount of air is breathed out in return."

This is what Empedocles says about respiration. But, as we have said, animals whose breathing can be seen breathe through the windpipe, whether they

Refutation of Empedocles.

---

[a] ῥινῶν here is evidently from ῥινός, " skin," as Diels pointed out. Empedocles is describing respiration through the pores. Aristotle (followed by many modern scholars) derived the word from ῥῖνες, " nostrils,"—thus missing the whole point.

474 a

ἀρτηρίας, διά τε τοῦ στόματος ἅμα καὶ διὰ τῶν
10 μυκτήρων· ὥστ᾽ εἰ μὲν περὶ ταύτης λέγει τῆς
ἀναπνοῆς, ἀναγκαῖον ζητεῖν πῶς ἐφαρμόσει ὁ
εἰρημένος λόγος τῆς αἰτίας· φαίνεται γὰρ τοὐναντίον συμβαῖνον. ἄραντες μὲν γὰρ τὸν τόπον,
καθάπερ τὰς φύσας ἐν τοῖς χαλκείοις, ἀναπνέουσιν
(αἴρειν δὲ τὸ θερμὸν εὔλογον, ἔχειν δὲ τὸ αἷμα τὴν
15 τοῦ θερμοῦ χώραν), συνιζάνοντες δὲ καὶ καταπλήτ-
τοντες,[1] ὥσπερ ἐκεῖ τὰς φύσας, ἐκπνέουσιν. πλὴν
ἐκεῖ μὲν οὐ κατὰ ταὐτὸν εἰσδέχονταί τε τὸν ἀέρα
καὶ πάλιν ἐξιᾶσιν, οἱ δ᾽ ἀναπνέοντες κατὰ ταὐτόν.
εἰ δὲ περὶ τῆς κατὰ τοὺς μυκτῆρας λέγει μόνης,
20 πολὺ διημάρτηκεν· οὐ γάρ ἐστιν ἀναπνοὴ μυκτήρων
ἴδιος, ἀλλὰ παρὰ τὸν αὐλῶνα τὸν περὶ τὸν γαρ-
γαρεῶνα, ᾗ τὸ ἔσχατον τοῦ ἐν τῷ στόματι οὐρανοῦ,
συντετρημένων τῶν μυκτήρων χωρεῖ τὸ μὲν ταύτῃ
τοῦ πνεύματος, τὸ δὲ διὰ τοῦ στόματος, ὁμοίως
εἰσίόν τε καὶ ἐξιόν. τὰ μὲν οὖν παρὰ τῶν ἄλλων
εἰρημένα περὶ τοῦ ἀναπνεῖν τοιαύτας καὶ τοσαύτας
ἔχει δυσχερείας.

25 VIII. Ἐπεὶ δὲ εἴρηται πρότερον ὅτι τὸ ζῆν καὶ
ἡ τῆς ψυχῆς ἕξις μετὰ θερμότητός τινός ἐστιν·
οὐδὲ γὰρ ἡ πέψις, δι᾽ ἧς ἡ τροφὴ γίνεται τοῖς
ζῴοις, οὔτ᾽ ἄνευ ψυχῆς οὔτ᾽ ἄνευ θερμότητός ἐστιν·
πυρὶ γὰρ ἐργάζεται πάντα· διόπερ ἐν ᾧ πρώτῳ
τόπῳ τοῦ σώματος καὶ ἐν ᾧ πρώτῳ τοῦ τόπου
30 τούτου μορίῳ τὴν ἀρχὴν ἀναγκαῖον εἶναι τὴν
τοιαύτην, ἐνταῦθα καὶ τὴν πρώτην θρεπτικὴν[2]
474 b ψυχὴν ἀναγκαῖον ὑπάρχειν. οὗτος δ᾽ ἐστὶν ὁ
μέσος τόπος τοῦ τε δεχομένου τὴν τροφὴν καὶ
καθ᾽ ὃν ἀφίησι τὸ περίττωμα. τοῖς μὲν οὖν
ἀναίμοις ἀνώνυμον, τοῖς δ᾽ ἐναίμοις ἡ καρδία

448

breathe through the mouth or the nostrils. Thus, if he is talking of breathing in this sense, we must consider how his explanation will fit the facts ; for it is clear that just the opposite happens. When men inhale they raise the region ⟨of the chest⟩, like the bellows in a forge (it is natural for heat to raise it, and for the blood to occupy the hot region) ; but they exhale by letting it collapse and settle down, like the bellows in the other case ; only in that case they do not admit and expel the air by the same passage, whereas those who breathe do. But if he is speaking only of breathing by the nostrils, he is guilty of a grave error. Respiration is not peculiar to the nostrils ; the breath passes through the channel near the uvula, where the roof of the mouth ends, and, as the nostrils are perforated, proceeds partly through them, and partly by the mouth, both when entering and when going out. Such is the nature and magnitude of the difficulties involved in the accounts given of breathing by other philosophers.

VIII. We have said before that life and the possession of soul depend upon some degree of heat ; for digestion, by which animals assimilate their food, cannot take place apart from the soul and heat ; for all food is rendered digestible by fire. Therefore the primary nutritive soul must reside in the region of the body, and in the part of that region, in which this principle directly resides. This is the region midway between that which receives the food and that by which the waste is discharged. It has no name in the bloodless animals, but in animals with

---

¹ καταπνίγοντες SP et plerique edd.
² θρεπτικὴν] τὴν θρεπτικὴν L.

474 b

τοῦτο τὸ μόριόν ἐστιν. ἡ τροφὴ μὲν γὰρ ἐξ ἧς
5 ἤδη γίνεται τὰ μόρια τοῖς ζῴοις ἡ τοῦ αἵματος
φύσις ἐστίν. τοῦ δ' αἵματος καὶ τῶν φλεβῶν τὴν
αὐτὴν ἀρχὴν ἀναγκαῖον εἶναι· θατέρου γὰρ ἕνεκα
θάτερόν ἐστιν, ὡς ἀγγεῖον καὶ δεκτικόν. ἀρχὴ δὲ
τῶν φλεβῶν ἡ καρδία τοῖς ἐναίμοις· οὐ γὰρ διὰ
ταύτης, ἀλλ' ἐκ ταύτης ἠρτημέναι πᾶσαι τυγχάνουσιν. δῆλον δ' ἡμῖν τοῦτο ἐκ τῶν ἀνατομῶν.
10 Τὰς μὲν οὖν ἄλλας δυνάμεις τῆς ψυχῆς ἀδύνατον
ὑπάρχειν ἄνευ τῆς θρεπτικῆς (δι' ἣν δ' αἰτίαν,
εἴρηται πρότερον ἐν τοῖς περὶ ψυχῆς), ταύτην δ'
ἄνευ τοῦ φυσικοῦ πυρός· ἐν τούτῳ γὰρ ἡ φύσις
ἐμπεπύρευκεν αὐτήν. φθορὰ δὲ πυρός, ὥσπερ
15 εἴρηται πρότερον, σβέσις καὶ μάρανσις. σβέσις
μὲν ἡ ὑπὸ τῶν ἐναντίων· διόπερ ἀθρόον τε ὑπὸ
τῆς τοῦ περιέχοντος ψυχρότητος, καὶ θᾶττον[1]
σβέννυται διασπώμενον. αὕτη μὲν οὖν ἡ φθορὰ
βίαιος ὁμοίως ἐπὶ τῶν ἐμψύχων καὶ τῶν ἀψύχων
ἐστίν· καὶ γὰρ ὀργάνοις διαιρουμένου τοῦ ζῴου,
καὶ πηγνυμένου διὰ ψύχους ὑπερβολήν, ἀποθνή-
20 σκουσιν. ἡ δὲ μάρανσις διὰ πλῆθος θερμότητος·
καὶ γὰρ ἂν ὑπερβάλλῃ τὸ πέριξ θερμόν, καὶ τροφὴν
ἐὰν μὴ λαμβάνῃ, φθείρεται τὸ πυρούμενον, οὐ
ψυχόμενον ἀλλὰ μαραινόμενον. ὥστ' ἀνάγκη γίνεσθαι κατάψυξιν, εἰ μέλλει τεύξεσθαι σωτηρίας·
τοῦτο γὰρ βοηθεῖ πρὸς ταύτην τὴν φθοράν.
25 IX. Ἐπεὶ δὲ τῶν ζῴων τὰ μὲν ἔνυδρα, τὰ δ'
ἐν τῇ γῇ ποιεῖται τὴν διατριβήν, τούτων τοῖς μὲν
μικροῖς πάμπαν καὶ τοῖς ἀναίμοις ἡ γινομένη ἐκ
τοῦ περιέχοντος ἢ ὕδατος ἢ ἀέρος ψύξις ἱκανὴ
πρὸς τὴν βοήθειαν τῆς φθορᾶς ταύτης· μικρὸν γὰρ
ἔχοντα τὸ θερμὸν μικρᾶς δέονται τῆς βοηθείας.

---

[1] θᾶττον ὅτι LP et plerique edd.

blood this part is the heart. The food from which the parts of animals directly derive their growth is the natural substance blood. But the blood and the veins must have the same source ; for the latter exist for the sake of the former, as its vessel and receptacle. But in sanguineous animals the heart is the source of the veins ; for the veins do not go through the heart, but all actually proceed from it. We can see this from dissections.

The other faculties of the soul cannot exist without the nutritive (the reason for this has been discussed in my work *On the Soul*),[a] nor can that exist without the natural fire in which nature has kindled it. But the destruction of fire, as has been said before, is either quenching or dying out. Quenching is destruction by contraries ; thus it is quenched by the coldness of the surrounding air both when it is a mass and (even more quickly) when scattered. Now this form of destruction is due to violence equally in creatures with and without soul ; for an animal dies just the same whether it is cut up by instruments or frozen by excess of cold. But dying out is due to excess of heat ; for if the surrounding heat is excessive the burning object, unless it receives food, dies, not because it grows cold, but by dying out. So then, if it is to survive, a cooling must take place ; for this protects it against destruction in this way.

IX. Since animals either belong to the water or spend their time on land, in the case of those which are bloodless and very small the cooling due to the surrounding envelope—whether water or air—is sufficient to protect them against destruction of this kind ; for as they contain little heat they need but

Breathing and life.

---

[a] *De An.* 411 b 18, 413 b 1.

451

474 b

30 διὸ καὶ βραχύβια σχεδὸν πάντα τὰ τοιαῦτ᾽ ἐστίν·
ἐπ᾽ ἀμφότερα γὰρ μικρὰ ὄντα μικρᾶς[1] τυγχάνει

475 a ῥοπῆς. ὅσα δὲ μακροβιώτερα τῶν ἐντόμων (ἄναιμα
γάρ ἐστι πάντα τὰ ἔντομα), τούτοις ὑπὸ τὸ διάζωμα
διέσχισται, ὅπως διὰ λεπτοτέρου ὄντος τοῦ ὑμένος
ψύχηται· μᾶλλον γὰρ ὄντα θερμὰ πλείονος δεῖται
τῆς καταψύξεως, οἷον αἱ μέλιτται (τῶν γὰρ μελιτ-
5 τῶν ἔνιαι ζῶσι καὶ ἑπτὰ ἔτη) καὶ τἆλλα δὲ ὅσα βομ-
βεῖ, οἷον σφῆκες καὶ μηλολόνθαι καὶ τέττιγες. καὶ
γὰρ τὸν ψόφον ποιοῦσι πνεύματι, οἷον τὰ[2] ἀσθμαί-
νοντα· ἐν αὐτῷ γὰρ τῷ ὑποζώματι, τῷ ἐμφύτῳ
πνεύματι αἴροντι καὶ συνίζοντι, συμβαίνει πρὸς
10 τὸν ὑμένα γίνεσθαι τρίψιν· κινοῦσι γὰρ τὸν τόπον
τοῦτον, ὥσπερ τὰ ἀναπνέοντα ἔξωθεν τῷ πλεύμονι
καὶ οἱ ἰχθύες τοῖς βραγχίοις. παραπλήσιον γὰρ
συμβαίνει κἂν εἴ τίς τινα τῶν ἀναπνεόντων πνίγοι,
τὸ στόμα κατασχών· καὶ γὰρ ταῦτα ποιήσει τῷ
πλεύμονι τὴν ἄρσιν ταύτην. ἀλλὰ τούτοις μὲν
15 οὐχ ἱκανὴν ἡ τοιαύτη ποιεῖ κίνησις κατάψυξιν,
ἐκείνοις δ᾽ ἱκανήν. καὶ τῇ τρίψει τῇ πρὸς τὸν
ὑμένα ποιοῦσι τὸν βόμβον, ὥσπερ λέγομεν, οἷον
διὰ τῶν καλάμων τῶν τετρυπημένων τὰ παιδία,
ὅταν ἐπιθῶσιν ὑμένα λεπτόν. διὰ γὰρ τοῦτο καὶ
τῶν τεττίγων οἱ ᾄδοντες ᾄδουσιν· θερμότεροι γάρ
20 εἰσι, καὶ ἔσχισται αὐτοῖς ὑπὸ τὸ ὑπόζωμα· τοῖς
δὲ μὴ ᾄδουσι τοῦτ᾽ ἐστὶν ἄσχιστον.

Καὶ τῶν ἐναίμων δὲ καὶ πλεύμονα ἐχόντων,
ὀλίγαιμον δ᾽ ἐχόντων καὶ σομφόν, ἔνια διὰ τοῦτο
πολὺν χρόνον δύνανται ἀπνευστὶ ζῆν, ὅτι ὁ πλεύμων
ἄρσιν ἔχει πολλήν, ὀλίγον ἔχων τὸ αἷμα καὶ τὸ
25 ὑγρόν· ἡ γὰρ οἰκεία κίνησις ἐπὶ πολὺν χρόνον

---

[1] μικρὰ ὄντα μικρᾶς] μικρᾶς ὄντα LPS et plerique edd.
[2] τὰ om. LSP.

little protection. Hence, too, nearly all such animals are short-lived ; for, being small, they have but little margin in either direction. But all the insects that are long-lived (for all insects are bloodless) are cleft at the waist, so that they may be cooled through the membrane, which is thinner there ; for being abnormally hot they need more cooling. Such are bees (for some bees live as long as seven years), and all other buzzing insects such as wasps and cockchafers and cicalas. They produce the buzzing sound by breath, like animals panting ; for just at the waist the breath inherent in them by rising and falling causes friction against the membrane ; for they move this region, just as those animals which draw their breath from the outside do with the lung, and fishes with their gills. It is similar to what would happen if one suffocated one of the respiring animals, by stopping its mouth ; for these too will make this heaving movement with the lungs. But while for them such movement does not produce sufficient cooling, for the others it does. It is by the friction against the membrane, as we said, that they make their buzzing, just as boys do through reeds pierced with holes, when they have put a thin membrane over them. This is how the cicalas which sing do so, for they are very warm creatures, and have a cleft at the waist ; but in those which do not sing there is no cleft.

Of sanguineous animals also which have a lung, but whose lung has little blood and is spongy, some can for this reason live for a long time without breathing, because the lung is capable of considerable expansion, having but little blood and moisture ; for its own movement is sufficient to keep it cool for a long time.

453

475 a

διαρκεῖ καταψύχουσα. τέλος δ' οὐ δύναται, ἀλλ'
ἀποπνίγεται μὴ ἀναπνεύσαντα, καθάπερ εἴρηται
καὶ πρότερον· τῆς γὰρ μαράνσεως ἡ διὰ τὸ μὴ
ψύχεσθαι φθορὰ καλεῖται πνίξις, καὶ τὰ οὕτω
φθειρόμενα ἀποπνίγεσθαί φαμεν.

30 Ὅτι δ' οὐκ ἀναπνεῖ τὰ ἔντομα τῶν ζῴων, εἴρηται μὲν καὶ πρότερον, φανερὸν δὲ καὶ ἐπὶ τῶν
μικρῶν ἐστι ζῴων, οἷον μυιῶν καὶ μελιττῶν· ἐν
475 b γὰρ τοῖς ὑγροῖς πολὺν χρόνον ἀνανήχεται, ἂν μὴ
λίαν ᾖ θερμὸν ἢ ψυχρόν (καίτοι τὰ μικρὰν ἔχοντα
δύναμιν πυκνότερον ζητεῖ ἀναπνεῖν)· ἀλλὰ φθείρεται ταῦτα καὶ λέγεται ἀποπνίγεσθαι πληρου-
5 μένης τῆς κοιλίας καὶ φθειρομένου τοῦ ἐν τῷ
ὑποζώματι θερμοῦ.[1]   διὸ καὶ ἐν τῇ τέφρα χρονι-
σθέντα ἀνίσταται. καὶ τῶν ἐν τῷ ὑγρῷ δὲ ζώντων
ὅσα ἄναιμα, πλείω χρόνον ζῇ ἐν τῷ ἀέρι τῶν ἐναίμων καὶ δεχομένων τὴν θάλατταν, οἷον τῶν ἰχθύων·
διὰ γὰρ τὸ ὀλίγον ἔχειν τὸ θερμὸν ὁ ἀὴρ ἱκανός
ἐστιν ἐπὶ πολὺν χρόνον καταψύχειν, οἷον τοῖς τε
10 μαλακοστράκοις καὶ τοῖς πολύποσιν. οὐ μὴν εἰς
τέλος γε διαρκεῖ πρὸς τὸ ζῆν, διὰ τὸ ὀλιγόθερμα
εἶναι, ἐπεὶ καὶ τῶν ἰχθύων οἱ πολλοὶ ζῶσιν ἐν τῇ
γῇ, ἀκινητίζοντες μέντοι, καὶ εὑρίσκονται ὀρυτ-
τόμενοι. ὅσα γὰρ ἢ μηδ' ὅλως ἔχει πλεύμονα ἢ
ἄναιμον, ἐλαττονάκις δεῖται καταψύξεως.

15 Χ. Περὶ μὲν οὖν τῶν ἀναίμων, ὅτι τοῖς μὲν ὁ
περιέχων ἀὴρ τοῖς δὲ τὸ ὑγρὸν βοηθεῖ πρὸς τὴν
ζωήν, εἴρηται· τοῖς δ' ἐναίμοις καὶ τοῖς ἔχουσι
καρδίαν, ὅσα μὲν ἔχει πλεύμονα, πάντα δέχεται
τὸν ἀέρα καὶ τὴν κατάψυξιν ποιεῖται διὰ τοῦ
ἀναπνεῖν καὶ ἐκπνεῖν. ἔχει δὲ πλεύμονα τά τε
20 ζῳοτοκοῦντα ἐν αὑτοῖς καὶ μὴ θύραζε μόνον (τὰ
454

But at last it is unable to go on ; the animal is choked if it does not respire, as has been said before. Decay which takes the form of destruction due to lack of cooling is called suffocation, and we say that animals which die in this way are suffocated.

We have stated before that among living creatures insects do not respire, and this is evident in the case of the small ones, such as flies and bees ; for they can swim in liquid for a long time, if it is not too hot or too cold (although animals which have but little strength try to breathe more frequently) ; but they perish and are said to be suffocated, when the belly is filled and the heat in their waist is exhausted. Hence, too, they revive after they have been for some time among ashes. Of water-animals too the blood-less can live in the air for a longer time than those which have blood and admit the water, *e.g.*, fishes ; for because they contain only a small quantity of heat, the air can keep them cool for a considerable time, especially the crustaceans and the cuttlefish. It does not enable them to live permanently thus because they contain but little heat, since most of the fishes too can live in the earth, though they cannot move and are found by digging. For animals which have no lung, or a lung without blood, require cooling less often.

X. As regards bloodless animals, then, we have stated that they are helped to live in some cases by the surrounding air and in others by moisture. As for animals with blood and a heart, all that have a lung admit the air and achieve cooling by breathing in and out. All viviparous animals have a lung— if, that is, they are viviparous internally and not

Breathing in warm-blooded animals.

---

¹ ὑγροῦ LSP, Bekker.

475 b

γὰρ σελάχη ζῳοτοκεῖ μέν, ἀλλ' οὐκ ἐν αὑτοῖς) καὶ
τῶν ᾠοτοκούντων τά τε πτερυγωτά, οἷον ὄρνιθες,
καὶ τὰ φολιδωτά, οἷον χελῶναι καὶ σαῦραι καὶ
ὄφεις. ἐκεῖνα μὲν οὖν ἔναιμον, τούτων δὲ τὰ
25 πλεῖστα τὸν πλεύμονα ἔχει σομφόν. διὸ καὶ τῇ
ἀναπνοῇ χρῆται μανότερον, ὥσπερ εἴρηται καὶ
πρότερον.

Χρῆται δὲ πάντα καὶ ὅσα διατρίβει καὶ ποιεῖται
τὸν βίον ἐν τοῖς ὕδασιν, οἷον τὸ τῶν ὕδρων γένος
καὶ βατράχων καὶ κροκοδείλων καὶ ἐμύδων καὶ
χελῶναι αἵ τε θαλάττιαι καὶ αἱ χερσαῖαι καὶ
30 φῶκαι· ταῦτα γὰρ πάντα καὶ τὰ τοιαῦτα καὶ τίκτει
ἐν τῷ ξηρῷ, καὶ καθεύδει ἢ ἐν τῷ ξηρῷ, ἢ ἐν τῷ
476 a ὑγρῷ ὑπερέχοντα τὸ στόμα διὰ τὴν ἀναπνοήν.
ὅσα δὲ βράγχια ἔχει, πάντα καταψύχεται δεχόμενα
τὸ ὕδωρ· ἔχει δὲ βράγχια τὸ τῶν καλουμένων σε-
λαχῶν γένος καὶ τῶν ἄλλων ἀπόδων. ἄποδες δ'
5 οἱ ἰχθύες πάντες· καὶ γὰρ ἃ ἔχει, καθ' ὁμοιότητα
τῶν πτερυγίων λέγουσιν.[1] τῶν δὲ πόδας ἐχόντων
ἓν ἔχει βράγχιον μόνον τῶν τεθεωρημένων ὁ καλούμενος κορδύλος. ἅμα δὲ πλεύμονα καὶ βράγχια
οὐδὲν ὦπταί πω ἔχον. αἴτιον δ' ὅτι ὁ μὲν πλεύμων
τῆς ὑπὸ τοῦ πνεύματος καταψύξεως ἕνεκέν ἐστιν
(ἔοικε δὲ καὶ τοὔνομα εἰληφέναι ὁ πνεύμων διὰ
10 τὴν τοῦ πνεύματος ὑποδοχήν), τὰ δὲ βράγχια πρὸς
τὴν ἀπὸ τοῦ ὕδατος κατάψυξιν· ἓν δ' ἐφ' ἓν χρήσιμον ὄργανον,[2] καὶ μία κατάψυξις ἱκανὴ πᾶσιν. ὥστ'
ἐπεὶ μάτην οὐδὲν ὁρῶμεν ποιοῦσαν τὴν φύσιν, δυοῖν
δ' ὄντοιν θάτερον ἂν ἦν μάτην, διὰ τοῦτο τὰ μὲν
15 ἔχει βράγχια τὰ δὲ πνεύμονα, ἄμφω δ' οὐδέν.

XI. Ἐπεὶ δὲ πρὸς μὲν τὸ εἶναι τροφῆς δεῖται
τῶν ζῴων ἕκαστον, πρὸς δὲ τὴν σωτηρίαν τῆς

externally only (for sharks are viviparous, but not internally) and among the oviparous such as are winged, like birds, or scaly, like tortoises, lizards, and snakes. The former class have a lung containing blood, but in the latter class the lung is usually spongy. Consequently they make less use of respiration, as has been said before.

Respiration is also employed by all animals which live and pass their time in the water, such as the genera watersnake, frog, crocodile, and freshwater tortoise ; also sea and land tortoises, and seals. All these and similar animals bring forth their young on dry land, and either sleep on dry land, or in water with their mouth above the surface for breathing. But all those that have gills cool themselves by admitting water ; such are the genus shark, and those of all other footless animals. All fishes are footless ; what they have in the place of feet get their name (πτερύγιον) from their similarity to wings (πτέρυξ). Of animals with feet one only, so far as we have observed, has gills, *viz.*, the water newt. But so far no creature has been observed with both lungs and gills. The reason is that the purpose of the lung is cooling by means of breath (its name—πνεύμων—seems due to its being a receptacle for breath—πνεῦμα) ; but the gills assist cooling by water ; one organ avails for one purpose, and one means of cooling is enough for every animal. Since, then, we see that nature does nothing in vain, and that if there were two organs one would be useless, for this reason some creatures have gills and some lungs, but none has both.

XI. Since every animal requires food for its existence, and cooling for its preservation, nature uses

Aquatic animals.

The mouth serves a

---

[1] ἔχουσιν Bekker.
[2] χρήσιμον ὄργανον] ὄργανον χρήσιμον P, Bekker.

476 a

καταψύξεως, τῷ αὐτῷ ὀργάνῳ χρῆται πρὸς ἄμφω
ταῦτα ἡ φύσις, καθάπερ ἐνίοις τῇ γλώττῃ πρός τε
20 τοὺς χυμοὺς καὶ πρὸς τὴν ἑρμηνείαν, οὕτω τοῖς
ἔχουσι τὸν πλεύμονα τῷ καλουμένῳ στόματι πρός
τε τὴν τῆς τροφῆς ἐργασίαν καὶ τὴν ἐκπνοὴν καὶ
τὴν ἀναπνοήν. τοῖς δὲ μὴ ἔχουσι πνεύμονα μηδ'
ἀναπνέουσι τὸ μὲν στόμα πρὸς τὴν ἐργασίαν τῆς
τροφῆς, πρὸς δὲ τὴν κατάψυξιν τοῖς δεομένοις
25 καταψύξεως ἡ τῶν βραγχίων ὑπάρχει φύσις. πῶς
μὲν οὖν ἡ τῶν εἰρημένων ὀργάνων δύναμις ποιεῖ
τὴν κατάψυξιν, ὕστερον ἐροῦμεν· πρὸς δὲ τὸ τὴν
τροφὴν μὴ διακωλύειν παραπλησίως τοῖς τ' ἀναπνέουσι συμβαίνει καὶ τοῖς δεχομένοις τὸ ὑγρόν·
οὔτε γὰρ ἀναπνέοντες ἅμα καταδέχονται τὴν τρο-
30 φήν· εἰ δὲ μή, συμβαίνει πνίγεσθαι παρεισιούσης
τῆς τροφῆς ἢ τῆς ὑγρᾶς ἢ τῆς ξηρᾶς ἐπὶ τὸν
πνεύμονα διὰ τῆς ἀρτηρίας· προτέρα[1] γὰρ κεῖται
ἡ ἀρτηρία τοῦ οἰσοφάγου, δι' οὗ ἡ τροφὴ πορεύεται
εἰς τὴν καλουμένην κοιλίαν. τοῖς μὲν οὖν τετράποσι
καὶ ἐναίμοις ἔχει ἡ ἀρτηρία οἷον πῶμα τὴν ἐπι-
476 b γλωττίδα· τοῖς δ' ὄρνισι καὶ τῶν τετραπόδων τοῖς
ᾠοτόκοις οὐκ ἔπεστιν, ἀλλὰ τῇ συναγωγῇ τὸ αὐτὸ
ποιοῦσιν· δεχόμενα γὰρ τὴν τροφὴν τὰ μὲν συνάγει,
τὰ δ' ἐπιτίθησι τὴν ἐπιγλωττίδα. προελθούσης δὲ
τὰ μὲν ἐπαίρει, τὰ δὲ διοίγει καὶ καταδέχεται τὸ
5 πνεῦμα πρὸς τὴν κατάψυξιν. τὰ δ' ἔχοντα βράγχια,
ἀφέντα διὰ τούτων τὸ ὑγρόν, διὰ τοῦ στόματος
καταδέχεται τὴν τροφήν· ἀρτηρίαν μὲν γὰρ οὐκ
ἔχουσιν, ὥστε ταύτῃ μὲν οὐθὲν ἂν βλάπτοιντο ὑπὸ
τῆς τοῦ ὑγροῦ παρεμπτώσεως, ἀλλ' εἰς τὴν κοιλίαν
10 εἰσιόντος. διὸ ταχεῖαν ποιεῖται τὴν ἄφεσιν καὶ
τὴν λῆψιν τῆς τροφῆς, καὶ τοὺς ὀδόντας ὀξεῖς

---

[1] προτέρα LZ : πρότερον.

the same organ for both these purposes ; just as in
some animals the tongue is used both to appreciate
flavours and for articulation, so in animals which have
a lung the mouth is used both for the mastication of
food and for breathing in and out. But in those which
have no lungs and do not breathe, the mouth serves
for mastication, while the gill-system supplies refrigeration as required. We shall explain later on [a]
how the aforesaid organs have the faculty to cause
refrigeration. To prevent interference by food, much
the same process occurs in creatures that breathe and
in those which admit moisture. They do not admit
food at the time of breathing ; if they did, they would
inevitably be choked by the food, whether dry or
wet, entering through the windpipe into the lung ;
for the windpipe lies in front of the oesophagus,
through which the food passes into what we call the
belly. In quadrupeds which have blood the windpipe
has a kind of lid called the epiglottis ; birds and
oviparous quadrupeds have no epiglottis, but they
achieve the same result by contraction of the windpipe ; for when taking in food the latter class contracts the windpipe, the former closes the epiglottis.
As the food continues on its way, the latter class
expands the windpipe again, the former opens the
epiglottis, and admits breath for the purpose of cooling. Creatures which have gills receive food through
the mouth as they discharge water through the gills ;
for they have no windpipe, so that they cannot
suffer harm from the water's falling into it, but only
from its entering the stomach. This is why they make
the discharge and the reception of food rapidly, and

[a] Ch. xiii.

ἔχουσι, καὶ καρχαρόδοντες σχεδὸν πάντες εἰσίν·
οὐ γὰρ ἐνδέχεται λεαίνειν τὴν τροφήν.

XII. Περὶ δὲ τὰ κητώδη τῶν ἐνύδρων ἀπορήσειεν
ἄν τις, ἔχει δὲ κἀκεῖνα κατὰ λόγον, οἷον περί τε
15 τοὺς δελφῖνας καὶ τὰς φαλαίνας, καὶ τῶν ἄλλων
ὅσα ἔχει τὸν καλούμενον αὐλόν. ταῦτα γὰρ ἄποδα
μέν ἐστιν, ἔχοντα δὲ πνεύμονα δέχεται τὴν θάλατταν. αἴτιον δὲ τούτου τὸ νῦν εἰρημένον· οὐ γὰρ
καταψύξεως ἕνεκεν δέχεται τὸ ὑγρόν. τοῦτο μὲν
γὰρ γίνεται αὐτοῖς ἀναπνέουσιν· ἔχουσι γὰρ πλεύ-
20 μονα. διὸ καὶ καθεύδουσιν ὑπερέχοντα τὸ στόμα,
καὶ ῥέγχουσιν οἵ γε δελφῖνες. ἔτι δὲ κἂν ληφθῶσι
τοῖς δικτύοις, ταχὺ ἀποπνίγονται διὰ τὸ μὴ ἀνα-
πνεῖν· καὶ ἐπιπολάζοντα φαίνεται τὰ τοιαῦτα ἐπὶ
τῆς θαλάττης διὰ τὴν ἀναπνοήν. ἀλλ᾽ ἐπειδὴ
ἀναγκαῖον ποιεῖσθαι τὴν τροφὴν ἐν ὑγρῷ, ἀναγκαῖον
25 δεχόμενα τὸ ὑγρὸν ἀφιέναι, καὶ διὰ τοῦτ᾽ ἔχουσι
πάντα τὸν αὐλόν· δεξάμενα γὰρ τὸ ὕδωρ, ὥσπερ
οἱ ἰχθύες κατὰ τὰ βράγχια, ταῦτα κατὰ τὸν αὐλὸν
ἀνασπᾷ τὸ ὕδωρ. σημεῖον δὲ καὶ ἡ θέσις τοῦ
αὐλοῦ· πρὸς οὐθὲν γὰρ περαίνει τῶν ἐναίμων, ἀλλὰ
30 πρὸ τοῦ ἐγκεφάλου τὴν θέσιν ἔχει, καὶ ἀφίησι τὸ
ὕδωρ. διὰ ταὐτὸ δὲ τοῦτο δέχεται καὶ τὰ μαλάκια τὸ ὕδωρ καὶ τὰ μαλακόστρακα, λέγω δ᾽
οἷον τοὺς καλουμένους καράβους καὶ τοὺς καρ-
κίνους. καταψύξεως μὲν γὰρ αὐτῶν οὐδὲν τυγχάνει δεόμενον· ὀλιγόθερμον γάρ ἐστι καὶ ἄναιμον
477 a ἕκαστον αὐτῶν, ὥσθ᾽ ἱκανῶς καταψύχεται ὑπὸ τοῦ
περιέχοντος ὑγροῦ· ἀλλὰ διὰ τὴν τροφήν, ὅπως μὴ
ἅμα δεχομένοις εἰσρέῃ τὸ ὑγρόν. τὰ μὲν οὖν
μαλακόστρακα, οἷον οἵ τε καρκίνοι καὶ οἱ κάραβοι,
παρὰ τὰ δασέα ἀφιᾶσι τὸ ὕδωρ διὰ τῶν ἐπιπτυγ-

have sharp teeth, set like a saw in most cases ; for it is impossible for them to chew their food.

XII. A difficulty might be raised about the cetaceans among water animals, although they too have a logical explanation—*e.g.*, about dolphins and whales, and all other creatures which have what is called a blowhole. For these have no feet, but have lungs, and yet admit the water. But the reason of this is the one already given ; for they do not admit the moisture for the purpose of cooling. This cooling takes place when they breathe, for they have lungs. This is why they sleep with their mouths above water, and the dolphins at any rate snore. Again, if they are caught in nets, they are quickly choked because they cannot breathe ; and they can be seen coming to the surface of the sea for the purpose of breathing. But since they have to do their feeding in water, they have to admit the water and then discharge it, and this is why they all have a blowhole ; for after having admitted the water they expel it again through the blowhole, just as fishes do through the gills. The position of the blowhole proves this ; for it leads to none of the parts with blood, but lies in front of the brain, and from there discharges the water. It is for the same reason that molluscs and crustacea admit water—I mean such creatures as lobsters and crabs. None of these happens to need cooling ; for each of these species is of low temperature and bloodless, so that it is sufficiently cooled by the surrounding water ; but they admit water in feeding, ⟨and so must expel it⟩ so that the water may not flow in as they are absorbing food. The crustacea, such as crabs and lobsters, discharge the water through the folds by

461

⁵ μάτων, σηπίαι δὲ καὶ πολύποδες διὰ τοῦ κοίλου
τοῦ ὑπὲρ τῆς καλουμένης κεφαλῆς. γέγραπται δὲ
περὶ αὐτῶν δι' ἀκριβείας μᾶλλον ἐν ταῖς περὶ τῶν
ζῴων ἱστορίαις. περὶ μὲν οὖν τοῦ δέχεσθαι τὸ
ὑγρόν, εἴρηται ὅτι συμβαίνει διὰ κατάψυξιν καὶ
διὰ τὸ δεῖν δέχεσθαι τὴν τροφὴν ἐκ τοῦ ὑγροῦ τὰ
¹⁰ τὴν φύσιν ὄντα τῶν ζῴων ἔνυδρα.

XIII. Περὶ δὲ τῆς καταψύξεως, τίνα γίνεται
τρόπον τοῖς τ' ἀναπνέουσι καὶ τοῖς ἔχουσι βράγχια,
μετὰ ταῦτα λεκτέον. ὅτι μὲν οὖν ἀναπνέουσιν ὅσα
πνεύμονα τῶν ζῴων ἔχουσι, πρότερον εἴρηται. διὰ
¹⁵ τί δὲ τοῦτο τὸ μόριον ἔχουσιν ἔνια, καὶ διὰ τί τὰ
ἔχοντα δεῖται τῆς ἀναπνοῆς, αἴτιον τοῦ μὲν ἔχειν
ὅτι τὰ τιμιώτερα τῶν ζῴων πλείονος τετύχηκε
θερμότητος· ἅμα γὰρ ἀνάγκη καὶ ψυχῆς τετυχη-
κέναι τιμιωτέρας· τιμιώτερα γὰρ τὰ τοιαῦτα τῆς
φύσεως τῆς τῶν ἰχθύων.¹ διὸ καὶ τὰ μάλιστα
²⁰ ἔναιμον ἔχοντα τὸν πνεύμονα καὶ θερμὸν μείζονά
τε τοῖς μεγέθεσι, καὶ τό γε καθαρωτάτῳ καὶ πλεί-
στῳ κεχρημένον αἵματι τῶν ζῴων ὀρθότατόν ἐστιν
ὁ ἄνθρωπος, καὶ τὸ ἄνω πρὸς τὸ τοῦ ὅλου ἄνω ἔχει
μόνον διὰ τὸ τοιοῦτον ἔχειν τοῦτο τὸ μόριον. ὥστε
²⁵ τῆς οὐσίας καὶ τούτῳ καὶ τοῖς ἄλλοις θετέον αἴτιον
αὐτό, καθάπερ ὁτιοῦν ἄλλο τῶν μορίων. ἔχει μὲν
οὖν ἕνεκα τούτου. τὴν δ' ἐξ ἀνάγκης καὶ τῆς
κινήσεως αἰτίαν καὶ τὰ τοιαῦτα δεῖ² νομίζειν συνεστάναι ζῷα, καθάπερ καὶ μὴ τοιαῦτα πολλὰ συνέ-
στηκεν· τὰ μὲν γὰρ ἐκ γῆς πλείονος γέγονεν, οἷον τὸ
τῶν φυτῶν γένος, τὰ δ' ἐξ ὕδατος, οἷον τὸ τῶν
³⁰ ἐνύδρων· τῶν δὲ πτηνῶν καὶ πεζῶν τὰ μὲν ἐξ ἀέρος

---

¹ ἰχθύων ci. Biehl: φυτῶν.

the hairy parts, but the cuttlefish and polypus through the hollow above the so-called head. I have given a more exact account of these in my *History of Animals*.[a] Concerning the admission of water, then, we have explained that it occurs for the purpose of cooling, and because those creatures which naturally live in water must derive their food from the water.

XIII. Next we must explain how this cooling takes place in creatures which breathe and in those which have gills. We have already stated that all living creatures that have lungs breathe. But two questions remain : why some creatures have this organ, and why those that have it need to breathe. The answer to the first is that animals higher in the scale of creation have more heat ; for they must at the same time have a higher form of soul ; for they have a higher nature than that of fishes. So the animals which have a lung with the most blood and heat are greater in size, and that whose blood is purest and in the greatest quantity of all living creatures is the most erect, that is to say man ; " up " in his case corresponds to " up " in the whole universe just because he has such a lung. So that the reason for its existence both in this and in other animals must be assumed, just as in the case of any other parts. It possesses it for this reason. One is bound to suppose that it is by necessity, and for the sake of motion that such creatures are so made, just as there are many that are not so made ; for some are made from a larger proportion of earth, such as the genus of plants, and others from water, such as the water animals ; but of the winged and land animals some

---

[a] *Hist. An.* 523 a 30, etc.

---

[2] δεῖ L : om. vulgo.

477 a

τὰ δ' ἐκ πυρός. ἕκαστα δ' ἐν τοῖς οἰκείοις τόποις
ἔχει τὴν τάξιν αὐτῶν.

XIV. Ἐμπεδοκλῆς δ' οὐ καλῶς τοῦτ' εἴρηκε,
477 b φάσκων τὰ θερμότατα καὶ πῦρ ἔχοντα πλεῖστον
τῶν ζῴων ἔνυδρα εἶναι, φεύγοντα τὴν ὑπερβολὴν
τῆς ἐν τῇ φύσει θερμότητος, ὅπως ἐπειδὴ τοῦ
ψυχροῦ καὶ τοῦ ὑγροῦ ἐλλείπει, κατὰ τὸν τόπον
ἀνασῴζηται ἐναντία ὄντα· θερμὸν γὰρ εἶναι τὸ
5 ὑγρὸν ἧττον τοῦ ἀέρος. ὅλως μὲν οὖν ἄτοπον πῶς
ἐνδέχεται γενόμενον ἕκαστον αὐτῶν ἐν τῷ ξηρῷ
μεταβάλλειν τὸν τόπον εἰς τὸ ὑγρόν· σχεδὸν γὰρ
καὶ ἄποδα τὰ πλεῖστα αὐτῶν ἐστιν. ὁ δὲ τὴν ἐξ
ἀρχῆς αὐτῶν σύστασιν λέγων γενέσθαι μὲν ἐν τῷ
ξηρῷ φησί, φεύγοντα δ' ἐλθεῖν εἰς τὸ ὑγρόν. ἔτι
10 δ' οὐδὲ[1] φαίνεται θερμότερα ὄντα τῶν πεζῶν· τὰ
μὲν γὰρ ἄναιμα πάμπαν, τὰ δ' ὀλίγαιμα αὐτῶν
ἐστιν.

Ἀλλὰ ποῖα μὲν δεῖ λέγειν θερμὰ καὶ ψυχρά, καθ'
αὑτὰ τὴν ἐπίσκεψιν εἴληφεν· περὶ δ' ἧς αἰτίας
εἴρηκεν Ἐμπεδοκλῆς, τῇ μὲν ἔχει τὸ ζητούμενον
λόγον, οὐ μὴν ὅ γε φησὶν ἐκεῖνος ἀληθές. τῶν μὲν
15 γὰρ ἕξεων τοὺς[2] τὰς ὑπερβολὰς ἔχοντας οἱ ἐναντίοι
τόποι καὶ ὧραι σῴζουσιν, ἡ δὲ φύσις ἐν τοῖς οἰκείοις
σῴζεται μάλιστα τόποις. οὐ γὰρ ταὐτὸν ἤ θ' ὕλη
τῶν ζῴων ἐξ ἧς ἐστιν ἕκαστον, καὶ αἱ ἕξεις καὶ
διαθέσεις αὐτῆς. λέγω δ' οἷον εἴ τι ἐκ κηροῦ
συστήσειεν ἡ φύσις, οὐκ ἂν ἐν θερμῷ θεῖσα δι-
20 έσωσεν, οὐδ' εἴ τι ἐκ κρυστάλλου· ἐφθάρη γὰρ ἂν
ταχὺ διὰ τοὐναντίον· τήκει γὰρ τὸ θερμὸν τὸ ὑπὸ
τοῦ ἐναντίου συστάν. οὐδ' εἴ τι ἐξ ἁλὸς ἢ νίτρου
σινέστησεν, οὐκ ἂν εἰς ὑγρὸν φέρουσα κατέθηκεν·

---

[1] οὐδὲ Christ : οὔτε.   [2] τοὺς om. LS et edd. plerique.

are made from air and some from fire. Each class has its sphere of life in the region appropriate to its preponderating element.

XIV. Empedocles is mistaken in saying that the creatures which contain most heat and fire live in the water, thereby escaping the excess of heat that lies in their nature, in order that, since they are short of coolness and fluid, they may be saved by the contrary character of their habitat ; for fluid is less hot than air. But it is quite absurd that every such animal should be born on dry land and then migrate to the water ; for most of them, one might say, have no feet. Yet he, describing how they are first formed, says that they are born on dry land, but that they escape from it and reach the water. Further it does not appear that they are warmer than the land animals ; for some of them are altogether bloodless, and others have only a little blood.

Empedocles on water animals.

However, the question which of them should be called hot and which cold has been dealt with separately. As for the explanation which Empedocles gives, in a sense what he tries to establish is reasonable, but his account is not correct. For while those who suffer from excess of any condition find relief in places or seasons of a contrary nature, their constitution is best preserved in the region corresponding to it ; for the matter of each individual animal is not the same thing as its states and dispositions. What I mean is this : if nature were to form anything out of wax, she would not preserve it by placing it in a hot atmosphere, nor if she had made a thing out of ice ; for it would be rapidly destroyed by its contrary ; for heat melts that which is constituted by its contrary. Nor if she had made a thing out of salt or nitre would she have taken it and placed it in water ; for

Empedocles refuted.

477 b

φθείρει γὰρ τὰ ὑπὸ θερμοῦ καὶ ξηροῦ συστάντα τὸ
ὑγρόν. εἰ οὖν ὕλη πᾶσι τοῖς σώμασι τὸ ὑγρὸν καὶ
25 τὸ ξηρόν, εὐλόγως τὰ μὲν ἐξ ὑγροῦ καὶ ψυχροῦ
συστάντα ἐν ὑγροῖς ἐστι, [καὶ εἰ ψυχρά, ἔσται ἐν
ψυχρῷ,]¹ τὰ δ' ἐκ ξηροῦ ἐν ξηρῷ. διὰ τοῦτο τὰ
δένδρα οὐκ ἐν ὕδατι φύεται, ἀλλ' ἐν τῇ γῇ. καίτοι
τοῦ αὐτοῦ λόγου ἐστὶν εἰς τὸ ὕδωρ, διὰ τὸ εἶναι
αὐτὰ ὑπέρξηρα, ὥσπερ τὰ ὑπέρπυρά φησιν ἐκεῖνος·
30 οὐ γὰρ διὰ τὸ ψυχρὸν ἦλθεν εἰς αὐτό, ἀλλ' ὅτι
ὑγρόν.

Αἱ μὲν οὖν φύσεις τῆς ὕλης, ἐν οἵῳπερ τόπῳ
εἰσί, τοιαῦται οὖσαι τυγχάνουσιν, αἱ μὲν ἐν ὕδατι
ὑγραί, αἱ δ' ἐν τῇ γῇ ξηραί, αἱ δ' ἐν τῷ ἀέρι θερ-
478 a μαί. αἱ μέντοι ἕξεις αἱ μὲν ὑπερβάλλουσαι θερμότητι ἐν ψυχρῷ, αἱ δὲ τῇ ψυχρότητι ἐν θερμῷ
τιθέμεναι σῴζονται μᾶλλον· ἐπανισοῖ γὰρ εἰς τὸ
μέτριον ὁ τόπος τὴν τῆς ἕξεως ὑπερβολήν. τοῦτο
5 μὲν οὖν δεῖ ζητεῖν ἐν τοῖς οἰκείοις τόποις ἑκάστης
ὕλης, καὶ κατὰ τὰς μεταβολὰς τῆς κοινῆς ὥρας·
τὰς μὲν γὰρ ἕξεις ἐνδέχεται τοῖς τόποις ἐναντίας
εἶναι, τὴν δ' ὕλην ἀδύνατον. ὅτι μὲν οὖν οὐ διὰ
θερμότητα τῆς φύσεως τὰ μὲν ἔνυδρα τὰ δὲ πεζὰ
τῶν ζῴων ἐστί, καθάπερ Ἐμπεδοκλῆς φησίν,
10 τοσαῦτ' εἰρήσθω, καὶ διότι τὰ μὲν οὐκ ἔχει πνεύμονα τὰ δὲ ἔχει.

XV. Διὰ τί δὲ τὰ ἔχοντα δέχεται τὸν ἀέρα καὶ
ἀναπνέουσι, καὶ μάλιστ' αὐτῶν ὅσα ἔχουσιν ἔναιμον, αἴτιον τοῦ μὲν ἀναπνεῖν ὁ πνεύμων σομφὸς
ὢν καὶ συρίγγων πλήρης. καὶ ἐναιμότατον δὴ
μάλιστα τοῦτο τὸ μόριον τῶν καλουμένων σπλάγ-
15 χνων. ὅσα δὴ ἔχει ἔναιμον αὐτό, ταχείας μὲν
δεῖται τῆς καταψύξεως διὰ τὸ μικρὰν εἶναι τὴν

¹ καὶ . . . ψυχρῷ susp. Christ, secl. Biehl.

water destroys that which is constituted by heat and dryness. If, then, the matter of which all bodies are composed is the wet and the dry, naturally that which is constituted of wet and cold lives in water [and if it is cold, will live in the cold], but what is constituted of the dry will live in the dry. For this reason trees do not grow in water, but in the earth. Yet on the same theory he would assign them to the water because they are too dry, just as he says of the too fiery. On this theory they would enter water not because it is cold, but because it is wet.

Thus the material constitution of anything corresponds in fact to its environment; in water live wet things, in earth dry, and in air hot. But the physical states which are excessively hot find greater relief in the cold, and those that are excessively cold in the warm; for their environment neutralizes the excess of their state. The means to this end must be sought in the regions appropriate to each kind of matter, and in the changes of the common seasons; for bodily states can be contrary to their environment, but matter cannot. Let this, then, suffice to show that it is not because of their natural heat, as Empedocles says, that some animals are aquatic and others terrestrial, and to explain why some have lungs and some have not.

XV. The reason why those that have lungs admit the air and breathe, and particularly those which have a lung charged with blood, is that the lung is spongy and full of tubes. This part contains more blood than any other of the internal organs. All creatures that have this part charged with blood need rapid cooling, because there is little margin for varia-

The function of the lung.

478 a

ῥοπὴν τοῦ ψυχικοῦ πυρός, εἴσω δ' εἰσιέναι διὰ
παντὸς διὰ τὸ πλῆθος τοῦ αἵματος καὶ τῆς θερμότητος. ταῦτα δ' ἀμφότερα ὁ μὲν ἀὴρ δύναται
ῥᾳδίως ποιεῖν· διὰ γὰρ τὸ λεπτὴν ἔχειν τὴν φύσιν
20 διὰ παντός τε καὶ ταχέως διαδυόμενος διαψύχει·
τὸ δ' ὕδωρ τοὐναντίον. καὶ διότι δὴ μάλιστ'
ἀναπνέουσι τὰ ἔχοντα τὸν πνεύμονα ἔναιμον, ἐκ
τούτων δῆλον· τό τε γὰρ θερμότερον πλείονος δεῖται
τῆς καταψύξεως, ἅμα δὲ καὶ πρὸς τὴν ἀρχὴν τῆς
θερμότητος τῆς ἐν τῇ καρδίᾳ πορεύεται τὸ πνεῦμα
25 ῥᾳδίως.

XVI. Ὃν δὲ τρόπον ἡ καρδία τὴν σύντρησιν ἔχει
πρὸς τὸν πλεύμονα, δεῖ θεωρεῖν ἔκ τε τῶν ἀνα-
τεμνομένων καὶ τῶν ἱστοριῶν τῶν περὶ τὰ ζῷα γε-
γραμμένων. καταψύξεως μὲν οὖν ὅλως ἡ τῶν ζῴων
30 δεῖται φύσις διὰ τὴν ἐν τῇ καρδίᾳ τῆς ψυχῆς ἐμπύ-
ρωσιν. ταύτην δὲ ποιεῖται διὰ τῆς ἀναπνοῆς, ὅσα
μὴ μόνον ἔχουσι καρδίαν ἀλλὰ καὶ πνεύμονα τῶν
ζῴων. τὰ δὲ καρδίαν μὲν ἔχοντα, πνεύμονα δὲ μή,
καθάπερ οἱ ἰχθύες διὰ τὸ ἔνυδρον αὐτῶν τὴν φύσιν
εἶναι, τῷ ὕδατι ποιοῦνται τὴν κατάψυξιν διὰ τῶν
35 βραγχίων. ὡς δ' ἡ θέσις ἔχει τῆς καρδίας πρὸς

478 b τὰ βράγχια, πρὸς μὲν τὴν ὄψιν ἐκ τῶν ἀνατομῶν
δεῖ θεωρεῖν, πρὸς δ' ἀκρίβειαν ἐκ τῶν ἱστοριῶν·
ὡς δ' ἐν κεφαλαίοις εἰπεῖν καὶ νῦν, ἔχει τόνδε τὸν
τρόπον. δόξειε μὲν γὰρ ἂν οὐχ ὡσαύτως ἔχειν
τὴν θέσιν ἡ καρδία τοῖς τε πεζοῖς τῶν ζῴων καὶ
5 τοῖς ἰχθύσιν, ἔχει δ' ὡσαύτως. ᾗ γὰρ νεύουσι τὰς
κεφαλάς, ἐνταῦθ' ἡ καρδία τὸ ὀξὺ ἔχει. ἐπεὶ δὲ
οὐχ ὡσαύτως αἱ κεφαλαὶ νεύουσι τοῖς τε πεζοῖς
τῶν ζῴων καὶ τοῖς ἰχθύσι, πρὸς τὸ στόμα ἡ καρδία
τὸ ὀξὺ ἔχει. τείνει δ' ἐξ ἄκρου τῆς καρδίας αὐλὸς

tion of their vital fire, and the air must penetrate the whole lung because of the quantity of blood and heat which it contains. Now air can easily fulfil both these functions ; for, because its nature is so rarefied, it rapidly pervades the whole and cools it. But water is just the opposite. From this it is obvious why animals which have blood in the lung breathe most ; for the warmer creature requires more cooling, and at the same time the breath passes easily to the source of heat, which lies in the heart.

XVI. How the heart communicates by passages with the lung should be studied from dissections, and by reference to the *History of Animals.*[a] Speaking generally, the nature of animals requires cooling owing to the fierce heat which the soul acquires in the heart. This cooling is achieved by breathing in the case of animals which have a lung as well as a heart ; but those which have a heart but no lung, such as the fishes, since their nature is aquatic, achieve this cooling by water through the gills. The position of the heart relatively to the gills should be studied visually from dissections, and in detail by reference to the *History* [b] ; but to summarize for our present purpose, the facts are as follows. One might suppose that the position of the heart is different in land animals and in fishes, but actually it is identical. For the apex of the heart lies in the direction in which their heads point. But since the heads of land animals and fishes do not point in the same direction, in the latter the heart has its apex directed towards the mouth. Now from the top of the heart a tube, like a sinewy

Connexion between heart and lung.

------

[a] *Hist. An.* 496 a, etc., 511 b, etc.
[b] *Hist. An.* 507 b 3.

478 b

φλεβονευρώδης εἰς τὸ μέσον, ᾗ συνάπτουσιν ἀλλή-
10 λοις πάντα τὰ βράγχια. μέγιστος μὲν οὖν οὗτός
ἐστιν, ἔνθεν δὲ καὶ ἔνθεν τῆς καρδίας καὶ ἕτεροι
τείνουσιν εἰς ἄκρον ἑκάστου τῶν βραγχίων, δι' ὧν
ἡ κατάψυξις γίνεται πρὸς τὴν καρδίαν, διαυλωνί-
ζοντος ἀεὶ τοῦ ὕδατος διὰ τῶν βραγχίων. ὡσαύτως
δὲ τοῖς ἀναπνέουσιν ὁ θώραξ ἄνω καὶ κάτω κινεῖται
15 πολλάκις δεχομένων τὸ πνεῦμα καὶ ἐξιέντων, ὡς τὰ
βράγχια τοῖς ἰχθύσιν. καὶ τὰ μὲν ἀναπνέοντα ἐν
ὀλίγῳ ἀέρι καὶ τῷ αὐτῷ ἀποπνίγονται· ταχέως γὰρ
ἑκάτερον αὐτῶν γίνεται θερμόν· θερμαίνει γὰρ ἡ
τοῦ αἵματος θίξις ἑκάτερον. θερμὸν δ' ὂν τὸ αἷμα
κωλύει τὴν κατάψυξιν· καὶ μὴ δυναμένων κινεῖν
20 τῶν μὲν ἀναπνεόντων τὸν πνεύμονα τῶν δ' ἐνύδρων
τὰ βράγχια διὰ πάθος ἢ διὰ γῆρας, τότε συμβαίνειν
τὴν τελευτήν.

XVII. Ἔστι μὲν οὖν πᾶσι τοῖς ζῴοις κοινὸν
γένεσις καὶ θάνατος, οἱ δὲ τρόποι διαφέρουσι τῷ
εἴδει· οὐ γὰρ ἀδιάφορος ἡ φθορά, ἀλλ' ἔχει τι κοι-
25 νόν. θάνατος δ' ἐστὶν ὁ μὲν βίαιος ὁ δὲ κατὰ
φύσιν, βίαιος μὲν ὅταν ἡ ἀρχὴ ἔξωθεν ᾖ, κατὰ
φύσιν δ' ὅταν ἐν αὐτῷ,[1] καὶ ἡ τοῦ μορίου σύστασις
ἐξ ἀρχῆς τοιαύτη, ἀλλὰ μὴ ἐπίκτητόν τι πάθος.
τοῖς μὲν οὖν φυτοῖς αὔανσις, ἐν δὲ τοῖς ζῴοις
καλεῖται τοῦτο γῆρας. ἔστι δὲ θάνατος καὶ ἡ
30 φθορὰ πᾶσιν ὁμοίως τοῖς μὴ ἀτελέσιν· τούτοις δὲ
παρομοίως μέν, ἄλλον δὲ τρόπον. ἀτελῆ δὲ λέγω
οἷον τά τε ᾠὰ καὶ τὰ σπέρματα τῶν φυτῶν, ὅσα
ἄρριζα. πᾶσι μὲν οὖν ἡ φθορὰ γίνεται διὰ θερμοῦ
τινος ἔκλειψιν, τοῖς δὲ τελείοις, ἐν ᾧ τῆς οὐσίας

[1] post αὐτῷ virgulam pro puncto scribendam monuit
G. R. T. Ross.

vein, runs to the junction at which all the gills meet. This is the largest tube, but from either side of the heart other tubes run to the extremity of each of the gills, and through these the cooling process reaches the heart, the water passing unceasingly through the channel of the gills. In the same way in respiring animals the chest moves frequently up and down as they admit and expel the breath, just as the gills of fishes move. Respiring animals are suffocated if the air is small in quantity and remains the same ; for in either case it quickly becomes hot, since the contact with the blood heats it. The heat of the blood checks the cooling ; and if, owing to disease or old age, respiring animals cannot move the lung, or aquatic animals the gills, then death supervenes.

XVII. Birth and death are common characteristics of all living creatures, but the manner in which they occur differs with the species ; for destruction does exhibit differences, although all its forms have a common element. Death may be either violent or natural ; violent when its origin is external, natural when it originates in the creature itself, and the constitution of the animal involved this end from the beginning, and it was no extraneous affection. This phenomenon is called withering in plants, and in animals old age. Death and destruction are the common fate of all animals alike which are not imperfect ; the latter have a similar end but in another way. By imperfect I mean, for instance, eggs and vegetable seeds before their roots appear. In all cases destruction occurs owing to a failure of heat, but in the perfect animal the failure lies in that part

Natural and unnatural death.

471

ἡ ἀρχή. αὕτη δ' ἐστίν, ὥσπερ εἴρηται πρότερον,
35 ἐν ᾧ τό τε ἄνω καὶ τὸ κάτω συνάπτει, τοῖς μὲν
φυτοῖς μέσον βλαστοῦ καὶ ῥίζης, τῶν δὲ ζῴων τοῖς
479 a μὲν ἐναίμοις ἡ καρδία, τοῖς δ' ἀναίμοις τὸ ἀνάλογον. τούτων δ' ἔνια δυνάμει πολλὰς ἀρχὰς
ἔχουσιν, οὐ μέντοι γε ἐνεργείᾳ. διὸ καὶ τῶν ἐντόμων ἔνια διαιρούμενα ζῇ, καὶ τῶν ἐναίμων ὅσα
5 μὴ ζωτικὰ λίαν εἰσί, πολὺν χρόνον ζῶσιν ἐξῃρη-
μένης τῆς καρδίας, οἷον αἱ χελῶναι καὶ κινοῦνται
τοῖς ποσίν, ἐπόντων τῶν χελωνίων, διὰ τὸ μὴ
συγκεῖσθαι τὴν φύσιν αὐτῶν εὖ, παραπλησίως δὲ
τοῖς ἐντόμοις.

Ἡ δ' ἀρχὴ τῆς ζωῆς ἐκλείπει τοῖς ἔχουσιν, ὅταν
10 μὴ καταψύχηται τὸ θερμὸν τὸ κοινωνοῦν αὐτῆς.
καθάπερ γὰρ εἴρηται πολλάκις, συντήκεται αὐτὸ
ὑφ' αὑτοῦ. ὅταν οὖν τοῖς μὲν ὁ πλεύμων τοῖς δὲ
τὰ βράγχια σκληρύνηται, διὰ χρόνου μῆκος ξηραι-
νομένων τοῖς μὲν τῶν βραγχίων τοῖς δὲ τοῦ πλεύμονος, καὶ γινομένων γεηρῶν, οὐ δύναται ταῦτα τὰ
μόρια κινεῖν οὐδ' αἴρειν καὶ συνάγειν. τέλος δὲ
15 γινομένης ἐπιτάσεως καταμαραίνεται τὸ πῦρ.

Διὸ καὶ μικρῶν παθημάτων ἐπιγινομένων ἐν τῷ
γήρᾳ ταχέως τελευτῶσιν· διὰ γὰρ τὸ ὀλίγον εἶναι
τὸ θερμόν, ἅτε τοῦ πλείστου διαπεπνευκότος ἐν τῷ
πλήθει τῆς ζωῆς, ἥτις ἂν ἐπίτασις γένηται τοῦ
μορίου, ταχέως ἀποσβέννυται· ὥσπερ γὰρ ἀκαριαίας
20 καὶ μικρᾶς ἐν αὐτῷ φλογὸς ἐνούσης διὰ μικρὰν
κίνησιν ἀποσβέννυται. διὸ καὶ ἄλυπός ἐστιν ὁ ἐν
τῷ γήρᾳ θάνατος· οὐδενὸς γὰρ βιαίου πάθους αὐτοῖς
συμβαίνοντος τελευτῶσιν, ἀλλ' ἀναίσθητος ἡ τῆς
ψυχῆς ἀπόλυσις γίνεται παντελῶς. καὶ τῶν νοση-
μάτων ὅσα ποιοῦσι τὸν πνεύμονα σκληρὸν ἢ φύ-

in which is the source of their being. This, as has been said before, lies at the point at which the upper and lower parts meet : in plants, between the shoot and the root ; in sanguineous animals, in the heart ; and in bloodless animals, in the corresponding organ. Some of these creatures have potentially (but not actually) many vital sources. This is why some of the insects continue to live when divided, and among sanguineous animals those which have not a very lively nature live for a considerable time after the heart is removed ; *e.g.*, tortoises move upon their feet so long as their shells are on, because their nature is of a low order of construction, on much the same level as that of insects.

The source of life fails its possessors when the heat which is associated with it is not moderated by cooling ; for then, as has been said several times, the heat is consumed by itself. When, then, in some animals the lung, and in others the gills, grow hard, the gills in the one case and the lung in the other drying through length of time and becoming earthy, the animal cannot move or expand and contract these parts. At last a crisis is reached and the fire dies out.

Consequently in old age they die rapidly, even when small ailments attack them ; for the heat in them being very little, since in their long life most of it has been breathed away, if any strain occurs in the part it is quickly extinguished ; just as though it contained a brief and tiny flame, it is extinguished by a slight movement. For this reason death in old age is painless ; for old men die without the occurrence of any violent disease, and the release of the soul occurs quite imperceptibly. All diseases which cause hardening of the lung by tumours or secretions

473

25 μασιν ἢ περιττώμασιν ἢ θερμότητος νοσηματικῆς
ὑπερβολῇ, καθάπερ ἐν τοῖς πυρετοῖς, πυκνὸν τὸ
πνεῦμα ποιοῦσι διὰ τὸ μὴ δύνασθαι τὸν πνεύμονα
μακρὰν αἴρειν ἄνω καὶ συνίζειν· τέλος δ᾽, ὅταν
μηκέτι δύνωνται κινεῖν, τελευτῶσιν ἀποπνεύσαντες.

XVIII. Γένεσις μὲν οὖν ἐστὶν ἡ πρώτη μέθεξις
30 ἐν τῷ θερμῷ τῆς θρεπτικῆς ψυχῆς, ζωὴ δ᾽ ἡ
μονὴ ταύτης. νεότης δ᾽ ἐστὶν ἡ τοῦ πρώτου κατα-
ψυκτικοῦ μορίου αὔξησις, γῆρας δ᾽ ἡ τούτου φθίσις,
ἀκμὴ δὲ τὸ τούτων μέσον. τελευτὴ δὲ καὶ φθορὰ
βίαιος μὲν ἡ τοῦ θερμοῦ σβέσις καὶ μάρανσις

479 b (φθαρείη γὰρ ἂν δι᾽ ἀμφοτέρας ταύτας τὰς αἰτίας),
ἡ δὲ κατὰ φύσιν τοῦ αὐτοῦ τούτου μάρανσις διὰ
χρόνου μῆκος γινομένη καὶ τελειοτάτη.[1] τοῖς μὲν
οὖν φυτοῖς αὔανσις, ἐν δὲ τοῖς ζῴοις καλεῖται
θάνατος. τούτου δ᾽ ὁ μὲν ἐν γήρᾳ θάνατος μάραν-
5 σις τοῦ μορίου δι᾽ ἀδυναμίαν τοῦ καταψύχειν ὑπὸ
γήρως. τί μὲν οὖν ἐστι γένεσις καὶ ζωὴ καὶ θάνατος, καὶ διὰ τίνας αἰτίας ὑπάρχουσι τοῖς ζῴοις,
εἴρηται.

XIX. Δῆλον δ᾽ ἐκ τούτων καὶ διὰ τίν᾽ αἰτίαν
τοῖς μὲν ἀναπνέουσι τῶν ζῴων ἀποπνίγεσθαι
10 συμβαίνει ἐν τῷ ὑγρῷ, τοῖς δ᾽ ἰχθύσιν ἐν τῷ ἀέρι·
τοῖς μὲν γὰρ διὰ τοῦ ὕδατος ἡ κατάψυξις γίνεται,
τοῖς δὲ διὰ τοῦ ἀέρος, ὧν ἑκάτερα στερίσκεται
μεταβάλλοντα τοὺς τόπους. ἡ δ᾽ αἰτία τῆς κινήσεως τοῖς μὲν τῶν βραγχίων τοῖς δὲ τοῦ πνεύμονος,
ὧν αἱρομένων καὶ συνιζόντων τὰ μὲν ἐκπνέουσι
15 καὶ εἰσπνέουσι τὰ δὲ δέχονται τὸ ὑγρὸν καὶ
ἐξιᾶσιν, ἔτι δ᾽ ἡ σύστασις τοῦ ὀργάνου, τόνδ᾽ ἔχει
τὸν τρόπον.

XX. Τρία ἐστὶ τὰ συμβαίνοντα περὶ τὴν καρ-

---

[1] τελειοτάτη MZ : τελειότητα.

or excess of morbid heat, as in the case of fevers, cause rapid breathing, because the lung cannot expand or contract far ; and at last, when they can no longer move the lung, they breathe their last and die.

XVIII. Birth is the first participation (in a warm medium) in the nutritive soul, and life is the continuance of this. Youth is the growth of the primary refrigerative organ, and old age is its destruction, the prime of life being between the two. Violent death or destruction is the extinction or waning of the heat (for destruction may occur from either of these causes), but natural death is the decay of the same due to lapse of time, and to its having reached its appointed end. In plants this is called withering, in animals death. Death in old age is the decay of the organ owing to its inability to cause refrigeration because of old age. Thus we have now defined birth and life and death, and explained why they occur among living creatures. The meaning of birth.

XIX. From these facts we can see why it is that respiring animals are suffocated in water, and fishes in air ; for the latter achieve cooling by means of water, and the former by means of air, and when they change their habitat each kind is deprived of its means. The reason for the movement in the one case of the gills, and in the other of the lung, by the expansion and contraction of which the former class exhale and inhale and the latter admit and expel water, and also the constitution of the organ, is to be explained as follows. Fish.

XX. There are three movements which take place Action of the heart.

ARISTOTLE

δίαν, ἃ δοκεῖ τὴν αὐτὴν φύσιν ἔχειν, ἔχει δ'
οὐ τὴν αὐτήν, πήδησις καὶ σφυγμὸς καὶ ἀναπνοή.

20 Πήδησις μὲν οὖν ἐστι σύνωσις τοῦ θερμοῦ τοῦ ἐν
αὐτῇ διὰ κατάψυξιν περιττωματικὴν ἢ συντηκτι-
κήν, οἷον ἐν τῇ νόσῳ τῇ καλουμένῃ παλμῷ, καὶ ἐν
ἄλλαις δὲ νόσοις, καὶ ἐν τοῖς φόβοις δέ· καὶ γὰρ οἱ
φοβούμενοι καταψύχονται τὰ ἄνω, τὸ δὲ θερμὸν
ὑποφεῦγον καὶ συστελλόμενον ποιεῖ τὴν πήδησιν,
25 εἰς μικρὸν συνωθούμενον οὕτως ὥστ' ἐνίοτ' ἀποσβέννυσθαι τὰ ζῷα καὶ ἀποθνήσκειν διὰ φόβον καὶ
διὰ πάθος νοσηματικόν.

Ἡ δὲ συμβαίνουσα σφύξις τῆς καρδίας, ἣν ἀεὶ
φαίνεται ποιουμένη συνεχῶς, ὁμοία φύμασίν ἐστιν,
ἣν ποιοῦνται κίνησιν μετ' ἀλγηδόνος διὰ τὸ παρὰ
30 φύσιν εἶναι τῷ αἵματι τὴν μεταβολήν. γίνεται δὲ
μέχρις οὗ ἂν πυωθῇ πεφθέν. ἔστι δ' ὅμοιον ζέσει
τοῦτο τὸ πάθος· ἡ γὰρ ζέσις γίνεται πνευματου-
μένου τοῦ ὑγροῦ ὑπὸ τοῦ θερμοῦ· αἴρεται γὰρ διὰ
τὸ πλείω γίνεσθαι τὸν ὄγκον. παῦλα δ' ἐν μὲν τοῖς
480 a φύμασιν, ἐὰν μὴ διαπνεύσῃ, παχυτέρου γινομένου
τοῦ ὑγροῦ, σῆψις, τῇ δὲ ζέσει ἡ ἔκπτωσις διὰ τῶν
ὁριζόντων. ἐν δὲ τῇ καρδίᾳ ἡ τοῦ ἀεὶ προσιόντος
ἐκ τῆς τροφῆς ὑγροῦ διὰ τῆς θερμότητος ὄγκωσις
ποιεῖ σφυγμόν, αἰρομένη πρὸς τὸν ἔσχατον χιτῶνα
5 τῆς καρδίας. καὶ τοῦτ' ἀεὶ γίνεται συνεχῶς·
ἐπιρρεῖ γὰρ ἀεὶ τὸ ὑγρὸν συνεχῶς, ἐξ οὗ γίνεται ἡ
τοῦ αἵματος φύσις· πρῶτον γὰρ ἐν τῇ καρδίᾳ
δημιουργεῖται. δῆλον δ' ἐν τῇ γενέσει ἐξ ἀρχῆς·
οὔπω γὰρ διωρισμένων τῶν φλεβῶν φαίνεται
ἔχουσα αἷμα. καὶ διὰ τοῦτο σφύζει μᾶλλον τοῖς
10 νεωτέροις τῶν πρεσβυτέρων· γίνεται γὰρ ἡ ἀνα-

in the region of the heart, which seem to be of the same character, but really are not : palpitation, pulsation, and respiration.

Palpitation is a violent concentration of the heat in the heart due to the chilling effect of waste products or secretions, as occurs in the disease called heart palpitation among others, and also in fear ; for those who are afraid grow cold in their upper parts, and the heat retreating and collecting produces palpitation, being forced into so small a space that sometimes animals suffer extinction and die through fear or malady.

The characteristic beating of the heart, which, as we can see, goes on continuously, is like the throbbing of an abscess, but the latter is accompanied by pain, because there is an unnatural change in the blood ; and this pain continues until pus is formed and discharged. This latter affection is like boiling ; for boiling takes place when liquid is aerated by heat : it expands because its bulk increases. In the case of abscesses, if there is no outlet, the liquid becomes thicker, and the end comes in the form of putrefaction, whereas boiling results in the overflowing of the container. But in the heart pulsation is due to the expansion by heat of the liquid food-product which continually enters it. It occurs as the fluid rises to the furthest point of the heart wall. This is a continuous process ; for there is a continuous influx of this fluid, of which the blood is constituted ; for it is in the heart that blood is first manufactured. This fact is clearly shown in the first stages of generation ; for the heart can be seen to contain blood before the veins are differentiated. For this reason there is more pulsation in the young than in the old ; for more

477

ARISTOTLE

θυμίασις πλείων τοῖς νεωτέροις. καὶ σφύζουσιν αἱ
φλέβες πᾶσαι, καὶ ἅμα ἀλλήλαις, διὰ τὸ ἠρτῆσθαι
ἐκ τῆς καρδίας. κινεῖ δ' ἀεί· ὥστε κἀκεῖναι αἰεί,
καὶ ἅμα ἀλλήλαις, ὅτε κινεῖ. ἀναπήδησις μὲν οὖν
ἐστὶν ἡ γινομένη ἄντωσις πρὸς τὴν τοῦ ψυχροῦ
15 σύνωσιν, σφύξις δ' ἡ τοῦ ὑγροῦ θερμαινομένου
πνευμάτωσις.

XXI. Ἡ δ' ἀναπνοὴ γίνεται αὐξανομένου τοῦ
θερμοῦ, ἐν ᾧ ἡ ἀρχὴ ἡ θρεπτική. καθάπερ γὰρ καὶ
τἆλλα δεῖται τροφῆς, κἀκεῖνο, καὶ τῶν ἄλλων μᾶλλον·
καὶ γὰρ τοῖς ἄλλοις ἐκεῖνο τῆς τροφῆς αἴτιόν ἐστιν.
20 ἀνάγκη δὴ πλέον γινόμενον αἴρειν τὸ ὄργανον. δεῖ
δ' ὑπολαβεῖν τὴν σύστασιν τοῦ ὀργάνου παρα-
πλησίαν μὲν εἶναι ταῖς φύσαις ταῖς ἐν τοῖς χαλκείοις·
οὐ πόρρω γὰρ οὔθ' ὁ πνεύμων οὔθ' ἡ καρδία πρὸς
τὸ δέξασθαι σχῆμα τοιοῦτον· διπλοῦν δ' εἶναι τὸ
τοιοῦτον· δεῖ γὰρ ἐν τῷ μέσῳ τὸ θρεπτικὸν εἶναι
25 τῆς φυσικῆς δυνάμεως. αἴρεται μὲν οὖν πλεῖον
γενόμενον, αἰρομένου δ' ἀναγκαῖον αἴρεσθαι καὶ τὸ
περιέχον αὐτὸ μόριον. ὅπερ φαίνονται ποιεῖν οἱ
ἀναπνέοντες· αἴρουσι γὰρ τὸν θώρακα διὰ τὸ τὴν
ἀρχὴν τὴν ἐνοῦσαν αὐτῷ τοῦ τοιούτου μορίου ταὐτὸ
τοῦτο ποιεῖν· αἰρομένου γάρ, καθάπερ εἰς τὰς
30 φύσας, ἀναγκαῖον εἰσρεῖν[1] τὸν ἀέρα τὸν θύραθεν
480 b καὶ ψυχρὸν ὄντα καὶ καταψύχοντα σβεννύναι τὴν
ὑπεροχὴν τὴν τοῦ πυρός. ὥσπερ δ' αὐξανομένου
ᾔρετο τοῦτο τὸ μόριον, καὶ φθίνοντος ἀναγκαῖον
συνίζειν, καὶ συνίζοντος ἐξιέναι τὸν ἀέρα τὸν
εἰσελθόντα πάλιν, εἰσιόντα μὲν ψυχρὸν ἐξιόντα δὲ
5 θερμὸν διὰ τὴν ἁφὴν τοῦ θερμοῦ τοῦ ἐνόντος ἐν τῷ
μορίῳ τούτῳ, καὶ μάλιστα τοῖς τὸν πνεύμονα
ἔναιμον ἔχουσιν· εἰς πολλοὺς γὰρ οἷον αὐλῶνας τὰς

---

[1] εἰσρεῖν scripsi : εἰσφέρειν.

evaporation takes place in the young. All the veins [a] throb, and throb simultaneously with each other, because they are connected with the heart. The heart beats always, and therefore so do the veins; and they beat simultaneously with each other when the heart beats. Palpitation, then, is the reaction which the heart makes to the pressure of the cold, but pulsation is the aeration of the fluid by the agency of heat.

XXI. Respiration occurs owing to the increase of the hot substance which contains the nutritive principle. For just as all the other parts need food, so does this, and even more than the other parts; for it is the cause of nourishment for the rest. Now as it increases it must cause the organ to rise. One may regard the structure of the organ as very like that of both the bellows in a forge; for heart and lung conform closely to this shape. An organ of this kind must be double; for the nutritive part must be in the middle of the natural force. As it [b] increases it expands, and as it expands the part which surrounds it must also expand. This is what men seem to do when they breathe; they expand their chest because the principle of the organ described above resides in the chest and causes this same expansion; for as the chest rises the air from outside must flow in, as it does into the bellows, and being cold and refrigerative, quench the excess of fire. Just as increase makes this part rise, so decrease must make it subside, and as it subsides the air which has entered must pass out again. It enters in cold and passes out hot, because of its contact with the heat which resides in this part, especially in those whose lung contains blood. For each of the many canal-like tubes in the lung, into

The function of breathing.

[a] Veins of course do not throb. Aristotle does not distinguish between veins and arteries. See the introduction to *On Breath*, p. 484.
[b] *Viz.*, the " hot substance."

which the air passes, has a blood-vessel alongside, so that the whole lung seems to be full of blood. The entry of the air is called inhalation, its exit exhalation. And this occurs continuously as long as the creature lives and keeps this part continuously moving. This is why life depends upon inhalation and exhalation. The movement of the gills in fishes takes place in the same way. When the hot substance in the blood rises through the parts of the body the gills also rise, and let the water through ; but when it grows cool and retires through its channels towards the heart, the gills contract and expel the water. As the heat in the heart is continually rising, so as it cools it is continually received back again. Thus, as in respiring animals the life and death depend entirely on respiration, so in fishes they depend on admitting water.

We have now virtually completed our inquiry into life, death, and kindred subjects. As for health and disease it is the business not only of the physician but also of the natural philosopher to discuss their causes up to a point. But the way in which these two classes of inquirers differ and consider different problems must not escape us, since the facts prove that up to a point their activities have the same scope ; for those physicians who have subtle and inquiring minds have something to say about natural science, and claim to derive their principles therefrom, and the most accomplished of those who deal with natural science tend to conclude with medical principles.

INTRODUCTION

THIS curious little treatise is obviously un-Aristotelian, although the author is familiar, up to a point, with many of Aristotle's views. It appears to be a philosophical rather than a medical document; its tentative manner makes a remarkable contrast with the extreme dogmatism of the pseudo-Hippocratic *On Breaths* (περὶ φυσῶν).

It has caused much difficulty to editors and translators. The text, though much improved by Jaeger and others, is very uncertain; there is a general lack of coherence in the thought; and there are two important and recurrent ambiguities of meaning. The first lies in the word πνεῦμα itself. The author seems to have grasped something of the Aristotelian theory of σύμφυτον πνεῦμα (see Introduction to *On Length and Shortness of Life*, p. 391), and frequently uses the phrase, varying it at times by the apparently equivalent expressions ἔμφυτον πνεῦμα and φυσικὸς ἀήρ; but it is often uncertain whether he is speaking of πνεῦμα in this technical sense or of ordinary breath, and he seems to confuse the two in his own mind. In the second place he makes frequent use of the word ἀρτηρία (translated " air-duct "). This word applies primarily (in Aristotle's genuine works always) to the windpipe, and in several passages here (*e.g.*, 481 a 22, b 13) quite clearly refers to it; else-

484

where (*e.g.*, 484 a 14) it seems to be used of any duct; elsewhere again (*e.g.*, 484 a 1) it apparently refers to the arteries, though without any realization that they are blood-vessels.

There are many other minor puzzles and inconsequentialities which still resist interpretation; and even to notice them lies beyond the scope of this series. It has seemed best to give the reader a fairly literal rendering of the best text as yet available, and to leave the problems open for discussion.

ΠΕΡΙ ΠΝΕΥΜΑΤΟΣ

I. Τίς ἡ τοῦ ἐμφύτου πνεύματος διαμονή, καὶ
τίς ἡ αὔξησις; ὁρῶμεν γὰρ ὅτι πλέον καὶ ἰσχυρό-
τερον γίνεται καὶ καθ᾽ ἡλικίας μεταβολὴν καὶ κατὰ
διάθεσιν σώματος. ἢ ὡς τἆλλα μέρη, προσγινο-
5 μένου τινός ;[1] προσγίνεται δὲ τροφὴ τοῖς ἐμψύχοις,
ὥστε ταύτην σκεπτέον ποία τε καὶ πόθεν. δύο δὴ
τρόποι δι᾽ ὧν γίνεται, ἢ διὰ τῆς ἀναπνοῆς ἢ διὰ
τῆς κατὰ τὴν τῆς τροφῆς προσφορὰν πέψεως,
καθάπερ τοῖς ἄλλοις. τούτων ἴσως οὐχ ἧττον ἄν[2]
δόξειεν διὰ τῆς τροφῆς· σῶμα γὰρ ὑπὸ σώματος
10 τρέφεται, τὸ δὲ πνεῦμα σῶμα. τίς οὖν ὁ τρόπος ;
ἢ δῆλον ὡς ἐκ τῆς φλεβὸς ὁλκῇ τινὶ καὶ πέψει·
τὸ γὰρ αἷμα ἡ ἐσχάτη τροφὴ καὶ ἡ αὐτὴ πᾶσιν.
ὥσπερ οὖν [καὶ][3] εἰς τὸ ἀγγεῖον αὐτοῦ καὶ εἰς τὸ
περιέχον[4] λαμβάνει τροφὴν εἰς τὸ θερμόν. ἄγει
δ᾽ ὁ ἀὴρ τὴν ἐνέργειαν ποιῶν, τήν τε πεπτικὴν
αὐτὸς αὑτῷ προστιθεὶς αὔξει καὶ τρέφει. οὐδὲν
15 δ᾽ ἴσως ἄτοπον αὐτό γε τοῦτο, ἀλλὰ γενέσθαι τὸ
πρῶτον ἐκ τῆς τροφῆς. καθαρώτερον γὰρ ὃ τῇ
ψυχῇ συμφυές, εἰ μὴ καὶ τὴν ψυχὴν ὕστερον λέγοι
γίνεσθαι, διακρινομένων τῶν σπερμάτων καὶ εἰς

---

[1] σώματος. ἢ . . . τινός; Bussemaker: σώματος, ἢ . . .
τινός.
[2] post ἧττον ἂν vulgo legitur οὐχ οὕτω : del. Furlan.
[3] καὶ secl. Ross.   [4] περιέχον Ross : περιεχόμενον.

ON BREATH

I. How can we account for the maintenance of the breath inherent in us, and for its increase ? For we can see that it grows in volume and strength both with advancing years and with the condition of the body. Probably it increases, like the other parts of the body, by some accession. Now the accession which comes to animate things is nutriment, so that we have to consider the nature and source of nutriment in the case of breath. Nutrition may occur in two ways, either by respiration, or, as in the case of the rest of the body, by digestion of food. Of these two the method by means of food seems more likely ; for body is nourished by body, and breath is a body. What then is this method ? Clearly by some extraction and assimilation from the veins ; for blood is the ultimate food for every part alike. Thus the breath draws nutriment into its hot substance, as into a container or envelope. The air is the agent as producing activity, and by employing the digestive faculty causes growth and nourishment. Probably there is nothing strange in this ; it would be strange if breath were derived from food in the first instance. For that which is naturally akin to the soul is purer ; unless one supposes that the soul is born after the body, as the particles are separated out and realize

481 a

20 φύσιν ἰόντων. εἴ τε¹ περίττωμα πάσης τροφῆς ἐστί,
ποίᾳ διαπέμπεται τοῦτο; κατὰ μὲν γὰρ τὴν
ἐκπνοὴν οὐκ εὔλογον· ἀντιλαμβάνει γὰρ εὐθύς.
λοιπὸν δὲ δῆλον ὅτι διὰ τῶν τῆς ἀρτηρίας πόρων.
τὸ δ' ἐκκρινόμενον ἤτοι λεπτότερον ἢ παχύτερον.
ἀμφοτέρως δ' ἄτοπον· εἰ ⟨γὰρ⟩² τοῦτο πάντων
25 ἔσται καθαρώτατον, ⟨πῶς λεπτότερον;⟩³ εἰ δὲ παχύτερον, ἔσονταί τινες πόροι μείζους. εἰ δ' ἄρα
κατὰ τοὺς αὐτοὺς λαμβάνει καὶ ἐκπέμπει, τοῦτ'
αὐτὸ παράλογον καὶ ἄτοπον. ἡ μὲν οὖν ἐκ τῆς
τροφῆς αὔξησις καὶ διαμονὴ σχεδὸν ταῦτα.

II. Ἡ δ' ἐκ τῆς ἀναπνοῆς, ὥσπερ Ἀριστογένης
οἴεται (τροφὴν γὰρ οἴεται καὶ τὸ πνεῦμα πεττο-
30 μένου⁴ τοῦ ἀέρος ἐν τῷ πνεύμονι⁵· τοῦτο δ' εἰς τὰ
481 b ἀγγεῖα διαδίδοσθαι, ⟨καὶ⟩⁶ τὸ περίττωμα πάλιν ἐκ-
πέμπεσθαι) πλείους ἔχει τὰς ἀπορίας. ἥ τε γὰρ πέψις ὑπὸ τίνος; εἰκὸς μὲν γὰρ ὑπ' αὐτοῦ, καθάπερ
καὶ τῶν ἄλλων. αὐτὸ δὲ τοῦτ' ἄτοπον, εἰ μὴ
διαφέρει τοῦ ἔξω ἀέρος· οὕτω δ' ἡ θερμότης ἂν
5 πέττοι. καὶ μὴν καὶ παχύτερον αὐτὸν εὔλογον εἶναι
μεθ' ὑγρότητος τῆς ἀπὸ τῶν ἀγγείων ὄντα καὶ
τῶν ὅλων ὄγκων, ὥσθ' ἡ πέψις ἂν εἰς τὸ σωματῶδες εἴη. τὸ δὲ περίττωμα, εἴπερ γίνεται λεπτότερον, οὐ πιθανόν. ἄλογος δὲ καὶ ἡ ταχυτὴς τῆς
πέψεως. εὐθὺς γὰρ μετὰ τὴν εἰσπνοὴν ἡ ἐκπνοή.
10 τί οὖν τὸ οὕτω ταχὺ μεταβάλλον καὶ ἀλλοιοῦν;
ὑπολάβοι γὰρ ἄν τις μάλιστα τὸ θερμόν, καὶ μαρτυρεῖ οὕτως ἡ αἴσθησις· ὁ γὰρ ἐκπνεόμενος θερμός.

---

¹ εἴ τε Dobson : εἴτε.   ² γὰρ supplevi.
³ πῶς λεπτότερον; e versione Latina supplevit Dobson.
⁴ πεττομένου Furlan : πεττόμενον οὐ.
⁵ πνεύμονι Furlan : πνεύματι.   ⁶ καὶ suppl. Dobson.

their true nature. Again, if all food has a residue, how is it expelled in the case of breath? It is not reasonable to suppose that it is by exhalation; for this directly follows inhalation. Clearly the alternative is that it takes place by the channels of the windpipe. Now that which is excreted is either finer or coarser; but either alternative is absurd. How can it be finer, if breath is to be the purest substance? and if it is coarser, some of the channels must be proportionately wider. And if it receives and discharges by the same channels, this in itself is improbable and strange. So much for the theory that the growth and maintenance of breath are due to food.

II. The theory that they are due to respiration, as Aristogenes supposes (for he thinks that breath too is a form of food, the air being digested in the lung; this is absorbed into the several receptacles, but the residue is expelled again), involves even more difficulties. What is the agent in this digestion? Presumably the breath, just as it is in other cases. But this is in itself improbable, unless it differs from the outside air; in which case its heat might be digestive. Again, it would seem probable that the air should be coarser when combined with the moisture from the vessels and from the solid parts in general; so that digestion would tend towards corporeality. But that the residue becomes finer cannot be believed. Again, the rapidity of the digestion is unreasonable; for exhalation occurs directly after inhalation. What then is it which causes such rapid change and alteration? One might suppose that it is heat, and the senses give evidence of this, for the air breathed out

489

ἔτι δ' εἰ μὲν ἐν τῷ πνεύμονι καὶ τῇ ἀρτηρίᾳ τὸ
πεττόμενον, ἡ τοῦ θερμοῦ δύναμις ἐν τούτοις·
ὅπερ οὔ φασιν, ἀλλ' ἐν τῇ κινήσει τῇ τοῦ πνεύματος
15 ἐκθερμαίνεσθαι τὴν τροφήν. εἰ δ' ἐξ ἑτέρου τινὸς
οἷον ἐπισπᾶται ἢ καὶ κινοῦντος δέχεται, τοῦτ' ἔτι[1]
θαυμασιώτερον. ἅμα δὲ καὶ οὐκ αὐτὸ τὸ πρῶτον
κινοῦν.

Ἔτι δ' ἡ μὲν ἀναπνοὴ μέχρι τοῦ πνεύμονος,
ὥσπερ λέγουσιν αὐτοί, τὸ δὲ πνεῦμα δι' ὅλου τὸ
20 σύμφυτον. εἰ δ' ἀπὸ τούτου διαδίδοται καὶ πρὸς
τὰ κάτω καὶ πρὸς τὰ ἄλλα, πῶς ἡ πέψις οὕτω
ταχεῖα; θαυμασιώτερον γὰρ τοῦτο καὶ μεῖζον· οὐ
γὰρ διαπέμπει τοῦτό γ' εὐθὺς πεττόμενον τὸν ἀέρα
τοῖς κάτω. καίτοι τὸ μὲν δόξειεν ἂν ἀναγκαῖον
εἶναι τοῦτο τῆς πέψεως γινομένης ἐν τῷ πνεύμονι,
25 τῆς τ' ἀναπνοῆς κοινωνούντων καὶ τῶν κάτω.

Μεῖζον δ' οὕτως ἔτι καὶ παραδοξότερον τὸ συμ-
βαῖνον· οἷον γὰρ διόδῳ καὶ θίξει γίνεται μόνον ἡ
πέψις.

Ἄλογον δὲ καὶ τουτὶ καὶ λογοδεέστερον,[2] εἰ ὁ
αὐτὸς λόγος τῆς τροφῆς καὶ τοῦ περιττώματος.
εἰ δὲ δι' ἄλλου τινὸς τῶν ἐντός, οἱ αὐτοὶ λόγοι οἳ
30 καὶ πρότερον, εἰ μὴ τοῦτο λέγοι τις, ὡς οὐ πάσης
482 a τῆς τροφῆς οὐδὲ πᾶσι γίνεται περίττωμα, καθάπερ
οὐδὲ τοῖς φυτοῖς, ἐπεὶ οὐδὲ τῶν τοῦ σώματος
μερῶν ἑκάστου λαβεῖν ἔστιν· εἰ δὲ μή, οὔτι γε
παντός. ἀλλ' ἄρα γε ἡ μὲν ἀγγείων αὔξησις ἡ
αὐτὴ καὶ τῶν ἄλλων μορίων, εὐρυνομένων δὲ καὶ
5 δισταμένων τούτων πλείων ὁ ἀὴρ ὁ εἰσρέων καὶ
ἐκρέων. εἰ δέ τι ἀναγκαῖον ἐνυπάρχει, τοῦτο αὐτὸ

---

[1] ἔτι Jaeger : ἐστί.   [2] λογοδεέστερον Dobson : λογοδέστερον.

is hot. Moreover, if that which is digested is in the lung and in the windpipe, the potency of heat must be in them too ; this they deny, saying that the food is heated by the movement of the breath. It is still more amazing if the breath draws in heat, as it were, from some other source or receives it through some external impulse. Besides, it is then not itself the prime cause of the movement.

Moreover, respiration only reaches as far as the lung, as they themselves admit, but the breath which is inherent in the creature pervades the whole. If it is distributed from the lung to the other (including the lower) parts, how is digestion so rapid ? This is more remarkable and a greater problem ; for the lung does not dispatch the air the moment it is digested to the parts below ; and yet in some sense this would seem to be inevitable if digestion takes place in the lung, and respiration extends to the parts below.

But in this case the consequence is still more serious and incredible : *viz.*, that digestion takes place merely by passage and contact.[a]

This too is unreasonable, and still more indefensible, that the same account should apply to nutriment and to excrement ; while if digestion is due to any other of the internal organs, the same arguments apply as before, unless one is prepared to say that excretion does not occur from all kinds of food, nor in all animals, any more than it does in plants, since one cannot find it in each of the parts of the body, or at any rate, not in every animal. But on this view the vessels grow just like the other parts, and, as they expand and open out, the volume of air flowing in and out increases. And if they must contain some-thing, the question which we are asking—what the

[a] Dobson refers this sentence to b 15 above.

482 a

⟨δ⟩[1] ζητεῖται, τίς ὁ φυσικὸς καὶ πῶς οὗτος πλείων
ὑγιῶς, ἐκ τούτου φανερὸν ἂν εἴη.[2] τοῖς δὲ δὴ μὴ
ἀναπνευστικοῖς τίς ἡ τροφὴ τοῦ συμφύτου καὶ τίς
ἡ αὔξησις; οὐ γὰρ ἔτι τούτοις ἀπὸ τοῦ ἔξωθεν.
10 εἰ δ' ἀπὸ τῶν ἐντὸς καὶ τῆς κοινῆς τροφῆς, εὔλογον
κἀκείνοις (ἀπὸ γὰρ τῶν αὐτῶν τὰ ὅμοια καὶ
ὡσαύτως)· εἰ μὴ ἄρα καὶ τούτοις ἀπὸ τοῦ ἐκτός,
ὥσπερ καὶ τῶν ὀσμῶν αἰσθάνονται· ἀλλ' οὕτω γ'
οἷον ἀναπνοὴ γίνεται. περὶ οὗ κἂν ἀπορήσειέ τις,
15 εἰ κατὰ ἀλήθειάν ἐστιν, αὐτό τε τοῦτο προφέρων
καὶ τὴν ἐπίσπασιν τῆς τροφῆς (ὁλκὴ γὰρ ἅμα
πνεύματος), ἔτι δ' ὑπὲρ τῆς καταψύξεως ἀντι-
λέγων, ὡς κἀκείνων δεομένων. εἰ δὲ διὰ τοῦ
ὑποζώματος αὐτοῖς γίνεται, ταύτῃ δῆλον ὅτι καὶ
ἡ τοῦ ἀέρος εἴσοδος· ὥσθ' ὅμοιόν τι τῇ ἀναπνοῇ.
πλὴν οὐκ ἀφορίζεται τίς ⟨ἡ⟩[3] ὁλκὴ καὶ ὑπὸ τίνος. ἢ
20 εἰ μὴ ὁλκή, πῶς ἡ εἴσοδος; εἰ μὴ ἄρα αὐτομάτως.
τοῦτο μὲν οὖν ἔχει καὶ αὐτὸ καθ' αὑτὸ σκέψιν.

Τοῖς δὲ δὴ ἐνύγροις τίς ἡ τροφὴ καὶ αὔξησις τοῦ
συμφύτου; χωρὶς γὰρ τοῦ μὴ ἀναπνεῖν οὐδ' ἐν-
υπάρχειν ὅλως ἐν τῷ ὑγρῷ φαμὲν ἀέρα. λοιπὸν
25 ἄρα διὰ τῆς τροφῆς, ὡς οὐχ ὁμοίως πᾶσιν, ἢ κἀκεῖνα διὰ τῆς τροφῆς τὰ ⟨μὴ⟩[4] ἔνυγρα· τριῶν γὰρ

---

[1] ὃ suppl. Ross. [2] εἴη Bussemaker : εἶεν.
[3] ἡ suppl. Jaeger. [4] μὴ suppl. Jaeger.

natural breath is and how it healthily increases in volume—would be clear from what we have just said. But in creatures which do not respire, how is their inherent breath nourished and increased ? In their case the nutriment cannot come from the outside. But if it comes from the parts inside, and from the common source of food, it is reasonable that the same should be true in the case of the former ; for similar results come from the same causes and in the same way. Unless, of course, the latter too obtain the nutriment from the outside, just as they apprehend smells. But this implies a sort of respiration. Here one might question whether they really do not respire —instancing both this argument and their ingestion of food (for they must draw in breath at the same time), and objecting also on the ground of refrigeration, which they must require like any other creatures. If, then, this takes place through the diaphragm, clearly the entry of the air must also be by this way ; so that in a sense it would be like breathing. But this does not define what this drawing in is, nor by what agency it takes place. Or if there is no drawing in, how does the air enter ?—unless it does so spontaneously. This subject requires an inquiry all by itself.

Again, how is the inherent breath nourished and increased in the case of water animals ? For apart from the fact that they do not respire, we do not admit that there is any air at all in water. So the only remaining supposition is that it comes through food, in which case, either the process is not the same for all creatures, or else the non-aquatic animals also receive it through food ; for one of these three things

482 a

τούτων ἀναγκαῖον ἕν. καὶ ταῦτα μὲν ὡς περὶ τὴν
αὔξησιν καὶ τροφὴν τοῦ πνεύματος.

III. Περὶ δὲ ἀναπνοῆς οἱ μὲν οὐ λέγουσι τίνος
χάριν, ἀλλὰ μόνον ὃν τρόπον γίνεται, καθάπερ
30 Ἐμπεδοκλῆς καὶ Δημόκριτος· οἱ δ' οὐδὲ τὸν τρόπον
ὅλως λέγουσιν, ἀλλ' ὡς φανερῷ χρῶνται. δεῖ δὲ
καὶ εἰ καταψύξεως χάριν, αὐτὸ τοῦτο διασαφῆσαι.
εἰ γὰρ ἐν τοῖς ἄνω τὸ θερμόν, οὐκ ἂν ἔτι δέοιτο
⟨τὰ⟩[1] κάτω. τὸ δὲ σύμφυτον πνεῦμα δι' ὅλου, καὶ
ἀρχὴ ἀπὸ τοῦ πνεύμονος. δοκεῖ δὲ καὶ τὸ τῆς ἀνα-
35 πνοῆς εἰς πάντα διαδίδοσθαι κατὰ συνέχειαν, ὥστε
τοῦτο δεικτέον ὡς οὐκ ἔστιν. ἄτοπον δὲ εἰ μὴ
482 b δεῖταί τινος κινήσεως καὶ οἷον τροφῆς. εἰ δὲ δια-
πνεῖ πρὸς πᾶν, οὐκ ⟨ἂν⟩[2] ἔτι καταψύξεως εἴη χάριν.
ἀλλὰ μὴν καὶ ἡ διάδοσις ἄλλως τ' ἀναίσθητος, καὶ
τὸ τάχος αὐτῆς. καὶ πάλιν τὸ τῆς παλιρροίας,
εἴπερ ἀπὸ πάντων, θαυμαστόν, πλὴν εἰ ἄλλον τρό-
5 πον ἀπὸ τῶν ἐσχάτων, τὸ δὲ πρώτως καὶ κυρίως
ἀπὸ τῶν περὶ τὴν καρδίαν. ἐν πολλοῖς δ' οὕτω τὸ
τῶν ἐνεργειῶν καὶ τῶν δυνάμεων. ἄτοπον γοῦν[3]
ὅμως εἰ καὶ εἰς τὸ ὀστοῦν διαδίδοται· καὶ γὰρ δὴ
τοῦτό φασιν ἐξ ἀρτηριῶν. διό, καθάπερ εἴρηται,
σκεπτέον περὶ ἀναπνοῆς, καὶ τίνος ἕνεκα καὶ ποίοις
10 μέρεσι καὶ πῶς. ἔτι οὐδ' ἐπιφορὰ τῆς τροφῆς
φαίνεται πᾶσι δι' ἀρτηριῶν, οἷον αὐτοῖς τε τοῖς
ἀγγείοις καὶ ἄλλοις τισὶ τῶν μερῶν· ζῇ δὲ τὰ
φυτὰ καὶ τρέφεται. ταῦτα μὲν οἰκειότερά πως
τοῖς περὶ τὰς τροφάς.

IV. Ἐπεὶ δὲ τρεῖς αἱ κινήσεις τοῦ ἐν τῇ ἀρτηρίᾳ

---

[1] τὰ suppl. Dobson.   [2] ἂν suppl. Jaeger.
[3] γοῦν Dobson : οὖν.

must be true. So much, then, for the increase and
nutriment of the breath.

III. As for respiration, some authorities, such as Breath must have a definite function.
Empedocles and Democritus, do not state why it
takes place, but only the method by which it occurs;
others again do not even deal with the method at
all, but treat it as obvious. But we must make it
quite clear whether its purpose is really refrigeration.
For if the heat resides in the upper parts, refrigeration would not be needed in the parts below. But
the inherent breath pervades the whole, and its
source is the lung. Also the inhaled breath seems
to be continuously distributed into every part; so
we should have to prove that this is untrue. It is
strange too if the lower parts need no stimulation
and nothing in the way of nutriment; while if
the breath is all-pervasive its purpose cannot be
refrigeration. Again, the manner of its reflux, if it
comes from all parts, is astonishing, unless it returns
from the extremities in some other way, but in the
primary and proper sense from the region of the heart.
Such discrepancy of functions and faculties is common.
At least, however, it is absurd that the distribution
should reach the bone—as they say that it does from
the air ducts. So, as has been said, we must consider
respiration, what it is for, in what parts it occurs, and
how. Again, it does not appear that food is carried by
the air ducts to all organs, e.g., to the vessels themselves, and certain other parts; and plants ⟨which
have no air ducts⟩ live and are nourished. Perhaps
these questions are more relevant to a discussion of
food.

IV. There are three distinct movements of the How does breath act?

482 b

15 πνεύματος, ἀναπνοή, σφυγμός, τρίτη δ' ἡ τὴν
τροφὴν ἐπάγουσα καὶ κατεργαζομένη, λεκτέον ὑπὲρ
ἑκάστης καὶ ποῦ καὶ πῶς καὶ τίνος χάριν. τούτων
δ' ἡ μὲν τοῦ σφυγμοῦ καὶ τῇ αἰσθήσει φανερὰ
καθ' ὁτιοῦν μέρος ἁπτομένοις, ἡ δὲ τῆς ἀναπνοῆς
μέχρι μέν του φανερά, τὸ δὲ πλέον κατὰ λόγον,
20 ἡ δὲ τῆς τροφῆς ἅπασα κατὰ λόγον ὡς εἰπεῖν, ὡς
ἐκ τῶν συμβαινόντων δὲ κατὰ τὴν αἴσθησιν. ἡ
μὲν οὖν ἀναπνοὴ δῆλον ὡς ἀπὸ τοῦ ἐντὸς ἔχει τὴν
ἀρχήν, εἴτε ψυχῆς δύναμιν εἴτε ψυχὴν δεῖ λέγειν
ταύτην, εἴτε καὶ ἄλλην τινὰ σωμάτων μίξιν, ἢ δι'
25 αὑτῶν ποιεῖ τὴν τοιαύτην ὁλκήν. ἡ δὲ θρεπτικὴ
δόξειεν ἂν ἀπὸ τῆς ἀναπνοῆς· αὕτη γὰρ ἀνταπο-
δίδοται, καὶ ὁμοία τῷ ἀληθεῖ. εἰ δὲ μὴ[1] πᾶν
ὁμαλίζει τοῖς χρόνοις τὸ σῶμα κατὰ τὴν τοιαύτην
κίνησιν,[2] ἢ εἰ μηδὲν διαφέρει τὸ ἅμα, πάντα τὰ
μέρη σκεπτέον. ὁ δὲ σφυγμὸς ἴδιός τις παρὰ
30 ταύτας,[3] τῇ μὲν ἂν δοκῶν εἶναι κατὰ συμβεβηκός,
εἴπερ, ὅταν ἐν ὑγρῷ πλῆθος ᾖ θερμότητος, ἀνάγκη
τὸ ἐκπνευματούμενον διὰ τὴν ἐναπόληψιν ποιεῖν
σφυγμόν, ἐν τῇ ἀρχῇ δὲ καὶ πρῶτον, εἴπερ τοῖς
πρώτοις σύμφυτον· ἐν γὰρ τῇ καρδίᾳ μάλιστα καὶ
πρῶτον, ἀφ' ἧς καὶ τοῖς ἄλλοις· τάχα δὲ πρὸς τὴν
35 ὑποκειμένην οὐσίαν τοῦ ζῴου τὴν ἐκ τῆς ἐνεργείας
ἀνάγκη τοῦτο παρακολουθεῖν.

Ὅτι δ' οὐδὲν πρὸς τὴν ἀναπνοὴν ὁ σφυγμός·
483 a σημεῖον· ἐάν τε γὰρ πυκνὸν ἐάν τε ὁμαλὸν ἐάν τε
σφοδρὸν ἢ πρᾶον ἀναπνέῃ τις, ὅ γε σφυγμὸς
ὅμοιος καὶ ὁ αὐτός, ἀλλ' ἡ ἀνωμαλία γίνεται καὶ

---

[1] μή, vulgo : corr. Dobson.
[2] κίνησιν. ἢ vulgo : corr. Dobson.
[3] παρὰ ταύτας LPQB[a] : παρ' αὐτάς.

breath in the windpipe, *viz.*, respiration, pulsation, and thirdly that which introduces and acts upon food. We must explain of each where, how, and why it takes place. Of these the movement of pulsation is perceptible to those who touch any part ; that of respiration is perceptible up to a point, but most of it is a question of theory ; while the movement which affects nutriment is almost entirely theoretical, but, in so far as it can be determined from its results, it is a matter of perception. It is clear that respiration has its source from within, whether we are right to describe it as a function of the soul, or the soul itself, or else some mixture of bodies which by their means causes this attraction. The nutritive movement would seem to proceed from respiration, which corresponds to and indeed resembles it. As to whether the time taken by this movement is not uniform throughout the body, or whether its simultaneity makes no difference, all the parts must be examined. The pulsation is quite distinct from the other two. In one way it would seem to be only incidental, since, when there is much heat in a liquid, that which is evaporated must cause pulsation due to the trapping of the air within ; but it is also original and primary, since it is inherent in what is primary ; for it is found chiefly and primarily in the heart, from which it is communicated to the other organs. Perhaps this is a necessary consequence of the animal's underlying essence, which is realized in activity.

But there is evidence that pulsation has no connexion with breathing ; for whether a man breathes rapidly or evenly, heavily or quietly, the pulse remains the same and unaltered, but irregularity and

483 a

ἐπίτασις ἔν τε σωματικοῖς τισὶ πάθεσι καὶ ἐν τοῖς
5 τῆς ψυχῆς φόβοις ἐλπίσιν ἀγωνίαις.

Εἰ δὲ καὶ ἐν ταῖς ἀρτηρίαις ὁ σφυγμός, καὶ ὁ
αὐτὸς ὢν ῥυθμῷ καὶ ὁμαλὸς [ᾖ],[1] σκεπτέον· οὐκ
ἔοικε δέ γε τοῖς μακρὰν ἀπηρτημένοις. ἥκιστα δ'
ἕνεκά του φαίνεται γίνεσθαι, καθάπερ εἴρηται. τὸ
γὰρ αὖ τῆς ἀναπνοῆς καὶ τῆς ἐπαγωγῆς, εἴθ' ὡς
10 ἕτερα πάμπαν ἀλλήλων εἴθ' ὡς θάτερον πρὸς θάτερον, ἕνεκά του φαίνεται καὶ ἔχει τινὰ λόγον. τριῶν
δ' οὐσῶν πρότερον[2] εὔλογον εἶναι τήν τε[3] σφυγμώδη
καὶ τὴν ἀναπνευστικήν· ἡ γὰρ τροφὴ προϋπάρ-
χοντος. ἢ οὔ; τὸ μὲν γὰρ ἀναπνεῖν, ὅταν ἀπολυθῇ
τῆς κυούσης, ἡ δ' ἐπιφορὰ καὶ ἡ τροφὴ καὶ ξυν-
15 ισταμένου καὶ ξυνεστηκότος, ὁ δὲ σφυγμὸς εὐθὺς
ἐν τῇ ἀρχῇ ξυνισταμένης τῆς καρδίας, καθάπερ ἐν
τοῖς ᾠοῖς γίνεται φανερόν. ὥστε αὕτη πρώτη,
καὶ ἔοικεν ἐνεργείᾳ τινὶ καὶ οὐκ ἐναπολήψει πνεύματος, εἰ μὴ ἄρα τοῦτο πρὸς τὴν ἐνέργειαν.

V. Τὸ δὲ πνεῦμα τὸ ἐκ τῆς ἀναπνοῆς φέρεσθαι
20 μὲν εἰς τὴν κοιλίαν, οὐ διὰ τοῦ στομάχου (τοῦτο
μὲν γὰρ ἀδύνατον), ἀλλὰ πόρον εἶναι παρὰ τὴν
ὀσφύν, δι' οὗ τὸ πνεῦμα τῇ ἀναπνοῇ φέρεσθαι ἐκ
τοῦ βρογχίου[4] εἰς τὴν κοιλίαν καὶ πάλιν ἔξω· τοῦτο
δὲ τῇ αἰσθήσει φανερόν.

Ἔχει δ' ἀπορίαν καὶ τὰ περὶ τὴν αἴσθησιν. εἰ
25 γὰρ ἡ ἀρτηρία μόνον αἰσθάνεται, πότερα τῷ πνεύματι τῷ δι' αὐτῆς, ἢ τῷ ὄγκῳ, ἢ τῷ σώματι;
ἢ εἴπερ ὁ ἀὴρ πρῶτον ὑπὸ τὴν ψυχήν, τῷ κυριω-

---

[1] ᾖ omittendum censuit Dobson.
[2] πρότερον Furlan : πότερον.
[3] τε Jaeger : γε.   [4] βραγχίου Z.

498

excitement occur during some bodily ailments and
in conditions of fear, expectation, and conflict in the
soul.

We must next consider whether the pulsation
occurs also in the arteries, and with the same even
rhythm. It does not seem so in those which are
remotely connected ; and, as has been said, it does
not seem to have any purpose at all. The acts of
respiration and ingestion, on the other hand, whether
they are regarded as independent or as correlative,
clearly have a purpose and a rational explanation.
Of the three functions it is reasonable to suppose that
pulsation and respiration are prior ; for nutrition
implies something pre-existent. Or is this wrong ?
For respiration starts as soon as the embryo is re-
leased from its mother, and ingestion and nutrition
belong to it both during and after its formation, but
pulsation begins at the very outset while the heart is
forming, as can be observed in eggs. So that pulsation is prior in origin, and resembles an activity, and
not an interception of the breath, except in so far
as this contributes to its activity.

V. It is said that in respiration the breath is not Physiology
conveyed to the belly through the oesophagus (this of breath
would be impossible), but that there is a passage body.
along the loin, through which the breath is carried
by respiration from the windpipe into the belly
and out again ; and that this is perceptible by the
senses.

But the circumstances of this perception present
difficulties. For if only the windpipe perceives, does
it do so by the breath passing through it, or by its
bulk or by its bodily nature ? Or if air is first after
the soul, does it perceive by this, as superior and

483 a

τέρῳ τε καὶ προτέρῳ; τί οὖν ἡ ψυχή; δύναμίν
φασι τὴν αἰτίαν τῆς κινήσεως τῆς τοιαύτης. ἢ
δῆλον ὡς οὐκ ὀρθῶς ἐπιτιμήσεις τοῖς τὸ λογιστικὸν
30 καὶ θυμικόν· καὶ γὰρ οὗτοι ὡς δυνάμεις λέγουσιν.
ἀλλ᾽ εἰ δὴ ἡ ψυχὴ ἐν τῷ ἀέρι τούτῳ, οὗτός γε
κοινός. ἢ πάσχων γέ τι καὶ ἀλλοιούμενος εὐλόγως,[1]
ἂν ἔμψυχον ἢ[2] ψυχή, πρὸς τὸ συγγενὲς φέρεται, καὶ
τῷ ὁμοίῳ τὸ ὅμοιον αὔξεται. ἢ οὔ; τὸ γὰρ ὅλον
οὐκ ἀήρ, ἀλλὰ συμβαλλόμενόν τι πρὸς ταύτην τὴν
35 δύναμιν ὁ ἀήρ, ἢ οὕτω[3] ταύτην ποιοῦν, καὶ τὸ
ποιῆσαν τοῦτ᾽ ἀρχὴ καὶ ὑπόθεσις.

483 b

Τοῖς δὲ μὴ ἀναπνέουσιν, ἵνα ἀνεπίμικτος τῷ
ἔξω—ἢ οὔ, ἀλλὰ κατ᾽ ἄλλον τρόπον μιγνύμενος ;[4]
—τίς οὖν ἡ διαφορὰ τοῦ ἐν τῇ ἀρτηρίᾳ πρὸς τὸν
ἔξω; διαφέρειν γὰρ εὔλογον, τάχα δὲ καὶ ἀναγ-
5 καῖον, λεπτότητι· ἀλλ᾽ ἔτι δὲ καθ᾽ αὑτὸν θερμὸς
ἢ ὑφ᾽ ἑτέρου; φαίνεται γὰρ ὁ ἔσω καθάπερ ὁ
ἔξω· βοηθεῖται δὲ τῇ καταψύξει. πότερα δέ; ἔξω
μὲν γὰρ πραΰς, ἐμπεριληφθεὶς δὲ πνεῦμα, καθάπερ
πυκνωθεὶς καὶ διαδοθεὶς πως. ἢ μίξιν τινὰ ἀνάγκη
10 λαμβάνειν, ἐν ὑγρότητί τε καὶ σωματικοῖς ὄγκοις
ἀναστρεφόμενον; οὐκ ἄρα λεπτότατος, εἴπερ μέμικται. καὶ μὴν εὔλογόν γε τὸ πρῶτον δεκτικὸν
ψυχῆς, εἰ μὴ ἄρα καὶ ἡ ψυχὴ τοιοῦτον, καὶ οὐ
καθαρόν τι καὶ ἀμιγές· τὴν ⟨δ᾽⟩[5] ἀρτηρίαν μόνον
εἶναι δεκτικὴν πνεύματος, τὸ δὲ νεῦρον οὔ. διαφέρει δὲ καὶ ὅτι τὸ μὲν νεῦρον ἔχει τάσιν, ἡ δ᾽
15 ἀρτηρία ταχὺ διαρρήγνυται, καθάπερ καὶ ἡ φλέψ.

---

[1] virgula interpunxit Jaeger.   [2] ἢ Jaeger : ἡ.
[3] οὕτω Dobson : οὐ τό.
[4] verba ἢ οὔ . . . μιγνύμενος pro parenthesi habenda censuit Ross.   [5] δ᾽ add. Dobson.

prior ? What then is the soul ? A potentiality, they say, which causes movement of this kind. Surely it is clear that one is not right to censure those who call it the seat of reasoning and the passions ; for they too speak of these as potentialities. But if the soul resides in this air, the air at least is common. Surely if by some affection or change it becomes animate or soul, it is naturally attracted to what is akin to it, and like is increased by like. Or is this not so ? For, it may be said, the air is not the whole of the soul, but is something which contributes to this potentiality, or that which makes it in this particular sense ; and that which has made it is its principle and basis.

But in creatures which do not respire, where there is no mixture with the air outside—or is this not so, but is it mixed in some other way ?—what is the difference between the air in the windpipe and the air outside ? It is natural and perhaps inevitable to assume that it is finer. But again, is it hot in itself, or heated by something else ? For the air inside seems to be like the air outside ; but it is assisted by refrigeration. Which is the true view ? Outside the air is mild, but when enclosed it is breath, being condensed and distributed in a certain way. Must it not admit some mixture, by moving about in liquid and the solid parts of the body ? Then if it is mixed, it is not the finest substance. Yet it is natural that the primary receptacle of the soul should be such (unless the soul too is of the above nature, not pure and unmixed) ; and that the air duct alone should admit breath, and the sinew not. There is also this difference, that sinew is elastic, but air ducts, like veins, are easily burst. The skin con-

483 b

τὸ δὲ δέρμα ἐκ φλεβὸς καὶ νεύρου καὶ ἀρτηρίας,
ἐκ φλεβὸς μὲν ὅτι κεντηθὲν αἷμα ἀναδίδωσιν, ἐκ
νεύρου δὲ ὅτι τάσιν ἔχει, ἐξ ἀρτηρίας δὲ ὅτι διαπνοὴν
ἔχει· μόνον γὰρ δεκτικὸν πνεύματος ἡ ἀρτηρία.
τὰς δὲ φλέβας ἔχειν πόρους, ἐν οἷς[1] τὸ θερμὸν ὂν
20 ὥσπερ ἐν χαλκείῳ θερμαίνειν τὸ αἷμα· φύσει γὰρ
οὐκ εἶναι θερμόν, ἀλλ᾽ ὥσπερ τὰ τηκτὰ καταδια-
χεῖσθαι· διὸ καὶ πήγνυσθαι τὴν ἀρτηρίαν, καὶ ἔχειν
ὑγρότητα καὶ ἐν αὑτῇ καὶ ἐν τοῖς χιτῶσι τοῖς
περιέχουσι τὸ κοίλωμα. φανερὸν δ᾽ ἔκ τε τῶν
25 ἀνατομῶν εἶναι, καὶ ὅτι εἰς τὸ ἔντερον καὶ εἰς τὴν
κοιλίαν αἵ τε φλέβες καὶ αἱ ἀρτηρίαι συνάπτουσιν,
ἃς εἰκὸς εἶναι τὴν τροφὴν ἕλκειν. ἐκ δὲ τῶν
φλεβῶν εἰς τὰς σάρκας διαδίδοσθαι τὴν τροφήν, οὐ
κατὰ τὰ πλάγια ἀλλὰ κατὰ τὸ στόμα, καθάπερ
σωλῆνας. ἀποτείνειν[2] γὰρ ἐκ τῶν πλαγίων φλεβῶν
30 φλέβια λεπτὰ ἐκ τῆς μεγάλης φλεβὸς καὶ τῆς
ἀρτηρίας παρ᾽ ἑκάστην πλευράν, καὶ ἀρτηρίαν καὶ
φλέβα παρακεῖσθαι· καὶ τὰ ὀστέα δὲ καθάπτειν τὰ
νεῦρα καὶ τὰς φλέβας καὶ εἰς μέσα καὶ εἰς τὰς
συμβολὰς τῶν κεφαλῶν, δι᾽ ὧν τὴν τροφὴν δέχεσθαι
τοὺς ἰχθύας καὶ ἀναπνεῖν· εἰ δὲ μὴ ἀνέπνεον,
35 ἐξαιρεθέντας ἂν ἐκ τοῦ ὑγροῦ εὐθὺς θνήσκειν.

484 a

Τὰς δὲ φλέβας καὶ τὰς ἀρτηρίας συνάπτειν εἰς
ἀλλήλας καὶ τῇ αἰσθήσει φανερὸν εἶναι. τοῦτο δ᾽
οὐκ ἂν συμβαίνειν, εἰ μὴ ἐδεῖτο καὶ τὸ ὑγρὸν
πνεύματος καὶ τὸ πνεῦμα ὑγροῦ, τῷ θερμὸν εἶναι
ἐν νεύρῳ καὶ ἀρτηρίᾳ καὶ φλεβί, θερμότατον δὲ
5 καὶ οἷον φλεβωδέστατον τὸ ἐν τῷ νεύρῳ.

Ἄτοπον οὖν τῇ τοῦ πνεύματος χώρᾳ τὸ θερμόν,
ἄλλως τε καὶ καταψύξεως χάριν. εἰ δὲ ποιεῖ καὶ
οἷον ἀναζωπυρεῖ θερμῷ τὸ θερμόν, γίγνοιτ᾽ ἄν.

---

[1] οἷς Dobson : αἷς.   [2] ἀποτείνειν Jaeger : ἀποτείνει.

sists of veins, sinews, and air ducts ; of veins because it emits blood when pricked, of sinews because it is elastic, and of air ducts because it allows the passage of air ; for the air duct alone is receptive of air. The veins have passages, in which the heat resides and warms the blood, as though in a crucible ; for it is not naturally hot, but is diffused by heat, just as molten metals are. Hence, too, the air duct becomes hard, and has moisture in it, and in the coverings which surround the cavity. This is evident both from dissection and from the fact that the veins and air ducts, which are probably the vehicles of nutriment, are connected with the intestines and the belly. The nutriment is distributed to the flesh from the veins, not through their sides but through their mouth, as through pipes. For from the side veins fine little veins extend from the great vein and air duct along each rib, air duct and vein lying side by side. The bones, too, are connected by sinews and veins, both to the middle parts and to the junction of the head, through which fishes admit food and breathe : if they did not breathe, when taken out of the water they would die at once.

That the veins and air ducts are attached to each other is evident even to sense-perception. This would not be the case if the liquid did not need breath and the breath liquid, because there is heat in sinew, air duct, and vein, that in the sinew being the greatest and most similar to that in the vein.

Heat is not suitable to the place of the breath, especially in view of refrigeration. But if the animal generates and rekindles its heat by other heat, this would account for its presence there. Again, the

ἔτι πάντων τῶν ἐχόντων θερμότητα σύμφυτόν πως
ἡ διαμονή, μηδενὸς ἀντικειμένου μηδὲ καταψύχον-
10 τος· ὅτι γὰρ πάντα δεῖται καταψύξεως, σχεδὸν
φανερὸν τῷ ⟨τὸ⟩[1] αἷμα κατέχειν ἐν τῇ φλεβὶ τὸ
θερμὸν οἷον ἀποστέγον· διὸ καὶ ὅταν ἐκρυῇ, μεθ-
ιέναι τε καὶ ἀποθνήσκειν, τῷ τὸ ἧπαρ οὐκ ἔχειν
οὐδεμίαν ἀρτηρίαν.

VI. Πότερον δὲ τὸ σπέρμα διὰ τῆς ἀρτηρίας ὡς
15 καὶ συνθλιβόμενον, καὶ ἐν τῇ προέσει μόνον; ἐν
οἷς δὴ φαίνεται καὶ ἡ ἐξ αἵματος μεταβολὴ τῷ τὰ
νεῦρα ἀπὸ τῶν ὀστῶν τρέφεσθαι· καθάπτει γὰρ
αὐτά. ἢ οὐδὲ[2] τοῦτ᾽ ἀληθές· καὶ γὰρ ἐν τῇ καρδίᾳ
νεῦρον, καὶ νεῦρα δὲ ἐκ τῶν ὀστῶν ἠρτημένα. οὐ
20 συνάπτει δὲ ἐν ἑτέρῳ, ἀλλ᾽ εἰς σάρκα ἀποτελευτᾷ.
ἢ τοῦτό γ᾽ οὐθέν· εἴη γὰρ ἂν οὐθὲν ἧττον ἀπὸ τοῦ
ὀστοῦ ἡ τροφή. αὐτοῖς δ᾽ ἀπὸ τοῦ νεύρου τοῖς
ὀστοῖς μᾶλλον τὴν τροφήν. ἄτοπον γὰρ καὶ τοῦτο·
ξηρὸν ⟨γὰρ⟩[3] φύσει καὶ οὐκ ἔχον πόρους ὑγροῦ[4]· ἡ
τροφὴ δ᾽ ὑγρόν. σκεπτέον δὲ πρότερον,[5] εἴπερ
ἀπὸ τῶν ὀστῶν, τίς ἡ τοῦ ὀστοῦ τροφή, ἢ φέρουσι
25 πόροι καὶ ἐκ τῆς φλεβὸς καὶ ἐκ τῆς ἀρτηρίας εἰς
αὐτό.[6] καὶ ἐν πολλοῖς μὲν εὔδηλοι, μάλιστα δ᾽
εἰς τὴν ῥάχιν. τοὺς[7] δ᾽ ἀπὸ τῶν ὀστῶν γίνεσθαι
συνεχεῖς, ὥσπερ ταῖς πλευραῖς· τούτους δ᾽ ἀπὸ
τῆς κοιλίας τίνα τρόπον, ἢ πῶς τῆς ὁλκῆς γινομένης; ἢ τὰ πολλὰ ἄχονδρα, καθάπερ ἡ ῥάχις· ἀλλ᾽
30 οὔτι πρὸς τὴν κίνησιν. ἢ συνάψεως χάριν; δεῖ δέ,
καὶ εἰ ἀπὸ τοῦ νεύρου τὸ ὀστοῦν τρέφεται, τὴν
τοῦ νεύρου τροφὴν εἰδέναι. ἡμεῖς δέ φαμεν ἐκ τῆς

---

[1] τὸ add. Jaeger. 　　[2] οὐδὲ Jaeger : οὔτε.
[3] γὰρ add. Jaeger. 　　[4] ὑγροῦ Dobson : ὑγρούς.
[5] πρότερον Bussemaker : πότερον.
[6] αὐτό Bussemaker : αὐτόν. 　　[7] τοὺς Dobson : τὰς.

persistence of all things containing heat is in a sense a natural quality, provided that nothing counteracts or cools it. That everything needs cooling is almost obvious from the fact that the blood retains the heat in the veins as though sheltering it. So too when it flows out, the animal loses its heat and dies, because the liver has no air duct.

VI. Is the semen's passage through the air duct due to pressure also, and does it occur only in emission? Those facts also exhibit the change from blood (into flesh), due to the sinews being nourished from the bones ; for they knit them together. Possibly even this is not true ; for there is sinew in the heart too, and sinews attached to the bones, but they do not connect with anything else, but end in flesh. Possibly this is of no account ; for none the less the nourishment of the sinew would come from the bone. Yet this too is awkward, and the bones themselves would be nourished more naturally from the sinews ; for bone is naturally dry, and has no passages for liquid ; and the nutriment is liquid. We must consider first, if it comes from the bones, what the nutriment of bone is ; and whether the passages carry it from both vein and air duct to the bone itself. In many places these passages are visible, especially those leading to the spine. And those leading from the bones are continuous, as they are in the ribs ; but how do they lead from the belly, and how does the extraction happen? Most bones have no cartilage, like the spine ; they are not at all suited to movement. Are they for connexion? Again, if the bone is fed from the sinew, one must know what feeds the sinew. We ourselves say that it is from the liquid

Nerves.

505

484 a

ὑγρότητος γλίσχρας οὔσης τῆς περὶ τὸ αὐτό.
πόθεν δ' αὐτὴ καὶ πῶς, λεκτέον. τὸ ἐκ φλεβὸς
καὶ ἀρτηρίας τὴν σάρκα, ὅτι πανταχόθεν αἷμα τῇ
35 κεντήσει, ψεῦδος ἐπί γε τῶν ἄλλων ζώων, οἷον
ὀρνίθων καὶ ὄφεων καὶ ἰχθύων ἢ ὅλως τῶν ᾠοτόκων·
ἀλλὰ τῶν πολυαίμων τοῦτ' ἴδιον, ἐπεὶ τῶν ὀρνιθίων
γε καὶ τεμνομένων τὸ στῆθος ἰχώρ, οὐχ αἷμα. Ἐμ-

484 b

πεδοκλῆς δὲ ἐκ νεύρου τὸν ὄνυχα τῇ πήξει. ἆρ'
οὖν οὕτω καὶ δέρμα πρὸς σάρκα; ἀλλὰ τοῖς ὀστρα-
κοδέρμοις καὶ μαλακοστράκοις πῶς ἀπὸ τῶν ἐκτὸς
ἡ τροφή; τοὐναντίον γὰρ δοκεῖ μᾶλλον ἀπὸ τῶν
ἐντὸς ἢ τῶν ἐκτός. ἔτι δὲ ποία καὶ διὰ τίνων ἡ
5 ἐκ τῆς κοιλίας δίοδος; καὶ πάλιν ἡ ἐκείνων ἀνα-
στροφὴ πρὸς τὴν σάρκα, καίπερ ἄλογος οὖσα.
πολὺ γάρ τι θαυμαστὸν φαίνεται καὶ ἀδύνατον
ὅλως. ἆρά γε ἄλλοις ἄλλη τροφή, καὶ οὐ πᾶσι
τροφὴ τὸ αἷμα· πλὴν ἐκ τούτου τἆλλα.

VII. Τὴν τῶν ὀστῶν φύσιν ἆρα σκεπτέον εἰ[1] πρὸς
10 κίνησιν ἢ πρὸς ἔρεισμα, καὶ πρὸς τὸ στέγειν καὶ
περιέχειν, ἔτι δ' εἰ ὥσπερ ἀρχαὶ ἔνια, καθάπερ ὁ
πόλος. λέγω δὲ πρὸς μὲν κίνησιν, οἷον ποδὸς ἢ
χειρὸς ἢ σκέλους ἢ ἀγκῶνος, ὁμοίως τήν τε καμ-
πτικὴν καὶ τὴν κατὰ τόπον· οὐδὲ γὰρ τὴν τοπικὴν
οἷόν τε ἄνευ κάμψεως. σχεδὸν δὲ καὶ τὰ ἐρείσ-
15 ματα ἐν τούτοις. τὴν δὲ τοῦ στέγειν καὶ περιέχειν,
οἷον τὰ ἐν τῇ κεφαλῇ τὸν ἐγκέφαλον, καὶ ὅσοι δὴ
τὸν μυελὸν ἄρχειν. αἱ δὲ πλευραὶ τοῦ συγκλείειν.

---

[1] εἰ Jaeger : ἤ.

surrounding it, which is viscous. Then it must be explained whence this arises and how. The idea that flesh consists of vein and air duct, because blood issues from every point at which it is pricked, is false in the case of other animals such as birds, snakes, and fishes, and generally of oviparous animals. It is only in full-blooded animals that this happens, since, when the breast of a small bird is cut, it is not blood but serum that flows out. Empedocles thinks that nail is formed from sinew by a process of hardening. Is the relation between skin and flesh the same? But how is it possible for creatures with hard or soft shells to get their nutriment from outside? They seem on the contrary to derive it rather from inside than from outside. Again, how and by what course does the passage of foods from the belly occur? and again its return into flesh, unreasonable though it is? The process seems most remarkable and indeed quite impossible. Perhaps different animals have different forms of nutriment, and it is not blood in all cases; and yet the other forms are derived from blood.

VII. Now we have to consider the nature of the bones, whether they are designed for movement or for support, or as a covering and envelope, and whether some are the origin of movement, like the axis of the universe. By "for movement" I mean that of the foot or hand or leg or elbow, whether in bending or in movement from place to place; for this latter movement is impossible without bending. These are also the bones, one may say, which provide support. Covering and envelope I mean in the sense that the skull covers the brain, or as those who regard the marrow as a principle say that the bone covers the marrow. The ribs are to lock the body together.

The function of the bones.

The spine, from which the ribs radiate to lock the body together, is the fixed part which originates movement. There must be something of this character; for all movement depends on something stationary.

At the same time there must be a final cause; under which some class the motive principle, *i.e.*, the spinal marrow and the brain. Besides these there are bones at the junctions and for the purpose of locking together, *e.g.*, the collar-bone; perhaps its name (the key-bone) is derived from this. Each is well suited to its object. If, *e.g.*, the spine, foot and elbow, were not such as they are, there could be no bending either of the whole or of parts; for the bending of the elbow must take place inwards to achieve its purpose; and the same applies to the foot and the other limbs. All are for a purpose, as are their component bones; *e.g.*, the radius in the forearm for the purpose of bending the elbow and the hand. For without it we could not turn the palm up and down, nor raise and bend the feet, if there were not two radii *a* employed in the movement. In the same way we must consider other cases, such as the movement of the neck, whether there is one bone only concerned. We have also to consider bones which hold or link together, like the patella over the knee; and why other joints have no such bone. All bones which are concerned with movement, especially practical movement, *e.g.*, those of the elbow, legs, hands, and feet, are furnished with sinews. The rest have sinews for connexion, if they need them at all; for some, such as the spine, have little or no function, except bending. For what fastens the vertebrae together is serum and mucous

ἐστι καὶ ὑγρότης μυξώδης.   τὰ δὲ καὶ συνδεῖται
νεύροις, οἷον τὰ περὶ τὰ ἄρθρα.

VIII. Πάντων δ' ἐστὶ λόγος ὁ βελτίων ὡς καὶ
5 νῦν ζητεῖν· ἀλλὰ τὰς ἀρχὰς ἐφ' ἱκανόν, ὧν χάριν,
σκεπτέον. οὐκ ἂν δόξειε κινήσεως ἕνεκα τὰ ὀστᾶ,
ἀλλὰ μᾶλλον τὰ νεῦρα ἢ τὰ ἀνάλογον, ἐν ᾧ πρώτῳ
τὸ πνεῦμα τὸ κινητικόν, ἐπεὶ καὶ ἡ κοιλία κινεῖται
καὶ ἡ καρδία νεῦρα ἔχει· τὰ δ' οὐ πᾶσιν, ἀλλ'
10 ἐνίοις, ἀνάγκη καὶ πρὸς τὴν τοιαύτην κίνησιν νεῦρα
ἔχειν, ἢ εἰς τὸ . . . ὁ γὰρ πολύπους ἐπ' ὀλίγον
καὶ κακῶς βαδίζει. δεῖ γὰρ τοῦτο λαβεῖν ὥσπερ
ἀρχήν, ὅτι πᾶσιν [ἢ]¹ ἄλλου τινὸς χάριν ἄλλα² πρὸς
τὴν κίνησιν τὴν οἰκείαν, οἷον τοῖς μὲν πεζοῖς πόδας,
καὶ τοῦτο τοῖς μὲν ὀρθοῖς δύο, τοῖς δὲ παντελῶς
15 ἐπὶ τῆς γῆς πλείους, ὅσοις ἡ ὕλη γεωδεστέρα καὶ
ψυχροτέρα (τὰ δὲ καὶ ἄποδα ὅλως ἐγχωρεῖ· βίᾳ
γὰρ οὕτω κινεῖσθαι)· τοῖς ⟨δὲ⟩³ πτηνοῖς πτέρυγας,
καὶ τούτων τὴν μορφὴν οἰκείαν τῇ φύσει. διάφορα
δὲ τοῖς πτητικωτέροις καὶ βραδυτέροις.⁴ πόδας δὲ
τροφῆς χάριν καὶ ἀναστάσεως, πλὴν τῆς νυκτερίδος·
20 διὸ καὶ τὴν τροφὴν ἐκ τοῦ ἀέρος, καὶ μὴ δεῖσθαι
διαναπαύσεως· οὐ δέονται γὰρ δὴ ἄλλως.⁵ τὰ δὲ
ὀστρακόδερμα τῶν ἐνύδρων ὑπόποδα διὰ τὸ βάρος.
καὶ ταῦτα μὲν πρὸς τὴν κατὰ τόπον ἀλλαγήν· ὅσα
δὲ πρὸς τὴν ἄλλην χρείαν, ὥσπερ ὑφηγεῖται καὶ
ἑκάστου τὰ ἴδια, καὶ εἴ τι μὴ προφανές, οἷον διὰ

---

¹ ἢ del. Jaeger.   ² ἄλλα Jaeger : ἀλλά.   ³ δὲ add. Jaeger.
⁴ βραδυτέροις Bussemaker : βραχυτέροις.
⁵ δὴ ἄλλως Dobson : δι' ἄλλων.

ᵃ Five or six letters are missing in all mss.

fluid. Other bones, such as those about the joints, are connected by sinews.

VIII. The best account of anything is obtained by the present method; but we must give due consideration to the final causes. That which serves the purpose of movement would seem to be not the bones but rather the sinews, or what corresponds to them, *i.e.*, that in which the breath which causes movement primarily resides, since even the belly moves and the heart has sinews; but only some parts, not all, have bones; whereas sinews are necessary for movement of this kind or for . . .ᵃ For the polypus walks very little, and that little indifferently. We must assume as a starting-point that all have different organs for different purposes, to suit their own special form of movement; for instance, feet for animals that move on land, and, of these, two for those which stand erect, more for those which move entirely on the ground, whose matter is more earthy and colder. (It is possible for some creatures to be entirely footless: for in this case they can be moved by force.) Again, winged creatures have wings, whose shape is appropriate to their nature. The organs differ according as the creature flies faster or slower. They have feet for the purpose of getting food and for standing upright—except for the bat; hence the bat gets its food from the air, and does not need to rest for the purpose; certainly not for any other reason. The hard-shelled water creatures have feet to support their weight. These members serve for locomotion; all that serve any other need are as the special qualities of each animal dictate. This is true even if the reason is not obvious—for instance why many-

Bones (continued).

25 τί τὰ πολύποδα βραδύτατα (καίτοι τὰ τετράποδα
θάττω τῶν διπόδων)· πότερον ὅτι ἐπὶ γῆς ὅλα τὰ
σώματα; ἢ ὅτι φύσει ψυχρὰ καὶ δυσκίνητα; ἢ
δι' ἄλλην αἰτίαν;

IX. Οἱ ἀναιροῦντες ὡς οὐ τὸ θερμὸν τὸ ἐργαζόμενον ἐν τοῖς σώμασιν, ἢ ὅτι μία τις φορὰ καὶ
30 δύναμις ἡ τμητικὴ τοῦ πυρός, οὐ καλῶς λέγουσιν.
οὐδὲ γὰρ ὅλως[1] τοῖς ἀψύχοις ταὐτὸ ποιεῖ πᾶσιν,
ἀλλὰ τὰ μὲν πυκνοῖ, τὰ δὲ μανοῖ, καὶ τήκει, τὰ δὲ
πήγνυσιν. ἐν δὲ δὴ τοῖς ἐμψύχοις οὕτως ὑποληπτέον, ὥσπερ φύσεως πῦρ ζητοῦντα, καθάπερ τέχ-
νης· καὶ γὰρ ἐν ταῖς τέχναις ἕτερον τὸ χρυσοχοϊκὸν
35 καὶ τὸ χαλκευτικὸν καὶ τὸ τεκτονικὸν πῦρ ἀποτελεῖ,
καὶ τὸ μαγειρικόν. ἴσως δ' ἀληθέστερον ὅτι αἱ
485 b τέχναι· χρῶνται γὰρ ὥσπερ ὀργάνῳ μαλάττουσαι
καὶ τήκουσαι καὶ ξηραίνουσαι, ἔνια δὲ καὶ ῥυθμί-
ζουσαι.

Τὸ αὐτὸ δὴ[2] τοῦτο καὶ αἱ φύσεις· ὅθεν δὴ καὶ πρὸς
ἄλληλα διαφοραί. διὸ γελοῖον πρὸς τὸ ἔξω κρίνειν·
5 εἴτε γὰρ διακρῖνον εἴτε λεπτῦνον εἴθ' ὁτιδήποτ' ἐστὶ
τὸ θερμαίνεσθαι καὶ πυροῦσθαι, διάφορα[3] ἕξει τὰ
ἔργα τοῖς χρωμένοις. ἀλλ' αἱ μὲν τέχναι ὡς
ὀργάνῳ χρῶνται, ἡ δὲ φύσις ἅμα καὶ ὡς ὕλη.

Οὐ δὴ τοῦτο χαλεπόν, ἀλλὰ μᾶλλον τὸ τὴν φύσιν
αὐτὴν νοῆσαι τὴν χρωμένην, ἥτις ἅμα τοῖς αἰσθη-
10 τοῖς πάθεσι καὶ τὸν ῥυθμὸν ἀποδώσει. τοῦτο γὰρ
οὐκέτι πυρὸς οὐδὲ πνεύματος. τούτοις δὴ κατα-
μεμῖχθαι τοιαύτην δύναμιν θαυμαστόν. ἔτι δὲ

---

[1] ὅλως Dobson : ὅλα.   [2] δὲ Z, Bekker.
[3] διάφορα Neustadt : διαφορὰν.

footed creatures are the slowest movers, and yet
quadrupeds move faster than bipeds. Is it because
the whole body is on the ground (*i.e.*, in the case of
the many-footed creatures) ? Or because the others
are naturally cold and hard to move ? Or again is
it for some other reason ?

IX. Those who maintain that heat is not the
operative principle in bodies, or that fire has only
one motion and potentiality, *viz.*, for disintegration,
are inaccurate. For even in inanimate things it does
not produce the same effect universally, but makes
some denser and some rarer ; some it melts and
others it hardens. So in the case of creatures possessing soul we must assume that the results are the
same, and seek the effects of fire in nature, just as we
should in a craft ; for fire produces different results
in the craft of the goldsmith, the coppersmith, the
carpenter, and the cook. Perhaps more accurately
the crafts produce the different results ; for they use
the fire as an instrument for softening, melting, or
drying, and in some cases for tempering.

It is the same with individual natures ; hence the
differences they exhibit. To judge, then, from the
outside is absurd ; for whether we consider the action
of heat and fire as separating, refining, or anything
else, the results will vary according to the user. But
while the crafts use fire as an instrument, nature uses
it also as matter.

This presents no difficulty, but rather the difficulty
lies in the fact that nature, which uses the fire, should
herself be an intelligent agent, capable of assigning
to objects their proper form, as well as sensible affections. This is beyond the scope of fire or breath, and
so it is remarkable that such a faculty should be
combined with these substances. Moreover the same

Heat in the body.

513

τοῦτο θαυμαστὸν ταὐτὸν καὶ περὶ ψυχῆς· ἐν τούτοις
γὰρ ὑπάρχει. διόπερ οὐ κακῶς εἰς ταὐτόν, ἢ
ἁπλῶς ἢ μόριόν τι τὸ δημιουργοῦν, καὶ τὸ τὴν
κίνησιν ἀεὶ τὴν ὁμοίαν ὑπάρχειν ἐνέργειαν· καὶ γὰρ
15 ἡ φύσις, ἀφ᾽ ἧς καὶ ἡ γένεσις. ἀλλὰ δὴ τίς ἡ
διαφορὰ τοῦ καθ᾽ ἕκαστον θερμοῦ, εἶθ᾽ ὡς ὄργανον
εἴθ᾽ ὡς ὕλην εἴθ᾽ ὡς ἄμφω; πυρὸς γὰρ διαφοραὶ
κατὰ τὸ μᾶλλον καὶ ἧττον. τοῦτο δὲ σχεδὸν
ὥσπερ ἐν μίξει καὶ ἀμιξίᾳ· τὸ γὰρ καθαρώτερον
μᾶλλον. ὁ αὐτὸς δὲ λόγος καὶ ἐπὶ τῶν ἄλλων
20 ἁπλῶν. ἀνάγκη γάρ, ἐπείπερ ἕτερον ὀστοῦν καὶ
σάρξ ἡ ἵππου καὶ ἡ βοός, ἢ τῷ ἐξ ἑτέρων εἶναι ἢ
τῇ χρήσει διαφέρειν. εἰ μὲν οὖν ἕτερα, τίνες αἱ
διαφοραὶ ἑκάστου τῶν ἁπλῶν; καὶ τίς . . . ;[1]
ταύτας γὰρ ζητοῦμεν. εἰ δὲ ταὐτά, τοῖς λόγοις ἂν
διαφέροι. ἀνάγκη γὰρ δυεῖν θάτερον, καθάπερ ἐν
25 τοῖς ἄλλοις· οἴνου μὲν γὰρ καὶ μέλιτος κρᾶσις[2] διὰ
τὸ ὑποκείμενον, οἴνου δ᾽ αὐτοῦ, εἴπερ ἕτερα,[3] διὰ
τὸν λόγον. διὸ καὶ Ἐμπεδοκλῆς λίαν[4] ἁπλῶς τὴν
τοῦ ὀστοῦ φύσιν, ⟨ἐπεί⟩,[5] εἴπερ ἅπαντα τὸν αὐτὸν
λόγον ἔχει τῆς μίξεως, ἀδιάφορα ἐχρῆν ἵππου καὶ
λέοντος καὶ ἀνθρώπου εἶναι. νῦν δὲ διαφέρει σκλη-
30 ρότητι, μαλακότητι, πυκνότητι, τοῖς ἄλλοις. ὁμοίως
καὶ σὰρξ καὶ τὰ ἄλλα μόρια. ἔτι δὲ τὰ ἐν τῷ αὐτῷ
ζῴῳ διαφέρουσι πυκνότητι καὶ μανότητι καὶ τοῖς
ἄλλοις, ὥστ᾽ οὐχ ἡ αὐτὴ κρᾶσις. παχὺ[6] μὲν γὰρ

---

[1] lacunam ind. Jaeger.   [2] κρᾶσιν LPQB[a], Bekker.
[3] ἑτέρα Jaeger : ἕτερα.   [4] λίαν Ross : αἰτίαν.
[5] ἐπεί add. Ross.   [6] ταχὺ Bekker.

remarkable feature occurs with the soul; for it inheres in these substances. Therefore the fact that its motion always exerts a similar activity may reasonably be referred to the same agent, either absolutely or to some definite effective part; for nature, from which they are generated, remains the same. But what difference can there be between the forms of heat occurring in each individual; whether we consider the heat as an instrument, or as matter, or as both? For the only differences that heat can show are differences of degree. One might say that they are due to being mixed or unmixed; for the purer is more intense. The same argument applies to other simple substances. For, since the bone and flesh of the horse and ox differ, either they must be composed of different substances, or the substances must be used differently. If they are actually different, what are the differences of each of the simple substances, and what is . . . ? It is for these that we are looking. If they are the same, then they can only differ in proportion. One must be true, as in other cases; for mixtures of wine and honey differ because of their substance, but mixtures of wine differ, if at all, in proportion. Therefore Empedocles described the nature of bone too simply, since if all bones have the same proportion in their mixture, there should be no difference between the bones of horse, lion, and man. But they do differ in hardness, softness, density, etc. So also do flesh and other parts differ. They even differ in hardness and softness and other qualities in the same animal, so that the proportion of mixing cannot be the same. For granting that thickness and thinness, greatness and smallness are quantitative differences, hardness and density and their contraries are differences in the quality of the mixture. But those who argue in this way must know how the constitutive element may vary by being greater or less in volume, by being isolated or mixed, or by being heated in something else like things which are boiled or baked. Perhaps this is the true solution ; for it is in the mixing that it effects the purpose of nature. Then the same explanation may be given of flesh ; for the differences in it are the same. And the accounts of vein, air duct, and the rest are approximately the same. So of two things one is true : either the proportion of their mixture is not always the same, or the definitions must not be conceived in terms of hardness, density and their contraries.

INDEXES

I. INDEX NOMINUM

519

II. INDEX RERUM

INDEX RERUM

521

INDEX RERUM

INDEX RERUM

INDEX RERUM

527